Cell
physiology

Frontispiece (overleaf)   Section through an array of sperm cells of the moth
Bombex. The fan-like objects are projections from the plasma membrane. Note
the characteristic nine-fold symmetry of the flagella and the modified mitochondria
(single, dark objects) situated within the plasma membrane. The orientation of
cells comes from the fact that they were sectioned *in situ* within the insect testis.
Magnification is about × 150,000. Staining was with uranyl acetate and lead
citrate. (Courtesy of Dr. David M. Phillips.)

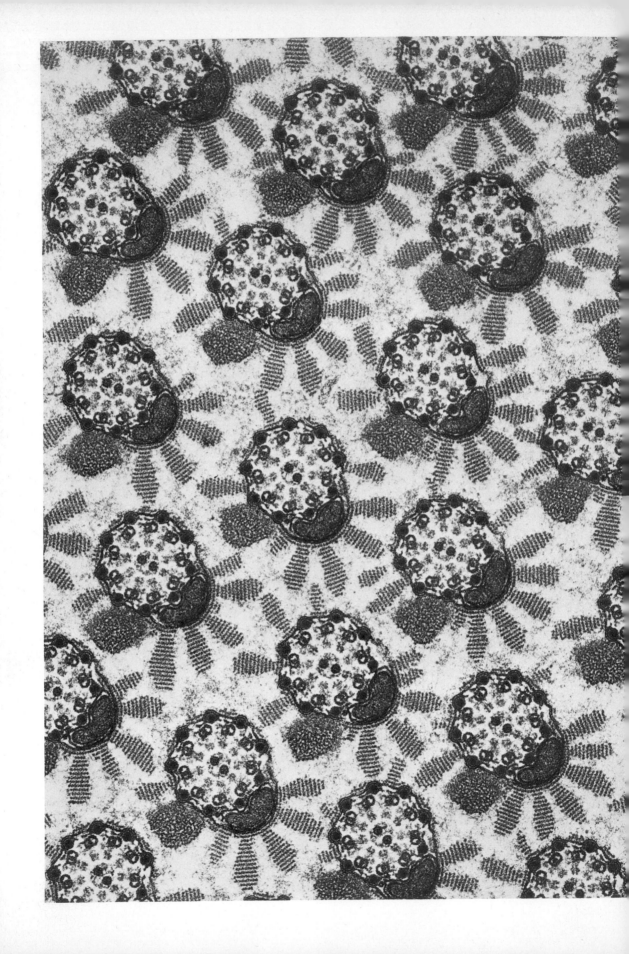

John L. Howland   Bowdoin College

Cell physiology

The Macmillan Company   NEW YORK

Collier-Macmillan Limited   London

Aan mijn vrouw
Aan mijn kinderen

The Macmillan Company
866 Third Avenue, New York, New York 10022

Collier-Macmillan Canada, Ltd.,
Toronto, Ontario

Library of Congress catalog card number: 76-187073

Printing: 1 2 3 4 5 6 7 8   Year: 3 4 5 6 7 8

Preface

This book is an introduction to cell physiology that places special emphasis on the regulation of cellular processes and on the fundamental role of membranes in the life of the cell. It has descended, in a number of ways, from an earlier book by this author,* which likewise stressed cellular regulation, but with a rather different organization and viewpoint. Work on the earlier book was begun when the author was about two years away from graduate study and was perhaps more enthusiastic about than experienced in teaching cell physiology. Since that time my ideas about what cell physiology is and how it can best be communicated have changed considerably. In writing this new text I wanted to reflect both this broader perspective of cell physiology and my continued enthusiasm for the subject. It is, of course, difficult to avoid enthusiasm when describing a field wherein so many miracles occur.

The present volume is somewhat unique in including discussion of a number of areas not usually found in such books. For instance, it might be noted that the present national concern with environmental matters has permitted inclusion of a chapter on environmental influences on cells (Chapter 19) that would have looked distinctly odd a few years ago. Since the author descended (or ascended) to the study of cell biology, biochemistry, and biophysics from an early interest in ornithology and a still earlier desire to climb mountains, inclusion of the chapter seems natural and appropriate.

In other respects, the book maintains an emphasis on matters of cellular regulation including sections on the control of enzyme activity and synthesis, regulation of growth, and the regulation of cell activities imposed by membrane systems. Part I of the book contains chapters dealing with cell structure, the special role of water in cells, energy, and the properties of proteins and enzymes. This section lays a foundation for subsequent discussion of cellular activities. Part II comprises five chapters dealing with different aspects of cellular metabolism—the

---

* *Introduction to Cell Physiology: Information and Control*, The Macmillan Co., New York, 1968.

chemical activities of cells. This, in turn, forms the basis for Part III, which discusses more complex aspects of cellular events: the general role of membranes, their involvement in excitability, and the basis of mechanical work by cells. Part IV deals with growth and genetic regulation. It includes chapters on growth, the role of nucleic acids in synthesis, the physiological effects of viruses, and the physiology of development. Part V, "The Cellular Environment," might be described as a discussion of physiological aspects of pollution with special attention to the effects of pesticides and other pollutants on cells.

A number of topics are emphasized in this book to a greater extent than usual. In addition to regulation, these include cellular energetics, the physiology of microorganisms, the action of viruses upon cells, and aspects of membrane structure and function. The virtue of emphasizing these areas resides in their central importance in contemporary cell study. Their inclusion also reflects the personal biases of the author, this prerogative being one of the very best reasons for writing books in the first place.

Work on the earlier book commenced while the author was enjoying the hospitality of the laboratory of Professor E. C. Slater in The Netherlands. Likewise, much of the work on the present one was carried out during a sabbatical year spent enjoying the equally generous hospitality of Professor Efraim Racker at Cornell University. The stimulating contact with both is here gratefully recorded. A number of colleagues have been most generous with their suggestions and assistance. They include Dr. G. J. Basbas, Dr. T. L. Bohan, Dr. S. S. Butcher, Dr. J. M. Calvo, Dr. R. R. Calvo, Dr. J. Enderson, Dr. W. T. Hughes, Dr. C. E. Huntington, Dr. Z. Kaniuga, Dr. E. B. Keller, Dr. J. B. Mathis, Dr. J. M. Moulton, Dr. E. Schwartz, Dr. C. T. Settlemire, and Dr. D. Der Vartanian. None of the above can be held responsible for any folly that persists in this book—without them there would have been a good deal more.

During the writing process, the author had the benefit of the wise and generous comments of Professor L. Kovac, who read much of the manuscript. It is not an exaggeration to say that, without his encouragement and help, the book would not and probably should not have been written. The author also wishes to record the skillful (and tolerant) secretarial assistance of Mrs. Virginia Richardson and Mrs. Jean Bohn. Finally, the first chapter was materially improved by having had an early version of it eaten (in manuscript form) by a particularly large Newfoundland dog, who appeared to enjoy, and certainly did not suffer from, the process. May the reader have an equally satisfactory experience.

J.L.H.

Harpswell, Maine

Contents

Contents

Part V   The cellular environment

Coda

Index

Part I

The structural
and chemical basis
of cell physiology

1 Cell organization

As cell study has progressed during recent years it has become increasingly obvious that there is no fundamental distinction to be made between structure and function. The study of cellular activity must always proceed with the cellular structure clearly in mind, and vice versa. Increasingly, the student of the cell is in a position to understand cell form as reflecting the underlying structure of subunit macromolecules, such as proteins and nucleic acids. In addition, he comes to understand the activity of living cells as highly regulated summations of many chemical and physical reactions in association with the action of enzymes, which are also proteins. Perhaps the ultimate fusion of morphology and activity is represented by those proteins that both contribute to the structure of cell membranes and are catalytically active enzymes.

This book will therefore proceed as if there were no line between cell morphology and activity, as in fact there is not. It is necessary, however, to begin somewhere, and a definition and brief description of the cell follow. The cell may be defined by what is called the *cell theory*, which, in turn, is really the history of cell study. Cell theory is little more than the observation that organisms tend to be made up of smaller, closed units, which were called *cells* by Robert Hooke in 1665 by analogy to the rooms in a nunnery. Hooke had, in fact, only seen the holes in cork where cells had been, but the name remains to describe the basic units of biological organization. About two centuries went by before microscopists acquired much understanding of the holes and of their place in the living scheme of things. The chief barrier to enlarged understanding of such matters was not any shortage of inquiring minds directed to cellular questions but rather the lack of optical apparatus capable of resolving the extremely small objects involved. The first completely functional *compound microscopes* (those allowing light to pass through more than a single lens) were developed during the 1830's; thus, few clear views of cellular structure were available much before that date. It should be added, however, that, over a century before that time, the great Dutch microscopist Anton van Leeuwenhoek made *simple microscopes* (i.e., not compound microscopes), which were constructed with such elegance and precision that they were of higher resolution than the compound instruments available in the middle of the 19th century. Anton van Leeuwenhoek was more than a brilliant deviser of optical instruments and observed the microscopic world through his microscopes with a remarkable mixture of towering insight and child-like wonder:

This was for me, among all the marvels that I have discovered in nature, the most marvelous of all.

Letter, October 9, 1676

Although we are not often persuaded that the establishment of new terminology represents an advance in understanding scientific matters, it is probably true that an important date in the pursuit of knowledge about the cell was 1839, when Purkinje spoke of the *protoplasm* as the material within the cellular chamber. Thus, the realization that protoplasm was a material stuff of living cells displaced the earlier notion that the cellular structure observed in plants, at least, was simply a structural matrix supporting, but not containing, the elements of living processes. In a conceptual way, the idea of protoplasm had to precede the notion that there could be such a thing as cell physiology, and the concerns that we now group under that name were, for many years, defined as the activities of living protoplasm.

Once cells became understood as the vessels of life, one of the first questions to be asked about them was that of their origin. During the last century, a number of brilliant speculations were made about this matter. For example, some proposed that small cells were formed by a process of budding (a view that turns out to be true in certain cases, such as the yeasts). Others maintained that a cell divided in two, and the halves then grew in size, and we know this proposal to be accurate in most cases. Finally, others (among whom were numbered Schleiden and Schwann) argued that the cell was formed from the nucleus, which, in turn, was formed from the nucleolus. This view must be called an imaginative speculation, which obviously grew out of the observed geometry of cells. Of course, in a literal sense, the proposal was simply wrong and should be placed on the heap of brilliant ideas that are, unfortunately, mistaken. However, we can note that there is even a germ of truth there; if one considers the flow of information in cells, it is true that the cell is formed from information contained in the nucleus (in the form of DNA). It happens that the nucleolus is under the control of the nucleus and not the other way around, so one must not press the analogy too far.

The final word on the origin of cells may be said to be that of Virchow in 1856 and, as perhaps befits the last word on anything, he said it in Latin:

Omnis cellula e cellula.

This idea that all cellular life must come from pre-existing cells found important support in the work of Pasteur later in the century when he employed the important approach of sterile culture technique. Pasteur demonstrated that even a favorable medium was unable to support a cell population until at least one cell was introduced. This concept forms the practical basis of the entire field of microbiology and tissue culture, where it is very important to know that cells are not created *de novo* by magical, or other, techniques. Of course, the one area where Virchow's dictum is of little help is in endeavoring to answer the interesting question of the origin of the *first cells*. Virchow lived in a time when this was not a particularly lively issue, owing to the rather different view of the Old Testament then prevailing. We now believe that the origin of life is an interesting question and one that should be asked in scientific terms. Unfortunately, we are not in a very good position at the moment to know how to answer it in any definite way, and are engaged largely in speculation.

The middle of the last century also provided increasingly accurate views of cell organelles, and one can only marvel at the highly detailed (and correct) information that was obtained. *Chloroplasts* were recognized at an early date, owing to their relatively large size, and in the 1880's Meyer observed that some of them contained smaller objects, which he called *grana*. Since grana were not seen by everyone, there was some question as to their existence, a question that was not settled in the case of some organisms until the availability of the electron microscope in the middle of the present century. Indeed, it could be said that Meyer carried microscopic observation of chloroplasts as far as it could go until the advent of the dramatically higher resolutions of electron microscopy. One sometimes wonders if certain of the questions presently frustrating us must await the development of new techniques whose nature we do not even suspect. Vacuoles, Golgi apparatus, endoplasmic reticulum, and mitochondria, all discussed later in this chapter, were observed prior to 1900, and considerable effort went into trying to discover their function. Chromosomes—including, luckily, the giant chromosomes of the fruit fly salivary glands—were known in the 1880's, although their role in the cell was not understood. It could be said that, by the turn of the century, most features of cellular structure were at least known, and it may not be too much of an oversimplification to add that the emphasis shifted then from *structure* to *structure in combination with function*, a concept that has occupied cell biologists since then.

Figure 1-1 Plant cell, illustrating the knowledge available in the 1890's. Modern names for the cell components are given as, in some cases, they differ from those in use at the time.

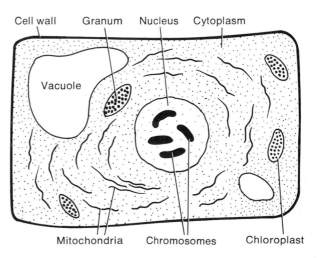

Cell wall   Granum   Nucleus   Cytoplasm

Vacuole

Mitochondria   Chromosomes   Chloroplast

Of course, one should not fail to note that purely morphological studies of cells have benefited from the increase in resolution obtained with better and better light microscopes, the development of ultraviolet microscopy,[1] and, finally, of electron microscopy. Clearly, our vision of cells has penetrated more deeply into levels of structure until, with the aid of very high resolution electron optics, we can even visualize individual (but large) molecules, such as DNA and proteins. It seems clear that, in the effort to improve resolution, the end is nowhere in sight.

Models of cell structure

The ability to view the interior of the cell in increasingly fine detail, as outlined above, has led to a succession of models describing the character of cells. The fact that these were models and not final and definitive views of cell structure is much more obvious in retrospect than when the various models were described. For example, Hooke's formulation of a model for cell structure was that cells were hollow rooms, a version not particularly consistent with modern electron microscopy but quite in keeping with Hooke's own observations.

Hooke's model was abandoned in the 1800's for the more complicated description of a cell as a closed sack containing a fairly liquid material (*cytoplasm*), in which float fairly solid objects (*organelles*). This was (and is) consistent with all observations possible in the pre-electron-microscope era and is at present the point of view of almost any layman who knows what a cell is. The sack-of-organelles model is quite satisfactory in many ways and tends to guide one in many of the procedures of analytical cytology described later in this chapter.

A number of refinements on the sack-of-organelles model became possible, based on advances in other fields. For example, the first half of the present century saw the rise and decline of cell models based on studies of colloidal phases of matter. One example of a colloidal system is the protein gelatin, which has looked to a number of people not unlike the material within a cell. Since cell material has been known for a long time to contain a great deal of protein, and since certain cell features are reminiscent of colloidal properties, much effort has been directed to the study of the cell as a colloidal system. However, the development of electron microscopy as well as modern protein chemistry has sent the colloid model into eclipse, because examination of protoplasm at high magnification reveals not so much a colloidal suspension as a complex mass of infolded membranes of a high degree of order and often of considerable beauty.

---

[1] The improvement of resolution obtained when ultraviolet light replaces visible wavelengths reflects the principle that one cannot expect to resolve objects with radiation of larger wavelength than the object. Ultraviolet light has a shorter wavelength than the visible region and high-energy electrons have still shorter wavelengths. The principle might be stated differently by noting that one cannot expect to probe the anatomy of a gnat with cannonballs.

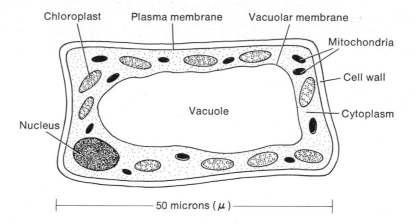

Chloroplast   Plasma membrane   Vacuolar membrane
Mitochondria
Cell wall
Vacuole
Nucleus
Cytoplasm

|— 50 microns (μ) —|

Figure 1-2   Cell from a green leaf. Note the perpherial location of cyto-plasm, the relative sizes of chloroplasts and mitochondria, and the nonidentity of the plasma membrane and cell wall.

Figure 1-3   General plan of an animal cell. This figure was drawn using several electron micrographs of liver cells as guides and illustrates such features as the complex plasma membrane, the double nature of the nuclear envelope, and the inclusions of granules, which, in this case, might well be composed of glycogen.

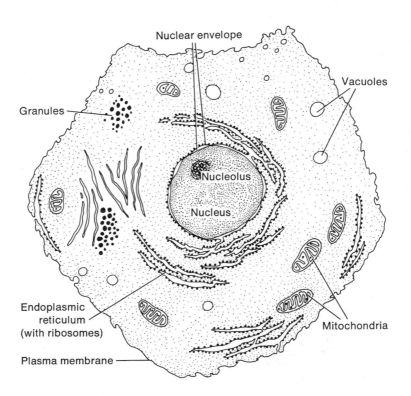

Nuclear envelope
Vacuoles
Granules
Nucleolus
Nucleus
Endoplasmic reticulum (with ribosomes)
Mitochondria
Plasma membrane

Cell organelles

Contemporary views of cell structure emphasize the close interrelations between parts, so that it has been impossible for a number of years to think of organelles as floating free in a homogeneous cytoplasm. Organelles should be viewed not so much as subcellular organs, complete in themselves, but as characteristic configurations of cytoplasm, usually involving membranes, which appear in a variety of cells investigated.

In the following sections, we shall describe a number of the more ubiquitous organelles, beginning with those found in cells from "higher organisms." In a later section, cells from "lower organisms" will be described. In the context of cell structure, "higher" and "lower" have a rather specific meaning that is worth mentioning here. A higher or *eukaryotic* cell is one possessed of a nucleus, bounded by a nuclear membrane. All organisms, save bacteria and blue-green algae, exhibit this sort of cell structure. On the other hand, the latter groups show a very different pattern of cell structure, termed *prokaryotic*, with much smaller cells, on the average. They have no membrane-bounded nucleus and lack many other organelles.

The fundamental membrane

Cells of all types are bounded by a membrane called the *plasma membrane*, and much of the internal structure of eukaryotic cells is composed of membranous organelles. All such cellular membranes have much in

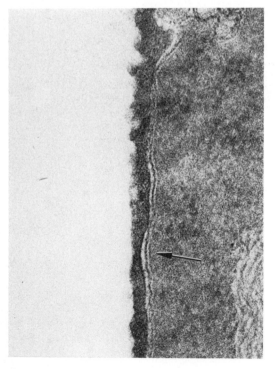

Figure 1-4 Thin section of the bacterium *Bacillus subtilis,* showing the double character of the plasma membrane (indicated by arrow). Fixation was with osmium tetroxide; magnification was × 180,000. From N. Nanninga, *J. Cell Biol.,* **42**, 733 (1969). (Courtesy of Dr. N. Nanninga.)

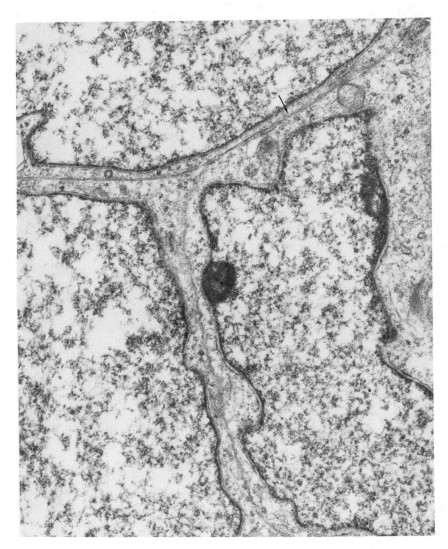

Figure 1-5 Junction between cells in a multicellular organism. This is a section of a portion of a tumor (lymphosarcoma), shown at a magnification of × 16,000. Cell membranes of the adjacent cells are pressed together, showing up as a double line (arrow). Note also the nuclei of the three cells in the field of vision. (Courtesy of L. Cohan.)

common, including their composition (which is largely protein and phospholipid) and their overall dimension of about 80–100 Å. Virtually all membranes present a similar appearance when viewed by high-resolution electron microscopy, exhibiting two dense parallel regions with a region of electron-transparent material in between (see Figure 1-4). This configuration is often called the *unit membrane*. Most membranes studied, whether they be bacterial plasma membranes or eukaryotic chloroplast membranes, exhibit specific patterns of permeability, admitting some molecules and excluding others from passage across their

Figure 1-6 Brush border of columnar epithelial cells. The brushes are evaginations of the plasma membrane that project into the lumen of the gut, giving the epithelial cells much greater surface area. The micrograph is of cells that have been obtained in suspension by treating the lining of a rat intestine with proteolytic enzymes under carefully controlled conditions. Note the profusion of mitochondria. The cells were fixed with glutaraldehyde and stained with osmium tetroxide. Magnification is × 16,000. (Courtesy of Dr. C. T. Settlemire.)

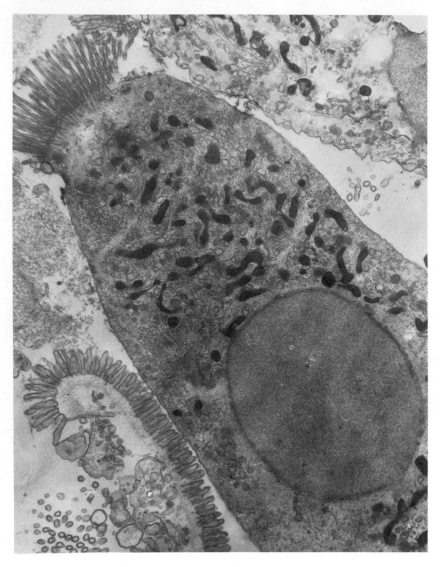

surface, and many membranes appear to play a central role in cellular energy coupling (see Chapter 6).

Plasma membranes often exhibit specialized structures in some regions of their surface, such as elongated projections, the *microvilli*, which may serve to increase their surface area. In the case of the striated border of the cells of the mammalian intestinal lining, the microvilli that compose it occur at a density of about 2×10^8 per square millimeter or more than 10^3 per cell, so that the increase in surface area must be impressive. Frequently, infolding of plasma membranes occurs, so that small regions of extracellular space become engulfed by the cell—a process called *pinocytosis*. This event enables the cell to incorporate extracellular medium without the necessity of its passage across the intact plasma membrane. In many cases, the inpocketing of plasma membrane is followed by pinching off of the pocket, thus producing an intracellular vesicle, the *pinocytotic vesicle*, which exists free in the cytoplasm for a time before finally disappearing.

Since the membrane around the engulfed material finally breaks down, the material does not encounter a permeability barrier in entering the cytoplasm, so that macromolecules that would be incapable of crossing the plasma directly nonetheless enter the cell and may contribute to cell nutrition.

There are also a number of other specialized regions of plasma membranes. For example, where adjacent cells are in contact, the plasma membrane may become thickened into discrete objects called *desmosomes*, which may have special significance in intercellular transfer of material. These are often observed in epithelial cells and often include very fine fibers (*tonofibrils*) radiating into the cytoplasm. Desmosomes were at one time thought to include a passage actually connecting the cytoplasm of the two cells, but even with careful electron microscopy no such connection has been discovered.

The nucleus

The name *nucleus* is consistent not only with its more or less central location in many cells but with the centrality of its role in regulation of cell activities. The nucleus, as we shall see, is the repository of genetic information of cells, that is, information enabling one generation to resemble preceding ones. Information for the synthesis of most cellular macromolecules is encoded in the nucleus in the form of the molecular structure of *deoxyribonucleic acid* or DNA. The central role of the nucleus is demonstrated by dramatic deleterious effects observed on removal of nuclei from otherwise intact cells. Enucleated cells at once lose the ability to divide and thus replicate themselves. Indeed, such cells, although continuing to respire and carry on certain metabolic activities, appear unable to maintain their structure and soon expire. The only example of a cell that is normally without a nucleus is the mammalian red blood cell, which is, however, derived from nucleated cells and only survives in the

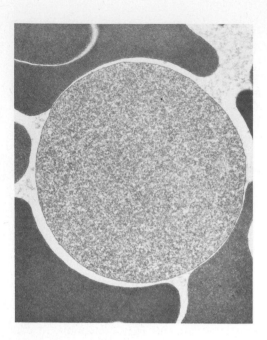

Figure 1-7 Human cell without a nucleus: a reticulocyte surrounded by mature red blood cells. This cell represents the final stage in the development of red blood cells; the nucleus has just been lost. Normally, such a cell would be found only in the blood-forming regions, but the conditions imposed by iron-deficiency anemia, in this instance, have permitted release into the bloodstream. These cells were stained with uranyl acetate; magnification is × 10,500. (Courtesy of L. Cohan.)

anuclear state for a matter of a few months. Even that span of survival is possible only because the red blood cell has such a specialized function, serving as a vessel for the oxygen-binding protein hemoglobin and possessing only limited metabolic capability.

The shape of the eukaryotic nucleus can be quite variable, depending on the cell in which it exists. In many cells, it is simply spherical, but in cells that are themselves flattened ("squamous cells") the nucleus may be deformed into a flattened disc. Many white blood cells contain irregular nuclei, which can be cup-shaped or quite randomly variable. Some plant nuclei are even filamentous. Moreover, while most cells contain a single nucleus, some are binucleate and still others contain as many as 100 nuclei. Examples of the latter category include a number of algae, molds, and certain osteoclasts (bone-eroding cells) from marrow.

The eukaryotic nucleus is surrounded by a two-layered *nuclear envelope*, of which each layer is, in turn, a typical double *unit membrane*. The envelope is typically pierced by a large number of *pores* (Figure 1-8), which may play a role in nuclear–cytoplasmic communication of material. It is likely that the pores represent something other than simple holes in the envelope since nuclei may be isolated from cells by means of differential centrifugation without loss of interior contents and while maintaining permeability barriers and osmotic properties that would be quite impossible if the envelope were a sieve.

Figure 1-8 (opposite) Nuclear pores revealed by the freeze-etching technique. The nucleus is above and exhibits profuse pores. Note also the mitochondria and other membranous structures in the cytoplasm. The cell wall and part of a neighboring cell are seen below. These cells are from an onion root tip; magnification is × 36,000. (Courtesy of Dr. Daniel Branton.)

13

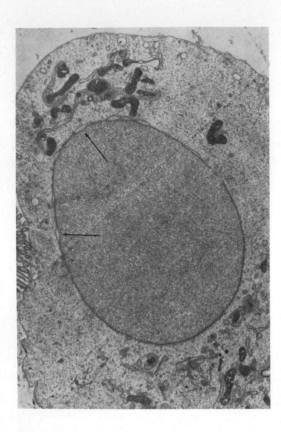

Figure 1-9 The nuclear envelope consists of two unit membranes. These are shown clearly in this electron micrograph of a columnar epithelial cell from the rat. Fixation was with glutaraldehyde, staining with osmium tetroxide. Magnification was × 8,000. (Courtesy of Dr. C. T. Settlemire.)

On the other hand, it is possible to inject colloidal gold into the cytoplasm of an amoeba and demonstrate its presence, finally, in the interior of the nuclei by means of electron microscopy. Thus, very large particles (up to about 100 Å) can pass through the nuclear envelope, most probably at the pores. It would seem that the pores are so modified as to permit the passage of such large objects, while preserving the osmotic integrity of the nucleus, a function that is also reminiscent of pinocytotic uptake of particulate material through the plasma membrane.

It should be added that the nuclear envelope enjoys a somewhat transitory existence in the life cycle of the eukaryotic cell, as it quite disappears at the time of cell division. When the final stage of division is completed, the envelope is seen to re-form, apparently through the coalescence of fragments of the cytoplasmic membrane system, the *endoplasmic reticulum*, discussed later in this chapter.

It has been recently shown that isolated nuclei are able to carry out respiration and that the responsible enzymes are associated with the nuclear envelope. It is likely that nuclear respiration is coupled to the synthesis of ATP, which, in turn, is utilized by the many energy-requiring processes that go on in the nucleus.

Apart from nucleic acids, nuclei contain large amounts of protein and some lipid, most probably complexed to protein. Included in the protein fraction are a number of enzymes, many of which are associated with DNA or RNA synthesis. As it has become increasingly possible to isolate nuclei that are uncontaminated with other cytoplasmic membrane systems, an increasing number of enzymes have been localized there. Thus, nuclear metabolism is becoming understood to include not only reactions involved in nucleic acid synthesis and energy transformations, but also a variety of pathways not as clearly related to the DNA–RNA aspect of nuclear function.

An important category of nuclear protein is the class of *histones*, which are highly basic proteins, existing in the nucleus in greater or lesser degree of complex with DNA. Histones are basic, owing to a large proportion of basic amino acids, including lysine, arginine, and histidine, all of which contain a positively charged nitrogen in the cellular pH range. The total DNA and histone contents of nuclei are about equal, as are the approximate number of histone-free amino groups (NH_4^+) and the number of negatively charged phosphate groups in DNA. This equivalence suggests that the function of the histones is to bind to DNA by electrostatic attraction between unlike charges, and such binding is readily demonstrated *in vitro*[2] as well as in the cell. The DNA–histone complex has long been regarded as playing a structural role in the nucleus, perhaps protecting DNA from mechanical stresses. It is also clear that the histone serves to neutralize the large net negative charge on the DNA molecule, thus preventing its binding of any small cations that may be in the nucleus. In the case of prokaryotic cells, there are no histones to fulfill such a role and the charge excess is largely neutralized by the presence of the divalent cation Mg^{2+}. In other words, in prokaryotic cells DNA exists as a magnesium salt, while in eukaryotic cells DNA contained in their true nucleus is in the form of a histone salt.

At the time of cell division in the eukaryotic cell, part of the interior contents of the nucleus condense to form the well-known chromosomes, which are described below. During the remainder of the cell's life cycle, the interior of the nucleus remains rather undifferentiated, showing a uniform granular appearance. The granularity (often called *chromatin*) probably represents aggregated material consisting of the complex of DNA and basic histones. Chromosomes, which materialize during division, likewise appear to be formed from the same two chief components.

The nucleolus

Within the nondividing nucleus, the most distinctive structural entity is a region of granular material called the *nucleolus*. There can be one or several nucleoli per nucleus, and staining reactions indicate that they are particularly rich in ribonucleic acid (RNA). Indeed, radioautographic experiments indicate that the nucleolus is the locus of extensive synthesis of RNA. Nucleoli may be isolated in a more or less purified state by

---

[2] The expression *in vitro* means "in glass," that is, under artificial conditions or outside the cell.

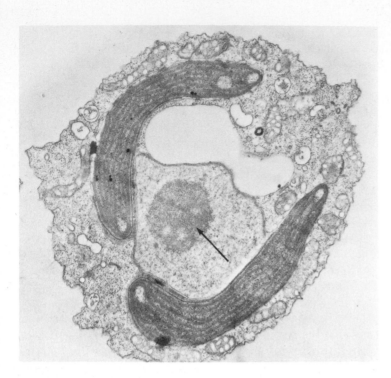

Figure 1-10 Nucleolus. This organelle is seen within the nucleus of the alga *Orchromonas danica*. Magnification is × 13,500. (Courtesy of Dr. Sarah P. Gibbs.)

disruption of nuclei and centrifugation at suitable speeds, and it appears that they contain large amounts of protein and RNA and rather small amounts of DNA. These organelles are unique in the cell in being relatively large objects that are *not* bounded by a membrane. Finally, electron microscopy reveals that there is some identifiable internal structure in nucleoli, including strands of what is probably a DNA–histone complex.

Without discussing the details of protein synthesis now, it is possible to note that nucleoli play an important role in that overall process and appear to be specifically associated with the production of *ribosomes*, which are the sites of protein formation in the cytoplasm. Because of this and because ribosomes exist as cytoplasmic particles readily seen with the electron microscope, it becomes much more clear why there should be pores in the nuclear envelope able to pass particulate matter. The evidence suggesting that ribosomes are manufactured in nucleoli is varied, but perhaps the most compelling line comes from studies on the development of the African toad *Xenopus*. There is a lethal mutation in this species that leads to a proportion of embryos failing to develop beyond a certain stage. The nuclei of these embryos do not contain nucleoli and it appears likely that the mutation represents a defect in their formation. It appears also that the reason for cessation of development and subsequent death in these cases is the inability of the embryos to manufacture ribosomes, so that when the demands of protein synthesis exceed the ability of the residual ribosomes from the maternal ovum, death ensues.

In addition, specific attachment between nucleoli and chromosomes have been noted during cell division; the regions of the chromosomes where this occurs are termed *nucleolar organizers*. Although evidence is

16

not very complete, it appears reasonable that the organizer loci represent the regions of DNA in the chromosome that are uniquely associated with the synthesis of ribosomal RNA; indeed, these regions form patterns on which ribosomal RNA is synthesized. During cell division, nucleoli disappear as recognizable objects under the electron microscope and, interestingly enough, during this period RNA synthesis is diminished. Finally, when nucleoli begin to reappear following division, the organizers seem to serve as centers for their formation.

Mitochondria

The portion of the cell within the plasma membrane but outside the nuclear envelope is called the *cytoplasm* and, viewed by electron microscopy, appears to contain several distinct, highly organized membranous entities. Of these *organelles*, the *mitochondria* are prominent in number and design (see Figure 1-11). Mitochondria are found in all aerobic higher cells and are involved in energy-transfer processes resulting from respiration. An animal cell might contain on the order of 1,000 mitochondria. They consist of a closed outer membrane and an inner membrane folded to form cristae, providing an increase in available surface. Since many of the processes of mitochondria take place on or in the membrane, the increased surface provides a greater number of sites of action and greater

Figure 1-11   Membrane structure of a mitochondrion. Note the cristae penetrating the interior. Magnification is × 62,400. (Courtesy of Dr. Stanley Bullivant.)

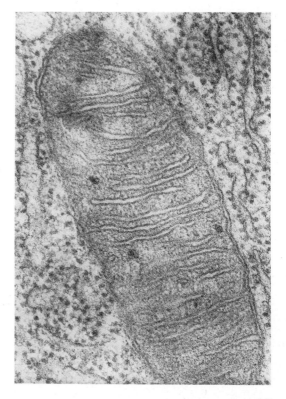

activity per mitochondrion. Very high resolution electron micrographs show the inner membrane to bear small knobs situated on short stalks. The knobs are said to be the loci of energy-transfer reactions, leading to the synthesis of the "high-energy" compound adenosine triphosphate, ATP. It appears that, of the two mitochondrial membranes, the outer membrane does not provide a barrier to the entry of any molecules, save very large ones such as proteins. On the other hand, the inner membrane accounts for most features of the specific permeability of mitochondria; many molecules and ions are quite excluded from the interior, while still others are only taken up by specific "pumps" at the expense of energy. The semipermeability of the mitochondrial inner membrane also gives the mitochondrion its osmotic properties whereby the transport of molecules or ions may be reflected in large volume changes of the mito-chondrion as a whole. Indeed, such swelling or shrinking is often monitored by electron microscopy or by means of light scattering and used as a measure of transport of material across the mitochondrial surface.

means of monitoring the volume changes of the mitochondria

Finally, mitochondria have been recently shown to contain DNA in the form of a single closed loop. As we shall see, chloroplasts contain a similar form of DNA, as do bacteria and a number of viruses. This similarity has encouraged those interested in the origins of cellular organelles to consider (but not necessarily accept) the hypothesis that mitochondria and chloroplasts are initially derived from microorganisms that invaded the ancestors of eukaryotic cells and established some sort of symbiosis. This point of view is consistent with the recent discovery of ribosomes within mitochondria and chloroplasts, again quite similar to those found in bacterial (and other) cells.

Lysosomes *(proteolytic enzymes)*

ref. "The Lysosome"

Another cytoplasmic organelle bounded by a closed membrane is the *lysosome*, which appears to contain enzymes, which catalyze the break-down of a number of molecules. Lysosomes, within their relatively un-complicated membrane, contain (1) a material, homogeneous in the electron microscope, that is certainly protein, (2) largely proteolytic enzymes (which hydrolyze protein), and (3) other enzymes that break down cell components, including ribonuclease, deoxyribonuclease, and phosphatases (see Table 1-1). In some instances, the protein within a lysosome may be observed to form large symmetrical crystals, reflecting the high concentration at which such protein must exist. Presumably, the function of the lysosomal membrane is to separate the lytic enzymes from the remainder of the cell in order to avoid destruction of components of the viable cell. When the cell becomes moribund, the membrane is observed to break down, releasing the lysosomal enzymes, which digest the remnants of the cell. It is unlikely, however, that this process rep-resents the major function of lysosomes. Rather, it is more probable that the enzymes sequestered within the lysosome play their central role in the living cell, catalyzing reactions that form part of the metabolic

moribund

18

Table 1-1   Some enzymes in lysosomes

Enzyme	Reaction*		
Proteases	Protein	→	Amino acids
Nucleases	Polynucleotides	→	Bases + phosphate + pentose
Phosphatases	Phosphate esters	→	Monophosphates
Glycosidases	Polysaccharides	→	Monosaccharides
Lipases	Lipids	→	Fragments thereof
Sulfatases	Sulfate esters	→	Fragments thereof

* All reactions are hydrolytic breakdowns of the general form $XY + H_2O \rightarrow XH + YOH$.

activity of the cell. If this is the case, the lysosomal membrane must play a regulatory role, since it ordinarily separates the enzymes inside from their substrates in the cytoplasm proper. Thus, its permeability properties, largely unstudied, would be expected to be of significance in control of the relevant portion of metabolism.

Lysosomes exist in a wide range of size varying from about 0.1 to 2 microns (1 micron = 10^{-4} centimeter). This variation is apparently related to one of function about which considerable information has

lysosome size

Figure 1-12   Role of lysosomes in phagocytosis. Lysosomal enzymes are synthesized at ribosomes and collected in the tubes of the Golgi apparatus (a). Lysosomes are formed there by a budding process. Meanwhile, phagocytotic vesicles are formed at the plasma membrane. These fuse with the lysosomes (b) to form a single vesicle, which contains both the "food" taken in and the enzymes to digest it. Finally, the vesicle breaks down (c) to release the products of digestion and (presumably) inactive enzymes into the cytoplasm.

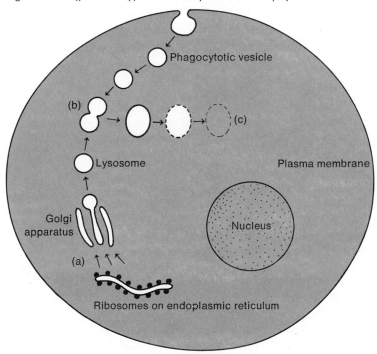

recently become available. Thus, lysosomes are all related to the hydrolytic digestion of intracellular molecules, but this digestion can occur in several contexts, including both reactions that are part of ordinary intracellular metabolism and those involved in the digestion of material taken in by phagocytosis (engulfment of solid nutrient by the plasma membrane). *phagocytosis* In the latter case, there appears to be a category of lysosomes that are produced by pinching off portions of the *Golgi apparatus* (see below) and which move in the cytoplasm until they are able to fuse with vesicles that have been produced by phago- or pinocytosis. Then the two vesicles fuse and the contents of the lysosome are placed into contact with the undigested material that has been incorporated by the cell (see Figure 1-12). Thus, the enzymes and substrates are mixed without the enzymes being allowed to enter the free cytoplasm where they might accomplish mischief by hydrolyzing essential portions of the cell. The ability of a lysosome to fuse with another vesicle might sound rather like magic, but it will become clear in a later chapter that fusion of membranes is by no means an uncommon event and that such plasticity is, indeed, an important feature of cellular membranes in general.

The Golgi apparatus

In most eukaryotic cells, a distinctive conformation of tubular membranes has been observed that is called the *Golgi apparatus* (formerly *complex*) after its discoverer. These membranes are most often located in the region of the nucleus and take the form of stacks of flattened vesicles (also called *dictyosomes*), as well as thin tubes or small sacs. The interrelation between the membranes of the Golgi apparatus is often highly complicated, with frequent interconnections. Indeed, there is so much variation and complexity in the picture presented by this class of organelle that the existence of the Golgi apparatus as a discrete organelle has been under question until fairly recent times. *dictyosomes*

The function of the Golgi apparatus has likewise been under discussion for a considerable period with important questions unresolved until quite recently. An important clue as to its role has been available for several decades, however, in the observation that cells with the capacity for active secretion (such as glandular cells) contain a highly developed Golgi apparatus when compared to cells with a less obvious secretory role.

A direct demonstration of the role of Golgi apparatus in cell secretion has, however, been rather difficult to obtain. For one thing, it would be most helpful to be able to isolate purified preparations of Golgi membrane systems so that its function might be studied *in vitro*. Unfortunately, this has not been easy, owing to the complexity and resultant fragility of the membranes, and the direct analysis of function has mostly depended upon the important technique of *radioautography*. When using this technique, one administers radioactive material to cells, cuts thin sections of them, and places the sections in close contact with fine-grained photographic emulsion. These are stored in the dark for a considerable period of time (often some months) and then developed. The locations of radioactive atoms are identified as the darkening of the developed emulsion, and when they are compared with photomicrographs taken of the same or adjacent sections, it becomes possible to localize the region where the tracer has ended up.

When this technique has been applied to the question of the function of Golgi apparatus, clear results have been obtained. For example, the introduction of radioactive amino acids to cells results in a rapid labeling of ribosomes, where the amino acids are incorporated into protein. Somewhat later, the label begins to be detectable in the Golgi apparatus, suggesting that proteins are synthesized in the ribosomes, as we shall see, and then collected in the Golgi membranes. There is no evidence that protein is actually synthesized at the Golgi membranes themselves, which, unlike the membranes of the endoplasmic reticulum, are devoid of ribosomes. On the other hand, there is radioautographic evidence that the Golgi membranes are the sites of the synthesis of polysaccharides, since they are rapidly labeled on addition of radioactive sugars. When proteins are collected in the Golgi apparatus, they are usually either secretory proteins, such as the digestive enzymes produced by the lining of the gut or the pancreas, or play a role in the intracellular digestion of material brought in by phagocytosis. When polysaccharides are synthesized and collected in the Golgi apparatus, they are likewise often secretory products, such as the carbohydrates of mucus or of the plant cell walls. Thus, the fundamental role of the Golgi apparatus in cellular secretory processes seems to be completely established.

Chloroplasts

Eukaryotic cells that are photosynthetic contain *chloroplasts*, in which are located the photosynthetic pigments and numerous enzymes involved in photosynthetic energy production and carbon assimilation. Although the exact form of chloroplasts is quite variable, especially as one compares flowering plants with the algae, in general the chloroplast is built up of layers of parallel membranes upon which the photosynthetic pigments, such as chlorophyll, are located. In many cases, there exist regions within the chloroplast where membranes are more closely applied than elsewhere ; such regions are termed *grana*, after their grain-like appearance in the light microscope. The whole chloroplast is bounded by an envelope formed from two unit membranes. Close examination of the internal chloroplast membranes reveals that a number of regions are in the form of closed sacs, greatly elongated or pressed together. These are called *thylakoids* and define regions of the chloroplast that are not continuous with the rest of the interior. Grana appear to be entirely composed of thylakoids and, when isolated from chloroplasts (they are then called *subchloroplast particles*), exhibit permeability properties consistent with the assumption that they contain an internal space bounded by a membrane. Electron

Figure 1-13 Tobacco chloroplast, exhibiting thylakoid membranes folded upon themselves to form grana. Fixation was with glutaraldehyde, staining with osmium tetroxide. Magnification is × 37,600. (Courtesy of Dr. Herbert W. Israel.)

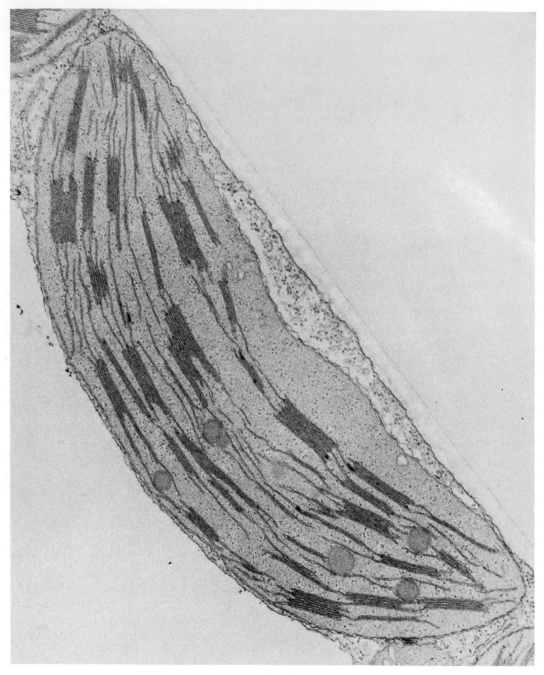

microscopic examination of chloroplast membranes at still higher resolution shows that they are composed of still smaller subunits often called *quantasomes*. It is tempting to imagine that these may represent basic units of photosynthetic function, although the evidence is presently inadequate.

quantasomes

As we have seen, both chloroplasts and mitochondria contain circular DNA and RNA. The presence of nucleic acids suggests that there must be information stored in these parts of the cell, and this storage, in turn, indicates a measure of autonomy from control by the nucleus. Indeed, there is a great deal of evidence for such autonomy, and both chloroplasts and mitochondria can be regarded as self-replicating organelles, under only partial regulation by the nuclear pool of genetic information. It will be seen in Chapter 16 that this capacity of self-propagation is reflected in active protein synthesis in both isolated mitochondria and chloroplasts.

self replicate

Figure 1-14 High-magnification (× 92,000) view of chloroplast membranes, showing configuration of grana. Preparation was as described in Figure 1-13. (Courtesy of Dr. Herbert W. Israel.)

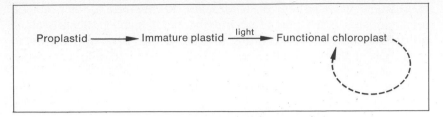

Figure 1-15   Outline of chloroplast development. Chloroplasts are either self-propagating (broken arrow) or develop from proplastids.

Such protein synthesis is similar to that taking place in the rest of the cytoplasm in that it occurs at the small particles called *ribosomes* (which we will define in the next section). The ribosomes of chloroplasts (and mitochondria), however, differ from those of the rest of the cytoplasm in being measurably smaller, and we shall see that the protein synthesis that occurs there is distinctive in having rather different sensitivity to certain antibiotics.

Not only do chloroplasts appear to be endowed with a considerable capacity for self-synthesis, but there exists also in many organisms, an alternative route for their formation. Thus, if one grows green plants in the dark, their color becomes progressively pale, owing to a very low concentration of the photosynthetic pigments. It is possible to germinate seeds in total darkness and they develop neither the color of normal plants nor exhibit the characteristic chloroplast structure. Instead, they do contain membrane-bound organelles called *proplastids*, which lack pigments, and the characteristic thylakoid configuration of developed chloroplasts. They do appear to contain DNA and to be capable of replication, so that one may think of proplastids as a self-perpetuating pool of potential chloroplasts. When these plants are placed in the light, the synthesis of chlorophyll and the other photosynthetic pigments occurs and their ultrastructure soon becomes normal. The relations between these organelles are summarized in Figure 1-15. It should be added that an analogous situation occurs in the case of mitochondria, where *promitochondria*, lacking the enzymes of the respiratory chain and the characteristic structure of mitochondria, have been demonstrated in yeast. When yeast cells are grown anaerobically, they are devoid of mitochondria and, when placed in air, rapidly manufacture them. It appears likely that promitochondria contain the DNA required for the synthesis of complete organelles, and it has also been demonstrated that they contain some, but not all, of the enzymes associated with energy metabolism.

The endoplasmic reticulum and ribosomes

Much of the cytoplasm in eukaryotic cells is filled with an intricate array of membranes, collectively called the *endoplasmic reticulum*. The endoplasmic reticulum (ER) appears to give rise to the nuclear envelope after

24

Figure 1-16 Rough endoplasmic reticulum. The membranes are here seen bearing a dense population of ribosomes. Magnification is × 80,400. (Courtesy of Dr. Stanley Bullivant.)

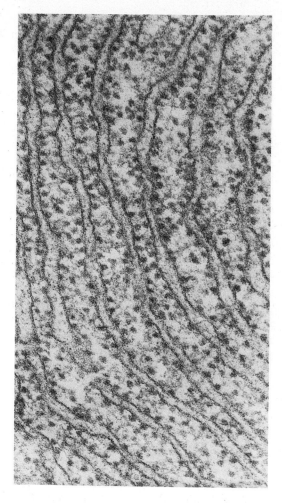

cell division. Much of the surface of the ER can be covered with ribosomes (see below), which give it a speckled appearance. Such regions are termed *rough endoplasmic reticulum*; *smooth endoplasmic reticulum* is devoid of attached ribosomes. Much of the ER is tubular and may be regarded as a channel for passage or storage of material. For example, endoplasmic reticulum, like the Golgi complex, is most highly developed in the case of cells that are active in the secretion of proteins.

Ribosomes are bound, in eukaryotic cells, to the endoplasmic reticulum and have a characteristic structure of their own. Examination in the electron microscope reveals them to consist of two subunits, one somewhat larger than the other. Since these subunits are generally characterized according to the rate at which they sediment in the ultracentrifuge (a very high speed centrifuge), they are usually described in terms of the *Svedberg unit* (S), which is a measure of their rate of sedimentation. For instance, the intact ribosomes of eukaryotic cells sediment with a constant of 80 S, while the two subunits sediment at about 40 and 60 S. Note that the S values are not additive. The details of ribosome structure, and especially their disaggregation into subunits, will be discussed further in the context

25

of protein synthesis. Each subunit contains both RNA and protein, the RNA of the large (60 S) subunit sedimenting at about 28 S and that of the small subunit sedimenting at about 18 S. There are a number of proteins present in ribosomes, some exhibiting enzymic activity, while other protein is quite basic and appears to bind to the RNA (ribonucleic acid) to form a salt in a manner analogous to the complex between DNA and histone. It now appears likely that the attachment between the ribosome and the endoplasmic reticulum links the large (60 S) subunit to the membrane.

When the cytoplasm of cells is disrupted so that the endoplasmic reticulum may be isolated, the ER itself breaks into small, closed fragments that are called *microsomes*. The microsome fraction of cells consists of closed vesicles bounded by an intact membrane that retains the ability to serve as an osmotic barrier and to carry on ion transport. When microsomes are formed from rough endoplasmic reticulum, the membranes bear ribosomes and such preparations are able to carry out protein synthesis. The microsomes (and, by extension, the endoplasmic reticulum in intact cells) contain a number of enzymes not related directly to protein synthesis, including some that function both in detoxification of certain drugs and in the synthesis of a number of fat-soluble substances, including fatty acids, triglycerides, steroids, and phospholipids. The endoplasmic reticulum also contains several proteins that appear associated with biological oxidations, including NADH-cytochrome c reductase and cytochrome b_5, a characteristic cytochrome of such membranes. The function of these components is unclear, but may be associated with active transport across the ER membrane as well as with the oxidation of drugs.

As we have said, the endoplasmic reticulum appears to function as an intracellular transport system and, as such, it exists as a very dynamic and fluid system of membranes. For instance, proteins synthesized at the ribosomes of rough endoplasmic reticulum pass into the interior (*cisternae*) of the ER tubule, where they are collected. The mechanism whereby the large newly manufactured proteins cross the ER membrane is unknown, but it is interesting that the part of the ribosome where the membrane is attached (the large 60-S subunit) is also the portion of the ribosome where the new protein is actually formed, so that the details of attachment may permit the protein to travel directly into the cisternae without actually requiring passage through the membrane.

Once the protein has entered the cisternae it may find its way to the Golgi apparatus, or it may be leaked into the cytoplasm outside the ER. There are electron micrographs of secretory cells where the membranes of the ER appear either to be in contact with those of the Golgi apparatus or, in some instances, to be in the process of giving rise to the closed membranes of that structure. In any case, we have seen that labeled protein, synthesized at the ribosomes, does gain access to the Golgi apparatus, where it appears to be stored.

On the other hand, there is also indication that the endoplasmic reticulum can give rise, by a process of "budding," to vesicles that contain protein. These are then seen to migrate in the cytoplasm until they either fuse with the Golgi membranes to empty their contents into the Golgi interior, or fuse with the plasma membrane, extruding their contents in the surrounding space. Whether protein, thus secreted, must necessarily be first collected in the Golgi apparatus is presently unclear, but there can be no doubt about the importance of the extraordinary fluidity of these membrane systems.

Indeed, the overall picture of ER, Golgi apparatus, lysosomes, and the plasma membrane is one of highly transitory structures, with continuous formation, fusion, and breakdown of membrane structures. We shall see in a later section that even such stable organelles as chloroplasts and mitochondria, when disrupted by detergents or mechanical means, are able to fuse again into closed vesicles, which in some instances retain the osmotic properties of the original organelle.

Peroxisomes

An additional membrane-bound organelle of wide occurrence in eukaryotic cells is the *peroxisome*, which is a vesicle containing several enzymes that either produce or decompose hydrogen peroxide (H_2O_2). These vesicles are about 1 micron in diameter and are similar to lysosomes, which are larger, in that they contain an apparently homogeneous solution of proteins. In some instances, peroxisomes contain a core that is a highly geometric array of protein, looking somewhat like a crystal.

Peroxisomes contain a number of enzymes that produce hydrogen peroxide as a product of an oxidation by molecular oxygen. These can be grouped under the general *oxidase* reaction,

$$AH_2 + O_2 \rightleftharpoons A + H_2O_2$$

and include D-amino acid oxidase, urate oxidase, and α-hydroxy acid oxidase. Since H_2O_2 is a powerful oxidizing agent and quite toxic to cells, being among other things a strong mutagen, its breakdown is catalyzed by another enzyme of peroxisomes, *catalase*, which catalyzes the following reaction:

$$2\,H_2O_2 \rightleftharpoons 2\,H_2O + O_2$$

Besides sequestering vesicles for hydrogen peroxide, peroxisomes often contain other enzymes not producing it. For example, in some instances they contain enzymes concerned with the breakdown of purines (a component of nucleic acids), and in others they contain the enzymes of the glyoxylate cycle, whose function is described in a later chapter.

Flagella

Most cells appear endowed with the power of motion, either moving with reference to their surroundings or at least circulating their cytoplasm about within the plasma membrane. Cells that are able either to propel themselves through a liquid medium or to move the medium across their surface do so by means of *cilia* (short and numerous) or *flagella* (longer and few). Even a cell fixed in its location within a multicellular organism may possess cilia in order to circulate extracellular fluid about its periphery. In eukaryotic cells, flagella and cilia are built on a universal pattern whereby the structure originates within a *basal granule* within the cytoplasm. In cross section, cilia and flagella are bounded by a sheath, which encloses an array of nine double tubules arranged in a radial fashion with two tubules in the center. A similar distinctive *nine-fold symmetry* is also present in the *centriole*, an organelle found in animal cells and associated with cell division, as we shall see.

Cilia and flagella are both of about the same diameter, about $\frac{1}{2}$ micron, and they range in length from about 5 microns (cilia) to 250 microns (a particularly gigantic flagellum). The basal granule has about the same diameter as the external flagella but differs in structure in not containing the central pair of tubules. The basal granule is presumed to function both in the synthesis of the external cilia and flagella and in the coordination of cilial beating. It is clear that the latter does exist, since waves of beating travel across the surface of a ciliated cell in a highly ordered fashion. What little is known of the mechanism of ciliary motion will be discussed in a later chapter in the context of cellular movement.

Centrioles have been described in most animal cells and in some lower plant cells as well. They are similar to the basal granules of cilia in exhibiting the characteristic nine-fold symmetry, but without the central pair of tubules. The close similarity between these organelles argues strongly for a common origin. Where they occur, centrioles are usually double and are found in the proximity of the nuclear envelope. The role of the centriole is somewhat clouded, but it appears to play a role in the formation of the spindle fibers involved in eukaryotic cell division. In this connection, the relationship between centriole and basal granule becomes apparent; both organelles seem involved in the formation of protein-containing objects that are, to at least some degree, tubular in nature. On the other hand, there are many cells entirely able to form spindle fibers, which, however, do not contain recognizable centrioles, so that it is impossible to be too dogmatic about their function.

The manner in which centrioles and basal granules are formed is not understood in any detail, but some evidence points to the supposition that

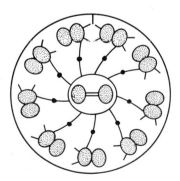

Figure 1-17 Example of the tubule arrangement within eukaryotic flagella. The diameter of the whole organelle is about 0.2 micron; the diameter of a single tubule is about 0.02 micron. Note the central sheath containing two central tubules, the small secondary fibers (seen end-on), and the outer ring of nine double tubules.

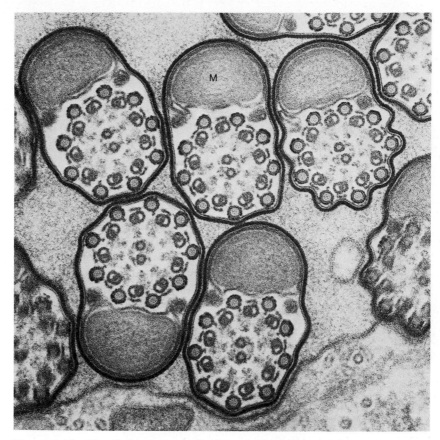

Figure 1-18 Flagella from sperm cells of the caddis fly (*Neuronia*). These sections are taken so as to include the mitochondrion (M) as well as the usual 2 + 9 arrangement. In addition to the nine double outer tubules, there is another set of nine still more peripheral tubules that appear to be characteristic of insect sperm flagella. Staining was with uranyl acetate and lead citrate. Magnification is about × 90,000. (Courtesy of Dr. David M. Phillips.)

they may be self-replicating. The appearance of a new centriole is normally in the immediate region of the pre-existing one, suggesting that the latter plays a part in its formation. Moreover, studies of cilia-less mutants of protozoans suggest that cilia are under a genetic control that is not wholly in the nucleus. This is supported by the cytochemical localization of a considerable amount of DNA in the peripheral part of ciliated cells, presumably associated with the basal bodies.

Cytoplasmic tubules and fibers

When high-resolution electron microscopy, together with relatively gentle fixation methods, became available, it was soon obvious that cytoplasm was often filled with a considerable array of extremely small tubules with diameters of about 0.02 micron (200 Å). These are not bounded by membranes and appear to be formed from protein subunits. *Microtubules* are commonly found in cellular structures that possess some rigidity, when compared to the remainder of the cytoplasm. For example, they are common in the long processes (axons and dendrites) of nerve cells where they would appear to provide a structural framework. They also compose the spindle fibers associated with cell division and are either the same as or similar to the tubules that run the length of flagella and cilia.

An obvious role for "microtubules" in the cell is as a structural matrix, although its being obvious does not necessarily imply its being true. Other functions have also been suggested and the small size of these objects has made it difficult to arrive at definitive answers. For instance, it is often argued that the tubules are channels for the movement of material within the cytoplasm and that they may play a special role in cytoplasmic streaming. In both instances, the main evidence is that the microtubules are often (but not always) aligned in the direction of net material transport in the cytoplasm, but it is clear that this might also represent only the passive alignment of the tubules in the face of bulk motion.

One possible probe for ascertaining the function of microtubules is provided by the observation that the drug *colchicine*

is able to lead to the destruction of tubular structures, notably the spindle fibers. Thus, the addition of colchicine to dividing plant cells, by preventing spindle formation, also prevents the partition of chromosomes into the daughter cells. Since chromosomes continue to replicate, this leads to the situation in which a cell may contain many times the usual number of chromosomes (*polyploidy*); the drug has therefore been of considerable use in the study of cell genetics. It is interesting that colchicine has been shown to bind to the protein of spindle fibers and it may also bind to that of other microtubular structures. The action of this material lends some support to the notion that the microtubules are largely of structural significance, since it can also lead to a loss of the characteristic shape of certain cell processes.

Cytoplasm often can be seen to contain a fabric of very small fibers, which, like microtubules, are probably of structural import. Fibers often are especially prominent in the region of contact between cells and they also often radiate from the basal granules of cilia. Perhaps the most dramatic occurrence of intracellular fibers (or filaments) is in muscle cells, where a highly ornate system of filaments forms the morphological basis of the contractile process, some of them being formed of the contractile protein actomyosin (see Chapter 14).

Extracellular material and intracellular inclusions

Plasma membranes are frequently surrounded by relatively inert material, which they have produced. This material usually has a mechanical role but does not normally have permeability properties like that of the membranes themselves. For example, removal of the cell wall of bacteria does not alter the pattern of permeability—that is, does not change the list of molecules that are either admitted or excluded. Some examples of such extracellular structures are given in Table 1-2.

In addition, many cells contain *inclusions*, which are particles or droplets of material that is not necessarily a universal component of cytoplasm but which is often a storage form of an energy-rich product of metabolism. Thus, oil droplets may be seen in many cells, notably bacterial and algal ones. Glycogen, a polymer of sugar, is a characteristic feature of many liver cells, whereas many higher plant cells can become distended with a heavy deposition of starch. A number of bacteria contain granules of poly-β-hydroxybutyrate, and others contain polymeric inorganic phosphate, both of which may be regarded as energy sources. Cells engaged in enzyme secretion often contain granules containing the enzymes that they produce (or precursors thereof); a good example is the *zymogen granules*, which contain digestive enzymes found in pancreatic cells. In those cases where the inclusion is a liquid that would be miscible with the cytoplasm, the inclusion is surrounded by a membrane and called a *vacuole*. The vacuoles of higher plant cells are often filled with a variety of compounds, including sugars, organic acids, and pigments, and may occupy a large fraction of the total volume of the cell. A final example of a cell inclusion is the rather special case where infection of the cell by a virus leads to formation of characteristic intracellular particles that are either the next generation of viruses or their precursors.

Table 1-2   Examples of extracellular structures

Structure	Location	Major constituents	Role
Protist cell wall	bacteria, blue-green algae	complex, often includes polymers of amino sugars, amino acids, lipids	mechanical: a rigid envelope for the plasma membrane
Eukaryotic cell wall	plants	polymer of sugars	mechanical: defines cell shape
Collagen	animals (fibrous tissue)	tropocollagen (a protein)	mechanical: forms important acellular structure in multicellular organism (e.g., bone, tendon, connective tissue)
Myelin sheath	animals (nerve fibers of vertebrates)	protein, lipids	electrical insulation about conducting nerve

The organization of prokaryotic cells

Prokaryotic cells exhibit a very different pattern from those eukaryotic cells discussed thus far, but they are far from simple. They are characterized by the lack of a nuclear envelope and absence of such relatively large and complicated organelles as mitochondria or chloroplasts. Indeed, these organelles are larger than many prokaryotic cells, and it might be argued that many of the special features of such cells are consequences of their small size. For example, a small spherical bacterium (coccus) might have a diameter of 1 micron, while a eukaryotic cell could be from 10 to 20 microns across.

We will take, as examples of prokaryotic cells, bacteria such as *Escherichia coli* (see Figure 1-19), but most comments would extend to blue-green algae as well. The cell is typically surrounded by a plasma membrane, which, in turn, is surrounded by a cell wall. It will be remembered from Table 1-2 that the prokaryotic cell wall is chemically unlike that of higher plants. The cell wall may merge with still more exterior layers of *capsule* material, which often are composed of sugar polymers and which can play an important role in the invasive properties of bacteria that live in multicellular organisms (that is, pathogenic bacteria). The wall and capsule are osmotically inert—which is to say that they do not present a barrier to free movement of molecules and ions, and they do not appear to contain any enzymes. On the other hand, the prokaryotic plasma membrane is the site of many enzymes, including those involved in energy metabolism. The plasma membrane also gives the cell its permeability properties and includes a number of specific carrier sites for specific molecules. For instance, the bacterium *Escherichia coli* is possessed of a specific protein in its plasma membrane that permits uptake of the sugar lactose, and

Figure 1-19 General plan of a prokaryotic cell. Note the lack of a nuclear envelope and, indeed, the absence of membrane systems within the cytoplasm. Vacuoles are also generally lacking and the cytoplasm does not exhibit the active streaming characteristic of eukaryotic cells. The storage granules could be composed of food stores such as glycogen, they could be droplets of fat, or they might be composed, in some instances, of polyphosphate.

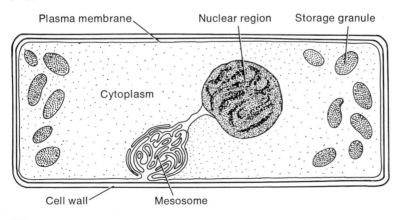

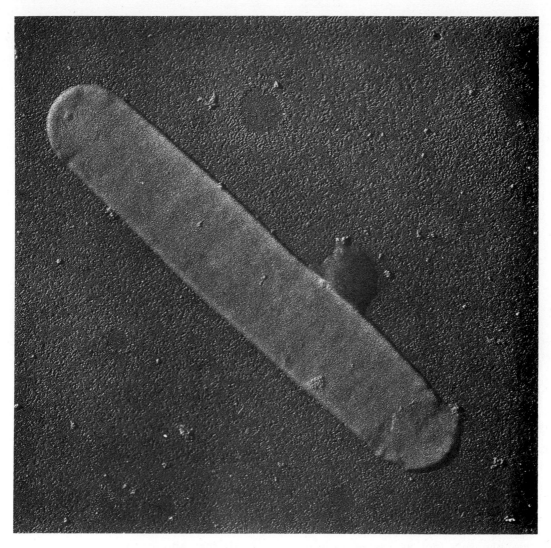

Figure 1-20 Isolated cell wall from *Bacillus lichenformis*. This is a shadowed preparation; magnification is × 37,000. From *Scientific American*, May 1969, with permission. (Courtesy of Dr. Nathan Sharon.)

which does not function to any measurable extent with most other sugars. Such *permeases* are discussed in a later chapter. It is interesting that bacterial cells shorn of their cell wall by treatment with a specific enzyme (lysozyme) differ from the intact cells by losing their characteristic shape, which appears to be dictated by the cell wall, but undergo no detectable change in enzymatic activity or in permeability. The cells are, however, unable to divide, suggesting that a certain mechanical rigidity is required for the latter process.

The plasma membrane of bacteria frequently forms a fold within the cytoplasm called a *mesosome*, which appears to play a role in replication of DNA when the cell divides. In photosynthetic prokaryotic cells,

mesosome

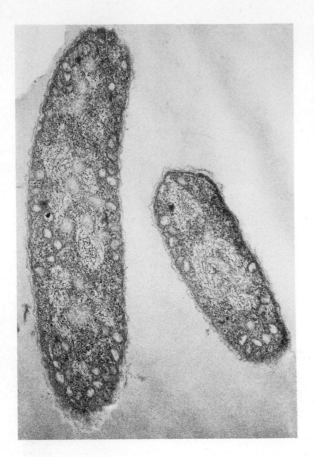

Figure 1-21 The photosynthetic bacterium *Rhodospirillum rubrum*. This thin section, magnified × 48,000, shows the nuclear region, the invaginations of the cell membrane that presumably bear the photosynthetic pigments, and the rather complex cell wall. The fuzzy gray areas in the center of the cell are probably deposite of glycogen. (Courtesy of Dr. G. Cohen-Bazire.)

infolded plasma membranes appear to be the site of photosynthetic pigments and associated enzymes, and take the form of closed vesicles in some photosynthetic bacteria called *chromatophores* or parallel internal membranes (in other photosynthetic bacteria and blue-green algae). Finally, the plasma membrane of prokaryotic cells often contains other pigments (not associated with photosynthesis) that appear to protect the cells from damage from visible and ultraviolet radiation.

The interior of prokaryotic cells contains no large closed membranous organelles. Ribosomes are evident and are not bound to membranes but rather are free in the cytoplasm. The content of ribosomes is dependent upon rate of growth, so that actively dividing bacteria are literally packed with them. There is a *nuclear region* unbounded by a membrane but containing DNA, often seen as fibrous material in the center of the cell. Often several nuclear regions are seen within a single plasma membrane, suggesting that, under some conditions, the nuclei divide more rapidly than the cell as a whole.

Many bacteria are motile and employ flagella to swim through liquid. The bacterial flagella are much smaller and less complicated than those of eukaryotic cells. Instead of the characteristic nine-fold symmetry of eukaryotic flagella, those of bacteria appear to be composed of 3 to 10 twisted filaments, which in turn are made up of smaller protein subunits.

34

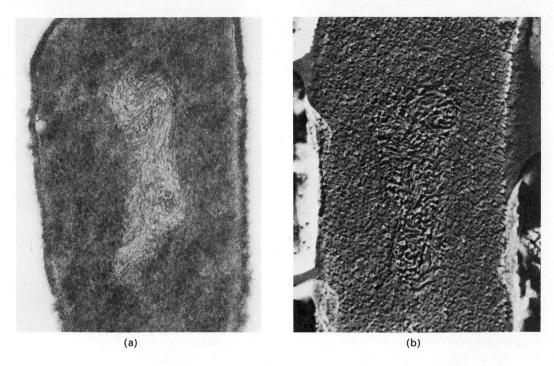

(a)   (b)

Figure 1-22   Nuclear region of bacteria. (a) A thin section of *Bacillus subtilis;*
magnification is × 76,000. (b) The same organism, prepared by freeze-etching.
Magnification is × 114,000. From N. Nanninga, *J. Cell Biol.,* **42**, 733 (1969).
(Courtesy of Dr. N. Nanninga.)

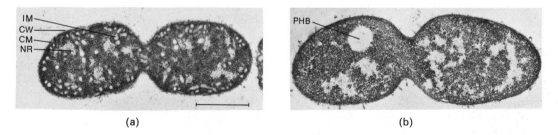

(a)   (b)

Figure 1-23   Thin sections of *Azotobacter vinelandii*. In (a) IM refers to a system
of internal membranes that is formed from invagination by the plasma membrane
under conditions when the bacterium is able to fix nitrogen (see Chapter 7). NR
stands for nuclear region, CM for cell (or plasma) membrane, and CW for cell wall.
In (b), PHB denotes a granule of poly-β-hydroxybutyrate.The bar indicates 1 micron.
Magnification is × 14,000. (Courtesy of Dr. Joel D. Oppenheim.)

The size of cells

Cells are commonly very small when compared to familiar objects, such as trees or human beings, and, since we are rather used to thinking in terms of objects of the latter scale, we find it difficult to contemplate the size of either bacteria or galaxies. Were we the size of bacteria, moreover, we would still experience difficulty looking (in the ordinary sense) at other bacteria (even if we were equipped to do so) since most details of bacterial structure are smaller than the wavelength of visible light. In any case, it is clear from the foregoing discussion that the feature of cells that most restricts their study (in both technical and practical ways) is their small size. We shall see in a later chapter some special techniques (other than microscopy) that are employed to obtain information from these small, but wonderful, objects.

In describing the size of cells, there are two possible approaches and we shall employ them both. In the first place, one can say that a given cell has dimensions that are a certain fraction of the size of some larger object in the human scale of things. For instance, a mammalian liver cell might be of the order of 10 microns across. To understand what this means, reflect on an object such as a large green pea (or, depending on personal taste, a rat's eyeball). These might be about 10 millimeters in diameter. If this distance were to be divided into 1,000 equal parts, each of them would be 0.01 millimeter long. One micron is 0.001 (one thousandth) millimeter, so that our 10-micron liver cell would also be 0.01 millimeter across. Thus, the liver cell would be said to be about one thousandth of the diameter of a pea (or rat's eye, or whatever). To put things somewhat differently, 100 microns is about the smallest object that one can resolve with the unaided human eye (in good working condition). The liver cell would span about one tenth of this, so that if one lined 10 liver cells end to end, they would be just visible as a tiny speck.

The figure of 10 microns is only an arbitrary one so as to give an idea of the magnitudes involved. In fact, some bacteria are one tenth this size, whereas the human ovum is over 100 microns and, therefore, just discernable to the eye. Table 1-3 gives additional information about the sizes of cells and parts thereof.

A second way to look at the dimensions of cells is to consider the relationship between size and the events that go on there. For example, molecules

Table 1-3   Dimensions in the cellular scale

Object	Range of length (or diameter) (microns)
Most plant cells	5–100
Most animal cells	5–50
Most bacterial cells	0.5–5
Chloroplasts (green plant)	2–10
Mitochondria	0.5–5
Ribosomes	0.02

require a finite interval of time to diffuse from one part of a cell to another, and it is often the case that they are produced at one location in a cell and subsequently react somewhere else. How long does this movement take? In fact, the rate at which molecules diffuse is a function of the temperature, the viscosity of the medium, and their molecular weight. It can be calculated that, under reasonable conditions, the amino acid glycine travels at a rate of about 100 microns per second, which means that it could traverse our 10-micron liver cell in about one tenth of a second. If cells were several orders of magnitude larger, the times required for molecules to diffuse across them would become very long, and one can see that there is probably a relationship between the rate at which biological events occur and the size of the cells in which they occur. Many of the processes in very large cells, for example, would be limited in rate by the process of diffusion; it appears, therefore, that evolution has proceeded with size and rate in suitable proportion.

Another way of considering the size of cells is with regard to volume and concentration. Consider a bacterial cell, which might have a volume of about 1 cubic micron. What does it mean to think of a solution of a certain molarity within this volume? In fact, many of the molecules that occur within a cell exist at concentrations lower than 1 micromolar (μM or 10^{-6} M). Knowing[3] that a 1-molar solution contains about 10^{24} molecules per liter, we can easily calculate that a cell with volume of 1 cubic micron will contain about 10^3 (or 1,000) molecules. A solution of 1 nanomolar concentration ($nM = 10^{-9}$ M), which is quite within the reasonable cellular range of concentration for certain molecules, will contain on the average one molecule per cell. Most of the chemical reactions and measurements with which we are familiar occur in large volumes with enormous numbers of molecules. Indeed, much of our available analysis of chemical events with regard to rate and energy require that there be a statistically meaningful number of molecules involved. For example, the field of thermodynamics, which is the analysis of energy exchange, is completely based on a statistical treatment of large numbers of atoms and molecules and loses its power when confronted with small numbers. It is clear that one molecule per cell (or even 1,000) is not a statistically meaningful collection and it is necessary to think very carefully in going from statements applicable to the larger scale of things to the cellular scale. More will be said about this problem in a later section.

There is yet another sense in which the size of cells may be more than fortuitous. It is often said that the reason that mitochondria and chloroplasts are not found in prokaryotic cells is that they are, themselves, larger than such cells and their inclusion would be topologically impossible. This reasoning sounds trivial at first, for it is quite possible to conceive of mitochondria that are, say, one tenth of the usual size, and these "micromitochondria" would fit nicely into even a small bacterium. While it is possible to conceive of this situation, such mitochondria do not appear to occur in nature, and it might be worth wondering why not.

---

[3] From Avogadro's number, 6.023×10^{23}.

After all, the White Queen pointed out the ease in imagining impossible things:

"There's no use trying," she said; "one *can't* believe impossible things."

"I daresay you haven't had much practice" said the Queen. "When I was your age, I always did it for half-an-hour a day. Why sometimes I've believed as many as six impossible things before breakfast."

An important aspect of the determination of the size of mitochondria and other cellular structures may be the minimal size of the macromolecules of which they are built. The inner membrane of mitochondria is probably the thickness of no more than a few respiratory enzymes, which, as we shall see later, are arranged in the membrane in a definite orientation. It is probably quite impossible to carry out the functions of energy coupling in such a membrane without the membrane's containing at least a few of these molecules, so that the width of the membrane is ordained by their size. One might wonder, then, if a micromitochondrion (or other organelle) might be built from minute, scaled-down, enzymes in order to accommodate them in the membrane. The answer to this is probably also "no," since the function of enzymes requires that they possess a fairly considerable surface in order to exhibit the specificity of binding that enzymes must possess. There are no enzymes with molecular weights of less than around 10,000 Daltons and the reason is probably that smaller ones would be unable to assume a conformation of high enough uniqueness. In any case, very small enzymes do not exist in nature and this fact determines, therefore, the size of the membranes that contain them and, indirectly, that of the organelles that the membranes form.

Cell-membrane topology

It is well known that the surface of a sphere divides all space into two components, inside and outside. Cells are often thought of as somewhat deformed spheres, with the world on one side of their closed cell membrane and cell contents on the other. Thus, all points within the cell would be separated from all points without by the cell membrane, and for a molecule (such as food) to pass from the outside to the interior, it would have to traverse the cell membrane. The same view might apply to the nucleus and the world (which would include the rest of the cell). Such ideas about cell topology, however, must take into consideration additional features of the membrane surfaces, such as the nuclear pores and continuity between different membrane systems. For instance, the apparent continuity of the endoplasmic reticulum with the nuclear membrane means that the topological relations between nucleus and surrounding cytoplasm are quite complicated (Figure 1-24). Such a relationship is obviously quite important with reference to the permeation of molecules to and from the nucleus.

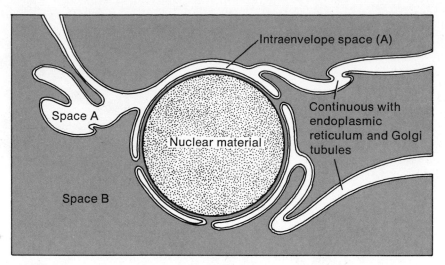

Figure 1-24 Continuity between nuclear envelope and endoplasmic reticulum. Space A is the space within the envelope and extends to the inside of the endoplasmic reticulum tubules and Golgi apparatus. Space B is the cytoplasm proper, and extends through the nuclear pores to the interior of the nucleus (provided the pores are really openings). See text for a discussion of this last point.

The remarks that have been made about cell structure in the preceding paragraphs are all of a general character, and it must be added that the details of cell structure are extremely rich, with great variation among cells of different organisms and within a single organism. The structure of a given cell must always be related to the particular function of that cell and its place in the evolutionary scheme.

As a compelling example of the relationship between the role of a cell and its structure, consider the red blood cell of mammals, which is modified exclusively for the transport of oxygen through the blood stream. Mammalian red blood cells, when circulating in the blood stream, have lost their nucleus and are devoid of the other cytoplasmic organelles, such as mitochondria, Golgi apparatus, and lysosomes. Indeed, they appear to retain only their plasma membrane, which contains a highly concentrated solution of the oxygen-binding protein hemoglobin, together with relatively few metabolic enzymes. As indicated earlier, the red blood cell survives with such a spartan morphology by (1) developing from a much more complex precursor, *reticulocytes*, which contain a full complement of organelles, and (2) having a relatively short half-life in the circulating condition.

One is frequently able to form opinions about the function or ecology of a cell from its microscopic appearance. We have noted that secretory cells often exhibit highly developed Golgi apparatus, where the secretory products are entrained before release. Cells that manufacture large amounts of protein, whether for external secretion as in the case of pancreatic cells or for internal use as in the case of rapidly dividing bacteria, contain relatively large numbers of ribosomes and, in the case of eukaryotic cells, a highly developed rough endoplasmic reticulum. Cells lining the

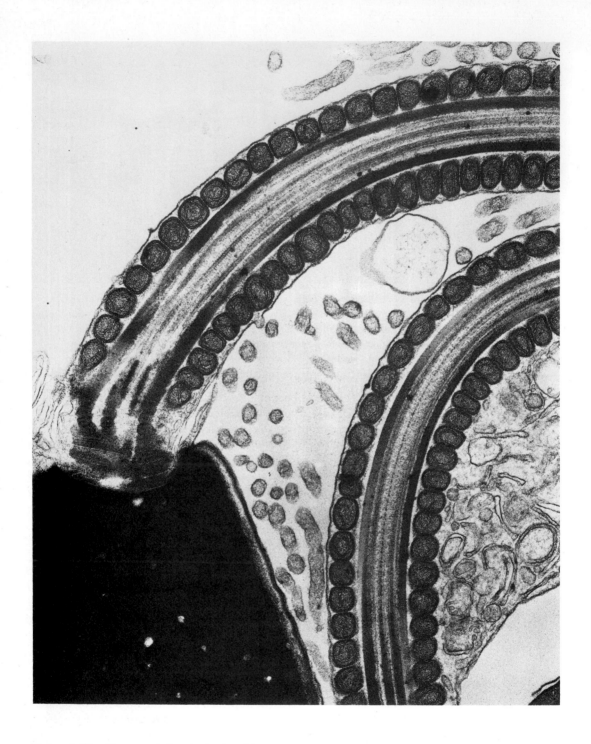

Figure 1-25 Close proximity of mitochondria to contractile fibers of sperm flagella. Magnification about × 6,000. (Courtesy of Dr. David M. Phillips.)

gut that have, as a major function, absorption of nutrients possess a plasma membrane bearing numerous microvilli that greatly increase the surface area in contact with the fluid to be absorbed.

It is likewise obvious that a central feature of the function of a sperm is rapid and sustained motion, and it is clear from Figure 1-25 that this function is reflected in the occurrence of a large flagellum that bears at its base—a midpiece containing modified mitochondria—the site of aerobic energy provision (in the form of ATP synthesis). Indeed, the spermatocyte may be called a packet of genetic information (the head) endowed with the power of motion.

Finally, in examining the structure of free-living protozoa one can gain a great deal of information about their ecological "life style." The pattern of cilia can often reflect their mode of feeding, enabling one to determine, for instance, whether they actively pursue their prey or, rather, pass particles to a passive "mouth" by the action of an array of cilia over their whole surface. Many protozoa contain, in addition to their ordinary nucleus, *micronuclei*, which contain smaller amounts of DNA. These appear to indicate that such organisms carry on sexual reproductive activities, so that they might be said to provide information, if not about the social mores of the beasts, then at least about their genetic capabilities.

Some techniques of analytical cytology

Although many techniques for cell study are available, and new ones are being developed and exploited with great rapidity, we shall be concerned here with only two approaches, which are of central importance. Both are directed to the general problem of correlating a cellular event (such as an enzymic reaction) with the fraction of the total cell structure involved.

An important area of technique is the powerful one of cytochemistry (selective staining of cells), where an event is localized microscopically in a cell by means of its ability to produce a visible alteration in the color of a dye. For example, a number of dyes are able to react with cellular oxidative enzymes, leading to a change in the oxidation-reduction state of the dye molecule and a consequent change in its color. Thus, the dye "stains" for the presence of the enzyme with which it reacts, and serves to locate that enzyme in a sectioned cell. Cytochemistry has yielded important results of this sort, and has frequently provided good correlation with results obtained in other ways. The chief limitation of the method has been the difficulty of demonstrating unambiguously that the dyes react only with the predicted enzyme, under conditions of cytological staining.

A second approach to the relationship between cell events and cell structure has been the physical separation of cells into their "subcellular" components. Thus, cells may be disrupted and fractionated into readily identifiable parts, including fairly pure nuclei, mitochondria, ribosomes, and other organelles (see Figure 1-26). Disruption may be brought about by grinding tissue with an abrasive such as alumina or frozen carbon

41

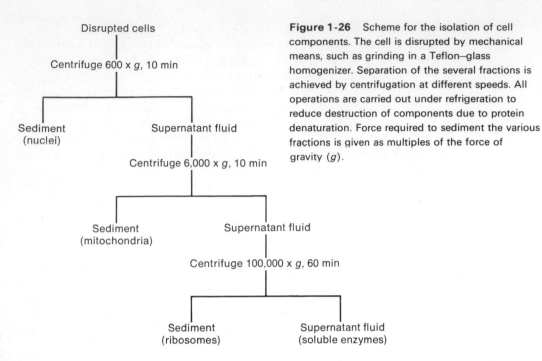

Disrupted cells

Centrifuge 600 x *g*, 10 min

Sediment
(nuclei)

Supernatant fluid

Centrifuge 6,000 x *g*, 10 min

Sediment
(mitochondria)

Supernatant fluid

Centrifuge 100,000 x *g*, 60 min

Sediment
(ribosomes)

Supernatant fluid
(soluble enzymes)

Figure 1-26 Scheme for the isolation of cell components. The cell is disrupted by mechanical means, such as grinding in a Teflon–glass homogenizer. Separation of the several fractions is achieved by centrifugation at different speeds. All operations are carried out under refrigeration to reduce destruction of components due to protein denaturation. Force required to sediment the various fractions is given as multiples of the force of gravity (*g*).

dioxide, by application of high-energy sound, or by extrusion at high pressure through a small orifice. Selection and careful use of the most appropriate method results in the breaking of the cell membrane and the liberation of reasonably intact organelles. One may then separate the organelles by differential centrifugation, where low-speed centrifugation sediments the larger organelles (such as nuclei); progressively greater forces are required for sedimentation of smaller ones (such as mitochondria and ribosomes). After centrifugation at a high speed (for example, 40,000 rpm[4]) for an hour, all organelles are sedimented, and the supernatant fluid contains much of the soluble material of cells, including the protein not bound to membranes. By careful selection of centrifugation speeds, it has been possible to isolate mitochondria, ribosomes, or chloroplasts, shown by visual and biochemical criteria to be reasonably free of contamination by other cell fractions. The central limitation of this method is probably uncertainty about the degree of purity as well as considerable uncertainty as to the degree of alteration in the properties of organelles, owing to the rigors of isolation. For example, conclusions about the properties of mitochondria in intact cells may only be drawn with extreme caution from studies using isolated mitochondria. The most convincing results are those inferred from evidence obtained using a variety of techniques.

---

[4] The rate at which particles sediment is related not to the rotation rate of a centrifuge rotor but to the centrifugal force produced. This force depends on the radius from the center of the rotor to the particle in the tube and is proportional to the first power of the radius and the square of the angular velocity. In centrifuges commonly used for cell fractionation, 40,000 rpm would correspond to about 100,000 times the force of gravity ($10^5 \times$ g).

Remarks on electron microscopy

Finally, since much of the information presented in this chapter on cell structure has come from electron microscopy, it appears useful to outline some features of the technique. The reader is probably aware that the essence of electron optics differs only from light microscopy in the replacement of electromagnetic radiation of visible wavelength (light) with a beam of electrons, and the consequent replacement of glass lenses with circular electromagnets. The advantage of electron over light optics is a measure of the much shorter wavelength of electrons at energies used (accelerated through a field of the order of 100,000 volts). Light microscopy at high magnification is limited by the wavelength of light, as it is quite impossible to obtain greater resolution than that wavelength. Since electrons have smaller wavelengths, they permit greater *resolution* (ability to discriminate between closer objects) and, hence, higher magnification is possible. The wavelength of electrons is inversely proportional to their energy, so that a desire for very high resolution has led to the development of microscopes with increasingly higher accelerating voltages (up to 10^6 volts, in a few instances), although, in practice, lens design is a more probable limitation. The basic construction of an electron microscope is shown in Figure 1-27. Since air is opaque to electrons, electron microscopes must pass the beam through high vacuum and the sample under study must be able to survive such conditions (which preclude examination of living cells).

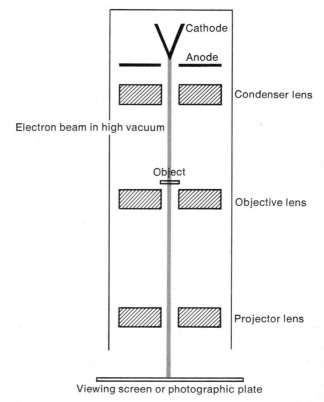

Figure 1-27 Electron microscope. Accelerating voltage of the order of 50–100 kilovolts is applied between the cathode and anode. The electron beam is shown undeflected; in practice, the electromagnetic lenses bend the beam, forming an image on the screen in a manner analogous to glass lenses bending the path of light in an ordinary light microscope.

Cathode

Anode

Condenser lens

Electron beam in high vacuum

Object

Objective lens

Projector lens

Viewing screen or photographic plate

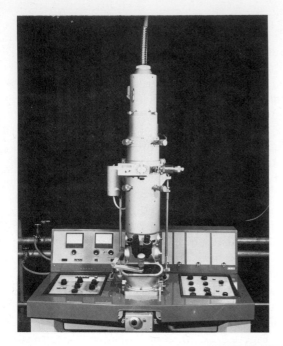

Figure 1-28 Electron microscope. (Courtesy of Picker Industrial Co.)

Figure 1-29 Section through the heads of sperm of caddis fly. This example of the thin-section technique shows the high levels of contrast that can be obtained. Note the dark nuclei of the sperm cells as well as the clearly defined membranes of the gonadal cells that contain them. (Courtesy of Dr. David M. Phillips.)

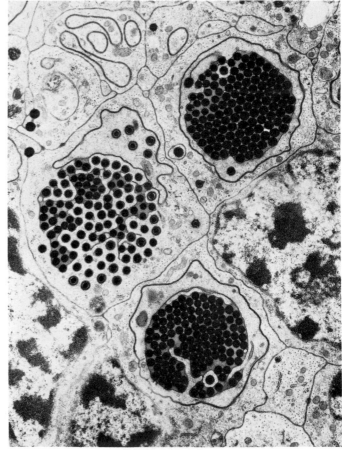

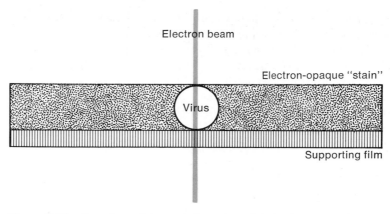

Electron beam

Electron-opaque "stain"

Virus

Supporting film

Figure 1-30 Negative staining technique. A particle that is relatively electron-transparent is embedded in a layer of electron-opaque material such as phosphotungstate. The particle (in this case, a virus) shows up in the electron micrograph as a hole in a dark field. See Figure 1-31.

There are several different techniques of sample preparation widely applicable to biological material, and each is suitable for disclosing different aspects of structure. For example, examination of whole cells is best carried out by preparing very thin stained sections. Refinement of the sectioning techniques for light microscopy has led to the ability to obtain sections as thin as 100 Å. These may be variously stained with material such as salts of uranium or osmium, which combine the properties of binding to cell structure and of being relatively opaque to electrons. (An example of such an electron micrograph is shown in Figure 1-29.)

Figure 1-31 Electron micrograph of *Bacillus subtilis* bacteriophage prepared by negative staining. Note that the stain brings out the surface details of the virus. Magnification is × 280,000. (Courtesy of Dr. A. K. Kleinschmidt.)

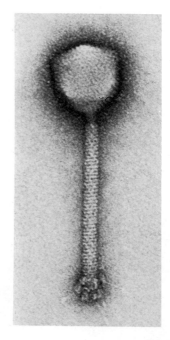

The visualization of small particles such as viruses, ribosomes, or membrane fragments may be accomplished by either of two means. It is possible to deposit the particles on a thin film of plastic and render them visible by *shadowing* with a heavy metal, such as platinum. The shadowing is done by evaporating metal from a hot wire in a vacuum chamber. Since the metal may come from one side only, it leaves a shadow behind the particle, which gives an indication of the three-dimensional structure of the material. The plating of metal on the particle also increases its size and, hence, its visibility in the electron micrograph (see Figure 1-20).

A second way to examine particles is by *negative staining*, in which particles are deposited on a film embedded in an electron-dense material (such as phosphotungstate). The particles then show up as "holes" in the dense film, since they are much more transparent to the electron beam than the surrounding stain (see Figure 1-30). An example of a negatively stained preparation is seen in Figure 1-31. The main advantage of this technique is its ability to reveal details of the surface structure of particles; a secondary advantage is the great ease and rapidity of preparing negatively stained material, so that only a few minutes need elapse between obtaining the material from its source and taking pictures.

In addition, the development of *scanning electron microscopes* during the last few years has led to increased ability to examine the surfaces of biological material. In such instruments, the electron beam scans the sample in a manner analogous to the electron beam in a television tube. Indeed, the picture thus obtained may be electronically displayed on a cathode-ray tube and has the advantage of presenting a highly three-dimensional view of the surface studied.

Figure 1-32 Use of scanning electron microscopy, employing red blood cells. (a) A normal red blood cell, magnification × 6,600. (b) A field of cells at the verge of hemolysis (osmotic lysis), magnification × 1,650. (Courtesy of Mr. Peter Morse.)

(a)   (b)

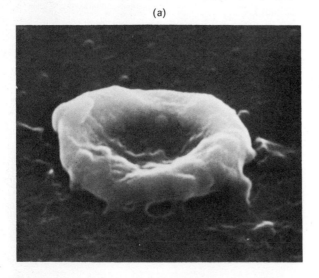
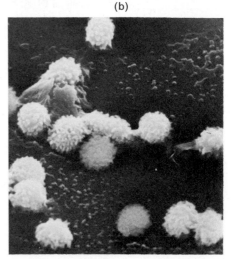

Figure 1-33   Electron micrograph of a yeast cell prepared by freeze-etching. Note that the nuclei exhibit a pitted appearance owing to the pores in the nuclear envelope. The arrows point to promitochondria, structures that precede mature mitochondria in yeast going from anaerobic to aerobic conditions. The bar indicates 1 micron. From R. S. Criddle et al., *J. Gen. Physiol.*, **54**, no. 1, pt. 2, 57 (1969). (Courtesy of Dr. G. Schatz.)

Finally, a technique called *freeze-etching* has recently been developed to provide unique opportunities to examine the surfaces of membranes. Material is rapidly frozen and then sectioned under conditions of extreme cold in a vacuum. Sectioning under these conditions cleaves cells along planes of inherent weakness, which usually means along the surfaces of membranes. In some instances, membranes can be shown to have been cleaved along the central plane of the bilayer—that is, to have been split down the middle. After sectioning, a mixture of platinum and carbon is evaporated onto the surface, and the original cellular material is digested away to leave a carbon–platinum "negative" of the original structure. This negative (or etching) may then be examined with the electron microscope, with resulting micrographs that are both revealing and remarkably beautiful (see Figure 1-33).

technique
cf. w'73
notes

47

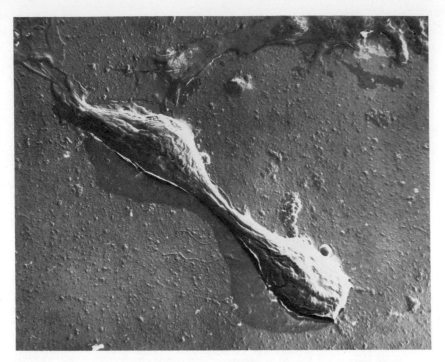

Figure 1-34 An additional approach in electron microscopy is to make a surface cast of an object and then to photograph the cast itself. In this case, a dividing myoblast, grown in culture, was prepared in this fashion. Magnification is about × 3000. (Courtesy of Martin Friedlander.)

Suggested Reading

Books

Brachet, J., *Biochemical Cytology*, Academic Press, New York, 1957.

Brachet, J., and A. E. Mirksy (eds.), *The Cell: Biochemistry, Physiology, Morphology*, vol. 2, Academic Press, New York, 1961.

DeRobertis, E. D. P., W. W. Nowinski, and F. A. Saez, *Cell Biology*, 4th ed., W. B. Saunders Co., Philadelphia, 1965.

Fawcett, D. W., *An Atlas of Fine Structure: The Cell, Its Organelles and Inclusions*, W. B. Saunders Co., Philadelphia, 1966.

Frey-Wyssling, A., and K. Mühlethaler, *Ultrastructural Plant Cytology*, American Elsevier Publishing Co., New York, 1965.

Grimstone, A. V., *The Electron Microscope in Biology*, St. Martin's Press, New York, 1968.

Gude, W. D., *Autoradiographic Techniques*, Prentice-Hall, Englewood Cliffs, N.J., 1967.

Haggis, G. H., *The Electron Microscope in Molecular Biology*, John Wiley & Sons, New York, 1967.

Haynes, R. H., and P. C. Hanawalt (eds.), *The Molecular Basis of Life: An Introduction to Molecular Biology*, W. H. Freeman and Co., San Francisco, 1968. (A collection of articles from *Scientific American*.)

Hurry, S. W., *The Microstructure of Cells*, Houghton Mifflin Co., Boston, 1966.

Jensen, W. A., *The Plant Cell*, Wadsworth Publishing Co., Belmont, Calif., 1964.

Kennedy, D. (ed.), *The Living Cell*, W. H. Freeman and Co., San Francisco, 1965. (A collection of articles from *Scientific American*.)

Ledbeter, M., and K. Porter, *Introduction to the Fine Structure of Plant Cells*, Springer-Verlag, New York, New York, 1970.

Novikoff, A. B., and E. Holtzman, *Cells and Organelles*, Holt, Rinehart and Winston, New York, 1970.

Pease, D. C. *Histological Techniques for Electron Microscopy*, 2nd ed., Academic Press, New York, 1964.

Porter, K., and M. A. Benneville, *An Introduction to the Fine Structure of Cells and Tissues*, Lea & Febiger, Philadelphia, 1963.

Rhodin, J. A., *An Atlas of Ultrastructure*, W. B. Saunders Co., Philadelphia, 1963.

Roodyn, D. B. (ed.), *Enzyme Cytology*, Academic Press, New York, 1967.

2 The role of water

In an age that is prepared to accept the reality of space exploration, consideration of possible life forms differing from familiar terrestrial ones has moved from the sphere of science fiction to that of serious consideration. Thus, considerable attention has been given to the proposition that, while life on this particular, possibly atypical, planet is based on a central role of water, such a basis is fortuitous. It is judged that water is only one of a number of compounds that could serve as solvent and substratum for evolving living systems and that one might well consider other possibilities such as ammonia and hydrogen sulfide. As our devices for direct measurement of living forms have yet to penetrate very remote regions of space and since the likelihood of life on other planets in this solar system does not appear to be high, there are, as yet, no data bearing on such possibilities. Such considerations should, however, stimulate us to reflect on the role of water in our world and, especially, to consider how its molecular properties render it unique and fit to serve as the basis for cellular (and all) life. The remainder of this chapter will be concerned with certain properties of water and it will be seen that this most familiar of chemical compounds is, in fact, not well understood at all. The poverty of our knowledge concerning the structure of water will be most dramatically underscored when we consider the possible occurrence of polymers of the water molecule, about which there is almost complete disagreement among workers in the field.

The special role of water in the processes of life has been taken seriously for a long time. Perhaps its full impact became apparent with the publication of *The Fitness of the Environment* by L. J. Henderson in 1925, a book of immense wisdom that is unique in science, in that it remains current (and exciting) almost a half-century after initial publication. In it, Henderson managed the almost impossible feat of looking at water as if it were not the most familiar substance in the world. He noted that many of its properties were anything but commonplace and that it appeared uniquely suited to its role in living systems. To put it more correctly, water appears to have been such a central determinant in the evolution of life on this earth that those of us who are products of that evolution find it difficult to conceive of life on any other basis. So great is our natural prejudice in this connection that it can be feared that if we ever encounter life on any other basis, we may fail to recognize it as life at all.

For the purposes of the present discussion, we may say that the role of water in life exists in three contexts. In the first place, water forms the natural environment of most cells, whether they be marine bacteria or human brain cells. This is not surprising, since water passes through biological membranes with ease, so that a cell, if not surrounded by water, would quickly dry out. Indeed, cells that really exist in a dry environment are highly modified for the purpose—a good example being a bacterial spore.

In the second place, cells contain water, which, to a first approximation, could be said to form the solvent for the chemical and physical events that go on there. The reasons why this is only a first approximation have to do with the importance of interfaces within cells, as we shall see. Finally, water forms an interfacial surface for certain exchanges that occur between living and nonliving regions of space. For instance, passage of oxygen from the atmosphere into a tissue of a multicellular animal (such as you) must occur through the "wet" interface provided by the lungs. Likewise, the entry of carbon dioxide into the photosynthetic tissue of green leaves is facilitated by the water constantly being passed into the leaves from below and evaporates (*transpires*) into the atmosphere. For all these reasons, it is not odd that water is, by far, the most abundant compound in living cells.

The structure of liquid water

Although, as we shall see, the detailed structure of liquid water is not yet completely understood, it is clear that the central property of the water molecule is its electrical polarity; that is, each water molecule possesses a region of electrical charge differing from the remainder. In fact, examination of Figure 2-1 reveals that the oxygen portion of water is negatively charged with respect to the two hydrogens. For this reason, water, although bearing no net electrical charge, behaves as an electrical dipole and therefore can interact both with other dipolar molecules and with an external electrical field. In pure water at temperatures from 0 to 100°C, as well as in dilute aqueous solutions, the most common interaction between dipolar molecules is that between two water molecules themselves. Thus, one can think of electrostatic attraction between the unlike charges of the oxygen of one water molecule and a hydrogen of another. Ignoring, for the time being, the geometry of this interaction, it should be clear that in a population of water molecules, at a given instant, many would be found to be associated together.

Many properties of water reflect the dipole association between molecules. For example, water exhibits a high surface tension, which indicates strong intermolecular forces. Likewise, when compared with other

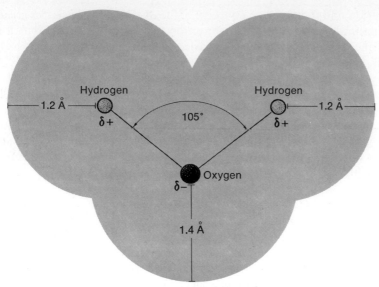

Figure 2-1 Construction of the water molecule. Note the van der Waals radii of oxygen (1.4 Å) and hydrogen (1.2 Å). The net result of the charge associated with the oxygen and hydrogen atoms is a dipole moment oriented in a vertical direction (in the figure).

compounds of generally similar size, water exhibits a very high boiling point, melting point, and heat of vaporization, all of which suggest that considerable heat input is required to overcome the strong forces between molecules.

Hydrogen bonds

The electrostatic binding that we have described between positive and negative regions of water molecules occurs with a specific geometry, which originates in the geometry of the water molecule itself. Thus, in a case where thermal motion does not act to randomize positions of the

Table 2-1 Heat of vaporization and surface tension of some liquids

Liquid	Surface tension* (ergs/cm²)	Heat of vaporization (cal/g)
Water	72	585
Acetic acid	28	97
Benzene	29	94
Ethanol	22	204
n-Hexane	18	79

* Measured at 20°C.

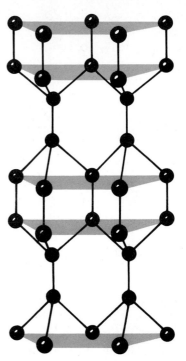

Figure 2-2 Configuration of water molecules in ice. The circles denote the locations of oxygen atoms; lines connecting them represent the axes of hydrogen bonds.

individual molecules, water molecules are related to each other in a precise lattice, or crystalline structure. This organization accounts for the crystalline nature of ice, whose structure is illustrated in Figure 2-2. Figure 2-3 indicates that the bond distances and angles between hydrogens and adjacent oxygen atoms are constant and, indeed, it turns out that there is a specific, measurable energy required to break these bonds of 4.5 kilocalories per mole. This is in contrast with bond energies of most covalent bonds (that is, bonds resulting from shared electrons), which are at least one order of magnitude higher. These relatively low-energy bonds between a positive hydrogen atom and some negative center, such as an oxygen atom, are called *hydrogen bonds*, and they not only account for the characteristic structure of ice but occur in a number of molecular systems of biological importance.

Hydrogen bonds are of great significance in liquid water, for they represent the intermolecular force that gives water many of its properties. Thus, while frozen water is a lattice of linked molecules, liquid water contains much of the same lattice structure, which, however, is continuously forming and breaking down. If one were able to obtain an instantaneous picture of the molecules in a sample of liquid water, one would find that over half the total number of molecules form very short-lived regions of pseudo-ice—with, however, very short half-lives. Such an association between water molecules forms not only in ice and liquid water but even in water vapor, although with a very much shorter half-life.

It should be added that hydrogen bonds are of great significance in a number of cases of biological importance in addition to that of water. Figure 2-4 illustrates several of these and it is clear that a hydrogen bond

53

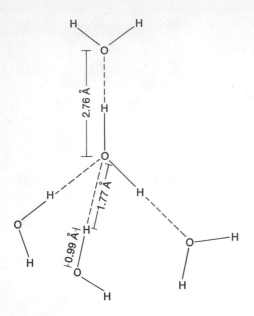

Figure 2-3 Hydrogen bonding between water molecules in ice. Note that the tetrahedral arrangement about the central molecule results in the characteristic lattice structure shown in Figure 2-2.

can form between hydrogen and one of several negative centers, notable examples of which are oxygen and nitrogen atoms. It will be seen in Chapter 3 that such hydrogen bonds are of great significance in determining the overall structure (conformation) of proteins and, in a later chapter, they will be seen to play a crucial role in genetics through the important phenomenon of specific pairing between the nucleotide bases of which DNA is constructed. In these cases, hydrogen bonds, although singly very weak, can contribute to structure in an additive fashion. Thus, a protein molecule or a nucleic acid molecule may include hundreds of

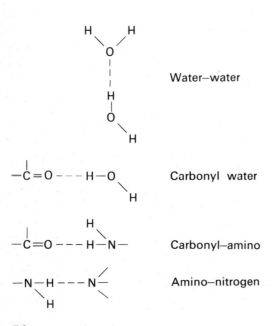

Figure 2-4 Examples of hydrogen bonds in biological systems.

Water–water

Carbonyl water

Carbonyl–amino

Amino–nitrogen

hydrogen bonds, which, together, may lead to the maintenance of a tight and stable configuration.

The structure of water near charged surfaces

Inasmuch as water molecules are, themselves, dipoles, they react to the presence of an external charge in a specific way, aligning themselves in its field. This behavior is of interest here in two connections. For one thing, a cation, such as the calcium ion, which bears a positive charge is, when in solution, surrounded by a sphere (or *hydration shell*) of water molecules, each oriented with its negative oxygen region directed toward the center (see Figure 2-5). Different species of ions possess hydration shells of differing radii that are of great importance in determining a variety of properties, including rate of diffusion and the ability to penetrate biological membranes. In general, large shells tend to retard transport if other factors are equal. Likewise, the size of hydration shells influences the ability of cations to bind to protein surfaces and to be bound to the circular molecules (ionophoretic antibiotics) that can serve as ion carriers.

A hydration layer of oriented water forms not only in the region surrounding a relatively small ion, such as potassium or calcium, but also in the region of a larger charged object, such as a protein molecule or, indeed, a cellular membrane. In the case of a protein molecule, the properties of the protein may reflect the differing degrees of hydration under different conditions. For instance, an enzyme (which is a protein) will bear a varying surface charge depending upon pH (as we shall see), and the charge will influence the degree of hydration. In turn, the amount of water bound to the enzyme surface in the region near the active, catalytic site can influence the ability of a small molecule to approach the active region, and, therefore, the activity of the enzyme.

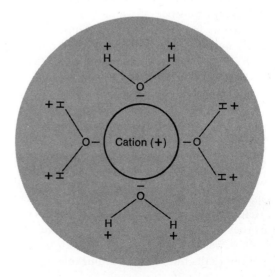

Figure 2-5 Hydration shell around a univalent cation. Note that if the central ion were an ion, the water molecules would be oriented with their dipoles in the opposite direction—that is, with the positive region directed toward the negative anion.

Table 2-2 Radii and degree of hydration of some biologically important ions

Ion	Crystal radius (Å)	Hydrated radius (Å)	Number of water molecules per ion
Cations			
Li^+	0.7	2.5	6.0
Na^+	1.0	2.3	4.6
K^+	1.3	1.9	2.3
Rb^+	1.5	1.8	1.7
Anions			
Cl^-	1.8	—	2.9
I^-	2.2	—	0

Likewise, by virtue of hydrogen bonding, water in the vicinity of a cellular membrane (which will be seen to have many charged centers) is thought to form a highly oriented layer, often a number of molecules thick. The formation of a sort of "ice" in the region of a membrane cannot fail to influence reactions and events that take place in or on the membrane surface. For example, passage of molecules across a membrane surface would be expected to require transit through the cohesive layer of highly oriented water, so that the properties of the molecule must somehow allow for its solubility in the relatively nonpolar membrane (containing much phospholipid) and also the highly polar water layer. Perhaps partly for this reason, ions with long side chains (such as salts of fatty acids) are often highly permanent; their ionic character would be expected to promote solubility in the water layer, whereas the presence of side chains contributes solubility in the membrane itself.

An additional consequence of the orientation of water about the membranes of cells and cell organelles is related to the structure of cells. That is, the interiors of cells are often found to be filled with closely packed membranes, so that one can estimate that a sizable fraction of a cell's total number of water molecules is situated within a few tens of angstroms from a membrane. Thus, a significant proportion of the cell water may possess a greater or lesser degree of oriented structure, so that such ice-like water would become significant in determining cell properties.

Water as a solvent

We have noted that the chemical and physical activities that take place within the cell occur either at the protein–lipid membranes or in solution spaces within them. To understand the events that occur in solution, it is useful to consider some of the properties of water as a solvent, especially as they apply to the cellular scale of events.

It is well known that water is an unusually good solvent, dissolving not only salts but many nonionizable organic molecules as well. Although nonionizable, such molecules are generally quite polar—that is, like water itself, they have a nonuniform distribution of charge. This charged character accounts for solubility in water of both salts (which are completely polar in that their positive and negative regions dissociate from one another) as well as such polar, but nondissociable molecules as alcohols and acetone. It will be recalled that a salt, such as potassium chloride, exists, in solution, as the separate anion and cation.

$$KCl \rightleftharpoons K^+ + Cl^-$$

This dissociation is favored by the presence of water molecules, which, because of *their* polarity, form hydration shells around the ions. These shells, in turn, stabilize the ionic (dissociated) form of the salt and ensure that the equilibrium position of the reaction will exist far in the direction of the individual ions. Thus, it is seen that the solvent capability of water is a direct consequence of its polarity. This may be expressed somewhat differently by saying that water has a high *dielectric constant* when compared to a nonpolar solvent such as a hydrocarbon. The dielectric constant expresses the ability of a material to oppose the attraction of unlike charges. The force F between two point charges of opposite polarity (of which two ions represent a close approximation) is given in terms of the magnitude of the charges, z_1 and z_2, the distance between them, r, and the dielectric constant, D:

$$F = \frac{z_1 z_2}{Dr^2} \tag{2-1}$$

Thus, the water's high value of D implies a relatively small force between the two ions and a relatively slight tendency for them to recombine. The ability of water to minimize the force between two ions is a direct consequence of its own polarity, since one influence of the hydration shell is to neutralize part of the ion's charge. In other words, a part of the ion's charge is occupied with attracting the part of the water molecule that bears the opposite charge and is, therefore, unavailable for interaction with other ions.

Table 2-3 Dielectric constants of some liquids

Liquid	Dielectric constant
Hexane	1.9
Benzene	2.3
Toluene	2.4
Ether	4.3
Acetone	21.4
Ethanol	24.0
Water	80.0
Hydrogen cyanide	116.0

The dissociation of water and pH

Not only does water promote the formation of ion pairs in other molecules, but it forms them itself. The dissociation actually involves two molecules of water and leads to the production of ions according to the equation

$$2\ HOH \rightleftharpoons H_3O^+ + OH^-$$

The fact that this reaction occurs, as opposed to a simple unimolecular dissociation into H and OH^-, indicates that there is a high degree of association between separate molecules and that the proton is quite mobile, so that the exchange may take place as shown:

$$\underset{H}{\overset{H}{O-H\cdots O}} \rightleftharpoons \underset{H}{\overset{H^+}{H-O^-\cdots H-O}}$$

where a dashed line indicates hydrogen bonding.

Since it was once believed that the dissociation of water was a simple unimolecular process, yielding H^+ and OH^-, it is quite common to refer to the hydrogen ion content of water when H_3O^+ is the actual species. In fact, this represents no real difficulty, as H_3O^+ is itself a good proton donor and a dissociable water molecule may be regarded as a weak acid. For this reason H_3O^+ and H^+ may be used interchangeably.

It will be remembered that ions in aqueous solution are surrounded by a hydration shell of water molecules and, since H_3O^+ obviously is to be found in aqueous solution, it is no exception. Figure 2-6 indicates that, on the average, three water molecules surround each hydronium ion, thus stabilizing it and forming the equilibrium in the direction of the dissociated form.

Since water is capable of ion formation and since many of the cell components form ions in the aqueous part of the cell, the degree of ionization of such compounds and the related H^+ (and H_3O^+) concentration

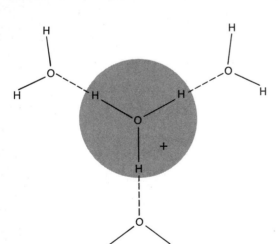

Figure 2-6 Hydronium ion surrounded by its hydration shell, consisting of (on the average) three water molecules. When hydration is taken into consideration, the formula of the ion becomes $(H_9O_4)^+$.

Table 2-4   The relation* between pH and
            concentration of H_3O^+ and OH^-

pH	(H_3O^+)	(OH^-)
1	10^{-1}	10^{-13}
3	10^{-3}	10^{-11}
5	10^{-5}	10^{-9}
7 (neutrality)	10^{-7}	10^{-7}
9	10^{-9}	10^{-5}
11	10^{-11}	10^{-3}
13	10^{-13}	10^{-1}

* Note that the values in this table reflect both the definition of pH (Equation 2-2) and an equilibrium constant of 1×10^{-14} for the dissociation of water at room temperature. Concentrations are molar.

of the cell are of considerable interest to the physiologist. Since the actual degree of ionization of water is quite small (the concentration of H_3O^+ in pure water is about 10^{-7} M), an exponential scale for expressing such concentration has been devised, according to which H_3O^+ concentration is expressed as pH:

$$pH = -\log_{10}(H_3O^+) \tag{2-2}$$

Thus, in a neutral solution at 25°C, the concentration of H_3O^+ equals that of OH^- and is 1×10^{-7}. The pH of this solution is given by

$$pH = -\log_{10}(1 \times 10^{-7}) = \log_{10}(10^7) = 7$$

The reader should satisfy himself that a solution at pH 6.0 has a H_3O^+ concentration 10 times as high as one at pH 7.0.

Acids and bases

The pH of a solution is the negative logarithm of the concentration of H_3O^+ and a measure of its acidity. (A low pH is that of an acidic solution, while a high pH reflects a basic one.) In the popular mind, acid is a matter of having a sour taste, a criterion that suffers somewhat for not being very quantitative and which has the additional defect of being rather impractical when applied to corrosive liquids, such as strong sulfuric acid. However, certain difficulties arise when one endeavors to be more explicit. For instance, the most useful practical definition of an acid for (biological) purposes is that an acid is a donor of H_3O^+ or, as we usually say, protons. A base is then a proton acceptor. An example of an acid would be hydrochloric acid, which donates protons on dissociation:

$$HCl \rightleftharpoons H^+ + Cl^-$$

In this case, HCl is a strong acid in that the reaction equilibrium favors virtually complete dissociation, so that the proton concentration is high.

At the risk of being somewhat tiresome, we should note that the H^+ is, as usual, equivalent to H_3O^+ by the reaction of H^+ with water. An example of a base would be ammonia, which is able to accept protons to become ammonium:

$$NH_3 + H^+ + OH^- \rightleftharpoons NH_4^+ + OH^-$$

Finally, a feature of this definition of acids and bases is that, if a proton is donated by one molecule (the acid), it must be accepted by something else (the base). Thus, in the reaction above, NH_3 is to serve as a base, and the acid is water. The relationship between acids and bases is perhaps more obvious in the case of acetic acid, which dissociates as follows:

$$CH_3COOH \rightleftharpoons H^+ + CH_3COO^-$$

In this instance, CH_3COOH is the acid, while CH_3COO^-, the acetate anion, is a base. The equilibrium constant for this reaction is given by the expression

$$K = \frac{(H^+)(CH_3COO^-)}{(CH_3COOH)} \tag{2-3}$$

and provides a measure of how readily the acid releases its protons. In the case of acetic acid the approximate value for K at $25°C$ is 1.7×10^{-5}. Since equilibrium constants for acids of interest range over many orders of magnitude, it has proved convenient to adopt a tactic similar to the logarithmic pH scale and to define the pK of an acid as

$$pK = -\log_{10} K \tag{2-4}$$

Thus, a strong acid is largely dissociated and therefore has a large equilibrium constant, K, and consequently a low pK. Examples of the pK's of several acids are given in Table 2-5.

Table 2-5   Some buffer systems of interest to biologists

Buffer	Acid form	Base form	pK	Remarks
Acetate	$-COOH$	$-COO^-$	4.73	
Borate	H_3BO_3	$B_4O_7^{2-}$	9.19	
Bicarbonate	H_2CO_3	HCO_3^-	6.1	Significant in regulation of pH of blood
Glycylglycine	$-COOH$	$-COO^-$	3.15	
Glycylglycine	$-NH_3^+$	$-NH_2$	8.25	
Phosphate	$H_2PO_4^-$	HPO_4^{2-}	6.7	Commonly used in biochemical studies
Tris(hydroxymethyl)-aminomethane	$\overset{NH_3^+}{HOH_2C\overset{\mid}{C}CH_2OH}$ $\underset{CH_2OH}{\mid}$	$\overset{NH_2}{HOH_2C\overset{\mid}{C}CH_2OH}$ $\underset{CH_2OH}{\mid}$	8.07	Commonly used in biochemical studies; basic solution does not absorb CO_2 from air, so that pH is stable; sometimes, however, toxic effects are noted

Table 2-6   The range of pH's found in the natural world*

Fluid	pH range
Human blood	7.3–7.5
Human stomach, contents	1–3
Intracellular fluid, human liver	6.9
Sea water	8.0
Peat bog	4.0
Cola	2.8

* The last entry should not be regarded as a recommendation.

An examination of Equation 2-3 will convince one that if the value for K is known for a particular compound, a certain relationship will exist between the concentration of protons and other components of the system. To be more precise, if one knows the value of K, which can be gotten from pK, the concentration of protons, and therefore the pH, can be obtained from the ratio of acid to base. Since one usually works in terms of pH and pK, it is of interest to relate these quantities; this may be done by using the *Henderson–Hasselbalch equation*:

$$pH = pK + \log_{10}\left[\frac{\text{proton acceptor}}{\text{proton donor}}\right] \qquad (2\text{-}5)$$

From this, one can obtain the pH of a solution if the pK and the ratio of base to acid are known. Likewise, one can obtain a value for the pK of a given acid–base system by measuring the pH at a known ratio of base to acid.

Since the Henderson–Hasselbalch equation is of great importance in the consideration of biological acid–base relationships, its derivation will now be given. Consider the dissociation of a weak acid,

$$HA \rightleftharpoons H + A^-$$

for which the equilibrium constant is

$$K = \frac{(H^+)(A^-)}{(HA)}$$

which is only a more general rendition of Equation 2-3. This may be solved for (H^+) to give

$$(H^+) = \frac{K(HA)}{(A^-)}$$

Then, if one takes the negative logarithm of both sides, we have

$$-\log_{10}(H^+) = -\log_{10} K - \log_{10}\frac{(HA)}{(A^-)}$$

61

From the definitions of pH and pK given earlier, this becomes

$$pH = pK - \log_{10}\frac{(HA)}{(A^-)}$$

which is usually written in the following equivalent form, obtained by changing the sign,

$$pH = pK + \log_{10}\frac{(A^-)}{(HA)} \tag{2-6}$$

where (A^-) and (HA) are the proton acceptors and donors, respectively, of Equation 2-5.

Buffers

Examination of Equation 2-6 shows that if one adds acid (or base) to a solution, the pH does not change in a linear manner, because it is proportional to the logarithm of the ratio of base to acid. If one (experimentally) adds acid to a solution and continuously records the pH, a curve such as that shown in Figure 2-7 is obtained. This is a titration curve and exhibits a shape similar to that shown for a variety of weak acids, differing, however, in the location of the midpoint. It is worth examining the shape of the curve in more detail. In the first place, there is a region where the curve is flat—that is, addition of more acid (or base) has relatively little effect on the pH. This is the region of greatest buffering capacity, buffering being simply the resistance of a solution to a change of pH on addition of acid or base. Since the pH at which a cellular process occurs is often critical, it is of interest that the interior of cells is highly buffered and, therefore, quite stable as to pH.

When one reflects on the nature of buffering, it becomes (or should become) clear that the main requirement for an effective buffer is that the molecule should be only partially ionized at a given pH. Thus, if there is

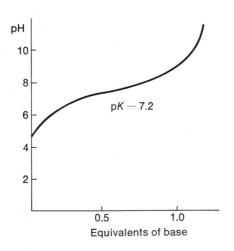

Figure 2-7   Example of an acid–base titration curve. In this instance, the acid is $H_2PO_4^-$, which is titrated to form HPO_4^{2-}. This is the system employed in phosphate buffers, often used in biological studies. Note that the curve is flattest in the region of the pK, where, therefore, the buffering power is greatest.

a reservoir of un-ionized acid, the addition of more protons, say in the form of a strong acid, will change the balance between the ionized and un-ionized forms but will produce relatively little alteration in the measured pH of the solution. This requirement for partial ionization at pH's where buffering occurs leads to the observation that buffers must be salts of weak acids, that is, mixtures of acids, protons, and the corresponding bases. Thus, a potassium phosphate buffer could be made up of a mixture of K_2HPO_4 and a strong acid, such as HCl, to bring it to the desired pH, but, whatever acid employed, the actual buffering system would be described by the following partial equation:

$$KH_2PO_4^- \rightleftharpoons H^+ + K_2HPO_4^{2-}$$

    (Acid)   (Base)

The ability of this system to absorb protons with relatively little change in pH reflects the lack of effect of protons on the equilibrium of this reaction. Added protons drive the reaction toward the left and are, themselves, lost as free entities, since they then form part of the un-ionized acid KH_2PO_4.

Further examination of the titration curve shown in Figure 2-7 indicates that the ability of a weak acid to exert a buffering effect on a solution varies with the pH. In fact, the buffering power is greatest where the curve is flattest, since that is where the balance between ionized and un-ionized forms is most equal. This region is of some special interest, since at its exact midpoint the pH can be shown to exactly equal the pK for that particular buffer. This is clear from the following argument. At the midpoint, the amount of acid and base are equal. Equation 2-6,

$$pH = pK + \log_{10} \frac{(A^-)}{(HA)}$$

becomes

$$pH = pK + \log_{10} 1$$

since $(A^-) = (HA)$. The log of 1 is, of course, 0 and so

$$pH = pK$$

Thus, the pK of a buffer system can be estimated from the titration curve, although the exact computation can be somewhat difficult, owing to difficulties in a precise location of the midpoint.

Buffers are of importance to the cell physiologist in two senses. In the first place, the cell is highly buffered and the mechanism by which this is accomplished is fundamental in understanding cell function. In fact, there is a wealth of weak acids available to the cell in the form of the dicarboxylic acids of metabolism, amino acids and their polymers, peptides and proteins, and a number of organic and inorganic phosphates. Since cells exist normally at a pH near 7 (neutrality), weak acids with pK's in the vicinity of 7 are of greatest utility. It should also be noted that a molecule with more than one ionizable group will exhibit different

63

pK's for each. We shall see later that this is the case with amino acids (and proteins).

Buffers are also important in a second, somewhat more practical sense. When one endeavors to measure a biological event outside the cell, control of pH is of great importance. For example, the rate of an enzyme-catalyzed reaction depends upon the pH at which it is measured, so that the use of buffers in reaction mixtures is imperative. The choice of buffers depends, among other things, on the pH to be maintained, so that a knowledge of the pK of buffer systems is of obvious interest. Table 2-5 gives a number of buffer systems that are of frequent use in biological studies, together with their pK's. The final selection of a buffer must include consideration of pK and of such specific matters as low toxicity and inability to react chemically with any of the components of the system being measured. For example, tris-HCl buffer, which is useful, in many cases, inhibits certain reactions, including some related to photosynthesis, possibly by binding certain required cations.

The Lewis definition of acids

In the preceding discussion of acids and bases we have regarded acids as proton (hydronium) donors and bases as acceptors. This view is of great practical value since it describes most acid-base events in dilute aqueous solutions and such solutions are of prime interest to physiologists. Moreover, the proton-donor theory of acids leads to the possibility of predictions and calculations [such as those arrived at through the Henderson–Hasselbalch equation (see Equation 2-5)], which are of the greatest importance to biologists. However, the only defect in this point of view seems to be its lack of generality; there are acid-base systems that do not fit into the proton donor-acceptor framework for the very good reason that they do not involve protons at all. For example, boron trichloride reacts readily with bases in nonaqueous solvents but clearly could not be said to serve as a proton donor:

$$BCl_3 + NH_3 \rightleftharpoons H_3N—BCl_3$$

To account for reactions of this general sort, G. N. Lewis devised a more general formulation of acids and bases based on the donation and acceptance of electrons rather than protons. In the example given, the *acid*, BCl_3, is an *electron acceptor*, while the *base*, ammonia, is the *donor*. The reader will note that this theory stresses a close similarity, almost an identity, between an acid and an oxidizing agent as well as between a base and a reducing agent. This view is more general than the proton donation–acceptance one but includes that view within itself. Note, for instance, that, as far as charge is concerned, the donation of a positive proton is quite the same as the acceptance of a negative electron. The choice as to whether we look at electrons or protons is one to be based on practicality and, to an extent, habit. Thus, most of the discussions to follow will consider acids as proton donors. However, at several points in the remainder of this book, the reader will note that there is fundamental disagreement among physiologists as to whether protons or electrons should be regarded as primary to certain processes, such as, for example, energy coupling in the vicinity of membranes. Perhaps the most important message of the Lewis theory, in this context, is the formal equivalence that may render some of the debate merely semantic.

Intracellular pH

Finally, it is worth considering how one might gain some idea of the intracellular pH or even the pH within an organelle, such as a mitochondrion or chloroplast. The technical aspects of such investigations are touched upon in a later chapter; here we shall merely consider how pH is measured *in principle* and whether such measurements are suitable to gross and cellular scales of events.

In the laboratory, pH is normally measured by means of an electrode (indicator dyes or dye-impregnated pH papers are rarely employed). Although the least ambiguous electrode system for measuring pH is the *hydrogen electrode*, which includes a platinum electrode and an atmosphere of hydrogen gas for rather obvious, practical reasons, the more common *glass electrode* is used. The glass electrode consists of a thin membrane of low-resistance glass, which is permeable to hydronium ions—but in the range of pH in which it operates, not to other ions that might interfere. Within the membrane is a solution of known pH in which is immersed a reference electrode. The electrode measures, in effect, the difference of pH on the two sides of the glass membrane, which is transformed into a potential (in millivolts) by suitable amplification. The main features of such a system are illustrated in Figure 2-8. Such pH meters are presently available, spanning a wide range of sensitivity, so that the full

Figure 2-8 Determination of pH with a glass electrode. A potential is developed across the glass membrane owing to different hydronium ion concentrations on either side. This potential is sensed by the Ag–AgCl electrode dipping into the inner solution. The use of the glass membrane depends on its selective permeability for hydronium ion and on the inclusion of a standard reference electrode (normally Hg–HgCl or "calomel"), which is in contact with the external solution by means of a salt bridge and which provides a stable reference potential against which the pH-generated potential may be measured.

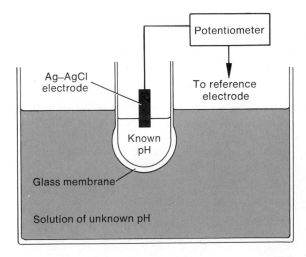

*Table 2-7   Properties of some useful pH
indicators*

Range	Indicator	pK of indicator
4.8–6.4	chlorphenol red	6.0
5.8–7.0	dimethyloxazolidinedione	6.1
5.8–6.8	bromocresol purple	6.3
6.0–7.6	bromothymol blue	7.0
6.8–8.0	phenol red	7.9
7.2–8.8	cresol red	8.3

measurement scale runs from 14 pH units in the simpler devices to as
little as 0.05 pH unit in the more sensitive ones.

While pH indicators are seldom of much use at the gross (laboratory)
scale, they are often used to estimate pH within cells and tissues, as they
can be injected into an isolated region and then observed with a visual
microscope or some other optical means. The essence of a pH indicator
is that it is a weak acid (buffer) whose ionized and un-ionized forms have
different colors and, therefore, differing absorbance spectra. Properties of
some indicators are given in Table 2-7. Since pH indicators are by necessity at least weak buffers, they must be used at sufficiently low concentrations so as to avoid altering the pH behavior of the system being studied.
For this reason, the practical limitation on indicators is that they must be
useful at low concentrations—that is, a small amount must produce a
large color change. In other words, they must exhibit a large *molar absorbancy index*,[1] which relates the absorbancy of a solution to its molar
concentration.

There is yet one additional consideration regarding the measurement of
cellular pH and this relates to the dimensions of cells and cell organelles.
The pH of a solution is a concept that springs directly from a consideration
of certain equilibria, such as those between water and its ionized forms,

$$2 \ H_2O \ \rightleftharpoons \ H_3O^+ + OH^-$$

or those involving certain other acids (water, in this instance, being a very
weak acid), such as acetic acid,

$$CH_3COOH \ \rightleftharpoons \ H^+ + CH_3COO^-$$

In such cases, one is invariably interested in the equilibrium constant, K,
for the reaction, and K, for reasons developed in Chapter 3, is a thermo-

---

[1] The molar absorbancy index, a, is defined as the ratio of absorbancy to concentration,
so that a 1 M solution measured across 1 centimeter will exhibit an absorbance equal to a.
To obtain concentration of a solution, divide the absorbancy by a since

$$a = \frac{A}{Ml}$$

where A is the measured absorbancy, M is the concentration in moles per liter, and l is the
length of the light path through the solution in centimeters.

dynamic quantity. Being thermodynamic in nature, K has meaning only when one considers a population of molecules large enough to be statistically manageable, for the simple reason that thermodynamics is a statistical description of how molecules and other small objects behave. For this reason, one cannot consider a thermodynamic description of a single molecule, or a few molecules, as thermodynamic arguments lose their force when this scale is reached. If one has a population of one or two molecules, the concept of equilibrium constant (or of equilibrium, itself) is devoid of meaning; the pH of a solution containing two protons would not only not be measurable, it would be meaningless. Our reason for laboring this point is that it becomes quite relevant when one speaks of the pH inside such a small chamber as a mitochondrion. It has been estimated that at a neutral pH a smallish mitochondrion would contain, on the average, something rather less than one free proton per mitochondrion. Clearly, if one is to think of intramitochondrial (or any intraorganelle) pH, and many, nowadays, have a strong desire to do so, then one, by necessity, means something quite different from the pH of a bucket of water. One is partially saved from conceptual disaster by thinking of a population of mitochondria large enough for the mitochondria themselves to be statistically significant, but, in any case, a good deal of caution is advised.

Suggested Reading

Books
Dick, D. A. T., *Cell Water*, Butterworth & Co., London, 1966.
Edsall, J. T., and J. Wyman, *Biophysical Chemistry*, vol. 1, Academic Press, New York, 1958.
Eisenberg, D., and W. Kauzmann, *The Structure and Properties of Water*, Oxford University Press, New York, 1969.
Hendrickson, J. B., *The Molecules of Nature*, W. A. Benjamin, New York, 1965.
Kavanau, J. L., *Structure and Function in Biological Membranes*, Holden-Day, San Francisco, 1965, p. 170.
Keosian, J., *The Origin of Life*, 2nd ed., Van Nostrand Reinhold Co., New York, 1968.
Speakman, J. C., *Molecules*, McGraw-Hill Book Co., New York, 1966. (Excellent paperback on the structure and properties of molecules in relation to biology.)

Articles
Klotz, I. M., "Water," in *Horizons in Biochemistry*, M. Kasha and B. Pullman (eds.), Academic Press, New York, 1962.
Palade, G. E., "The organization of living matter," in *The Scientific Endeavor*, Rockefeller University Press, New York, 1964, pp. 179–203.
Wald, G., "The origins of life," in *The Scientific Endeavor*, Rockefeller University Press, New York, 1964, pp. 113–34.
Whipple, H. E. (ed.), "Forms of water in biological systems," *Ann. N.Y. Acad. Sci.*, **125**, 249 (1965).

3 Energy and control

Cells do things. They move, either internally (the streaming of cytoplasm muscle contraction) or as a unit, through their environment. Cells also carry on a number of less obvious activities, such as the transfer of material through membranes, often against a gradient in concentration, the production of heat, the maintenance of an electrical potential, and the synthesis of an immense number of chemical compounds. Cells must exert themselves to get ahead in the world or, as we shall see, even to maintain the status quo. Exertion means the expenditure of energy, and it seems fair (and necessary) to ask where this energy comes from and how cells apply it to perform such wonders. However, to ask the question on anything but a silly level, it is first necessary to take a serious look at the whole problem of energy, to ask what it is and what forms it can take.

In the life of the cell, the type of energy that is most important is *chemical energy*. It is true that photosynthetic organisms (and therefore the whole living world) obtain energy in the form of radiation, but after the primary absorption event, radiant energy is quickly converted to a chemical form. There are a number of ways in which energy may be transferred, but the cell uses energy of chemical compounds, or, more properly, chemical bonds, as its currency. Therefore, we begin with a consideration of energy as related to chemical reactions and will expand the discussion to other forms of energy as the need arises.

Consider the chemical reaction

A $\rightleftharpoons$ B

To find out if the reaction is really possible, we begin by throwing in some A and either looking for B to appear or for A to be consumed. There are only two possibilities: A is converted to B or it is not. If it is, we say that the reaction (A $\rightleftharpoons$ B) is spontaneous. But, in deciding whether a reaction is spontaneous, we must include an additional dimension in the reaction scheme, energy.

A spontaneous reaction is an energy-yielding reaction in the sense that it may be used to drive one that is not by itself spontaneous. Energy that is yielded by a spontaneous reaction may also be partly in the form of heat, but this is by no means necessarily the case. Some reactions—including many biological ones—are able to drive other reactions but produce little heat change. Thus we might classify all reactions as energy-yielding and energy-requiring, which is to say spontaneous and not spontaneous. We would write the spontaneous equation involving A and B as

$$A \rightleftharpoons B + \Delta \text{ energy}$$

which is to say that in such a case energy is liberated. A reaction that is not spontaneous in a thermodynamic sense requires energy to make it go. Note that there are two sides to the coin of spontaneity: If a reaction is spontaneous in one direction, it most surely is not in the other. If energy is produced going in one direction, it must be consumed if the reaction proceeds in the other.

Note that we do not specify the form in which the energy might exist. In fact, it might consist of mechanical and heat components as well as in other forms. As an example of mechanical energy, one might cite the coupling of chemical reactions to the performance of work, such as gas volume changes against a given pressure or muscular contraction.

At this point it is necessary to observe that there are two senses in which a reaction may or may not proceed. It is possible to write a vast number of reactions that can never work, no matter how much energy may be available. These reactions are forbidden for the simple reason that no mechanism exists whereby the transformations may occur. Molecules must obey not only the laws of thermodynamics but also such additional restrictions as rules of valence and geometry. The alchemists were notoriously unlucky with such forbidden reactions, finding gold exceedingly hard to produce from base metals, even with the application of considerable energy in the form of heat or good will. Thus the study of energy relationships (thermodynamics) gives information about what is possible but can say nothing about mechanisms or about the rates at which things happen (which may very well be zero).

Free energy

Now, there are a number of ways to express energy, some of which are more applicable to a steam engine than to a reaction within the cell. We shall give the cell first priority and begin by defining a measure of energy that is most suitable to the study of ordinary chemical reactions, the *free energy* of a reaction. Free energy is usually denoted by G (for J. W. Gibbs, one of the founders of modern thermodynamics), and any change in free energy is given as ΔG. So our spontaneous reaction should be written

$$A \rightleftharpoons B \qquad \Delta G > 0$$

For the present, we can define free energy simply as that energy that determines spontaneity. The ordinary unit of free energy is the calorie, and just as chemical reactions are given in terms of moles, free energy is generally given in terms of calories per mole.

Cell biologists are fortunate in a number of respects (including the interesting nature of the field), and one additional bit of good luck comes to them when they consider energy and energy transformations. The fact is that most cellular reactions take place in solution under more or less constant conditions of temperature, pressure, and volume. Exceptions to this rule would tend to lead to the destruction of cells. This constancy of conditions enables us to ignore a great deal of complication in thermodynamics and make some very pleasant simplifications in our consideration of energy. The treatment that follows will be quite valid for reactions within cells but will be hopelessly inadequate for the study of reactions that might be found inside a volcano or within the cylinder of an automobile.

First, we might dissect the free-energy change of a reaction a bit and see how it is put together. We find, for example, that free energy may be expressed as follows:

Free energy = Heat component − order-disorder component

The precise nature of these two components may be stated in the more explicit equation

$$\Delta G = \Delta H - T \, \Delta S$$

In this equation ΔH represents the change in a quantity called *enthalpy*, which under conditions of constant pressure (which is all that we are interested in) is identical to the change in heat.

It is worth saying a bit more about *heat* (as represented by enthalpy in our discussion). Heat is one of the concepts of thermodynamics that suffers from being too familiar, an everyday idea that we are often unable to define because of our certainty about it. For our purposes, heat may be defined as that form of energy that is able to raise the temperature (the molecular kinetic energy) of material. Thus a unit of heat is the calorie, which is the amount of heat required to raise the temperature of 1 gram of water 1 degree Celsius, provided no other energy is exchanged. When one measures the heat absorbed or evolved in a chemical reaction, one detects only the ΔH term of the total free energy. Thus an energy-yielding (spontaneous) reaction might well produce a very low heat of reaction, provided the other term is sufficiently great. It is clearly impossible to predict the free energy of reactions by observations of ΔH alone.

The order-disorder component requires further discussion. First, in the equation above, T represents the absolute temperature at which the reaction occurs, and ΔS denotes the change in a quantity called *entropy*, which is of great interest to the biologist. Second, since entropy is not measured directly (there is no such thing as an entropy meter), its nature is not as obvious as that of, say, temperature or pressure, which may be sensed directly. Because entropy may have a somewhat unfamiliar ring,

70

we will define it in several ways and trust that the combination of them will be instructive.

Definition 1 *Entropy is a measure of the disorder in a given situation.* For example, a chemical or physical change might well result in a change in the orderliness of the reactants. When one dissolves salt in water, the ions, which were in an extremely orderly state in the precise crystal lattice of the solid, in solution begin flopping about in a quite random manner. In such a case, the event of dissolving would be attended by a large positive change in entropy, ΔS, and in fact any real process, when viewed in a sufficiently broad context, is seen to result in a greater or lesser increase in entropy, which is to say disorder. In this connection, we shall content ourselves (for the moment) with noting that the cell is an extremely orderly configuration of material, so that its formation (or dissolution) should likewise produce a significant ΔS.

Definition 2 *Entropy is a measure of the essential irreversibility of events in the real world.* At this point it is necessary to back down partially on the earlier statement to the effect that reactions tend to be reversible but not fully so. That is, one can never quite get back to the starting point. This may be illustrated by considering a reaction $A \rightleftharpoons B$ and its reverse, $B \rightleftharpoons A$, as a kind of loop:

A⟨⟩B

From everything we have said, it should be possible to go from A to B with, for example, energy production and then use that energy to go back to A by the other route. There should be no reason why, once started, the reaction could not cycle through A to B endlessly, but in fact this is not observed. At each stage of the cycle, one finds that a little energy becomes unavailable and that, as time runs on, more and more is lost to the system until the whole process grinds to a stop. The fact that energy becomes unavailable in any real process is another way of saying that any real process tends to move in the direction of disorder, and that energy loss in this connection is irrevocable. The energy change in any process related to the loss of order is given by the term $T \Delta S$, and the fact that its sign is minus in the definition of ΔG given above is a reflection of its character as a decline in available energy.

Since entropy is central to an understanding of many energy matters, it is vital to be able to obtain a measure of it. The definition that follows is both the most formal one and the one that makes its evaluation possible.

Definition 3 *In a reversible process taking place at constant pressure and absolute temperature T and involving an enthalpy change of dH, the entropy change is given by*

$$\Delta S = \int \frac{dH}{T}$$

71

The essential point is that the entropy contribution to an energy change may be obtained in reversible situations by the relatively simple expedient of measuring a heat change at a known temperature. Thus, as real processes approach reversibility, the entropy associated with them becomes increasingly accessible to precise mathematical treatment. In a truly reversible cyclic process (such as the hypothetical one described above), entropy for the entire pathway must be zero, while that associated with any segment of the path is given by the equation above. Clearly, in such a case, there must be regions where a decline in entropy occurs to balance increases found in others.

Definition 4 A last definition of entropy may be given in terms of probability. We have seen that real reactions are fundamentally irreversible and that disorder (and therefore, entropy) tends to increase in any real process. This is to say that, in the world, disorder is more probable than order and that, by the definition of "probable," things tend to approach the most probable situation. In the same way, the reason that reactions approach an equilibrium value is that it is the most probable configuration for the reactants and products to assume. A reaction near equilibrium (with its high probability) also becomes increasingly reversible and, for this reason, at equilibrium, entropy production ceases:

$$\frac{dS}{dt} = 0$$

Thus *entropy is closely related to probability*, and it is possible to define it in such terms.

A statistical approach shows that there is a very simple relationship between the probability of a certain state and the entropy change involved in getting there:

$$\Delta S = K \log_e \Delta P$$

where P is the probability of the state, $\log_e$ is the natural log (to the base e, or 2.718), and K a proportionality constant. In this relation it is implied that a very large change in probability, such as that from a structured crystal to a much more random solution, will lead to a large entropy change.

Anyone who has looked at electron micrographs of cells will conclude that the cell is highly improbable (orderly) and that the breakdown of the structure of the cell on death will be associated with a large increase in entropy. Since the disintegration of a cell involves a large energy loss to the $T\Delta S$ term of the free-energy expression, it follows that it takes a lot of free energy to make a cell to begin with.

Now, if there is a large energy requirement to achieve this orderly arrangement—the cell—it should also be said that energy is required just to keep a cell orderly and intact. In other words, cells require energy even when they are not growing or making more cells. They require a continuous influx of energy just to keep ahead. The reason for this should be clear from the preceding discussion. Whenever a cell does something, there is absorption of free energy, a portion of which is lost forever in the form of the $T\Delta S$ term (entropy). In a sense, order is being constantly siphoned off as cellular events take place, and the maintenance of the supremely orderly and therefore improbable cell requires a continuing uptake of energy throughout its life.

Free energy and equilibria

To see more clearly why free energy (and therefore entropy) is of interest to the biologist, let us return to the general reaction resulting in the transformation of A to B. We have said that the reaction is very nearly reversible and that one might study it going in either direction provided energy is available to drive it in the energetically unfavorable direction. This is a bit too simple, and to see what is going on we must introduce a new aspect of the reaction, its equilibrium. In this connection, it must be said that reactions seldom go all the way. A more likely situation would be for A to become 90 percent converted to B at the end of the reaction. If this were the case, we would also find that, by starting with B, the reaction would proceed only 10 percent of the way. In other words, there appears to be an equilibrium position, which a reaction will, in time, approach, no matter from which side. The position of the equilibrium is a property of any given reaction and has nothing to do with the rate at which the equilibrium is approached. In our simple case of $A \rightleftharpoons B$, the equilibrium position may be expressed by the equilibrium constant

$$K = \frac{(B)}{(A)}$$

where (A) and (B) denote the molar concentrations of A and B. If the reaction were more complicated, the equilibrium constant would arise in the same way as the concentrations of the products multiplied together divided by those of the reactants, so that the constant for the conversion of A + B to C + D would be

$$K = \frac{(C)(D)}{(A)(B)}$$

In the case where the stoichiometry differs from a ratio of one between reactants and products, such as in the equation $2\,A + B \rightleftharpoons 3\,C + D$, the equation becomes

$$K = \frac{(C)^3(D)}{(A)^2(B)}$$

It is obvious that if the equilibrium of this reaction is far in the direction of the products (C and D), the equilibrium constant will be a large number. It is also clear that when one defines K, it is necessary to specify the direction being considered, the choice being arbitrary. It is clear from the equations above that a value for K is transformed into its reciprocal by considering the reaction as proceeding in the opposite direction. We shall

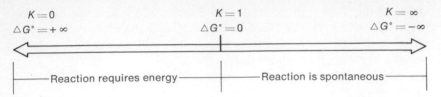

Figure 3-1 Relation between the equilibrium constant, K, and standard free-energy change, ΔG°.

see that the same transformation leads to a change in sign in the free-energy change. In other words, if a reaction is spontaneous in one direction, it cannot be so in the other.

Now, in the light of the equilibrium constant, it is clear that spontaneous reactions are those whose equilibrium lies far in the direction of the products. Under such conditions—when concentrations of reactants are far from their equilibrium values—reactions are able to yield energy rather than require it. This may be expressed by the sign of the free-energy change. In the reaction

$$A \rightleftharpoons B$$

if ΔG°, the free-energy change with the reactants in the standard state (at a $1\,M$ concentration at a specified temperature), is positive, the reaction requires added energy to make it go. A negative value, on the other hand, means that the reaction is far from equilibrium and, under those conditions, spontaneous and energy-yielding.

If the sign of ΔG° tells whether the reaction is spontaneous or not, the magnitude of ΔG° should tell us something about the actual position of the equilibrium. This is the case, and a simple equation gives the relationship between the two as

$$-\Delta G^\circ = RT \log_e K$$

where ΔG° is again the change in free energy under standard conditions, R is the gas constant,[1] T the absolute temperature, and K the equilibrium

[1] $R = 1.98$ calories per mole-degree.

Table 3-1   The relation between equilibrium constant and free-energy change: an example

K	ΔG° (kcal)
10^{-3}	$+4.1$
1	0
10^3	-4.1

constant of the reaction. Since R and T are both positive, it is clear that when $\Delta G°$ is positive, $\log_e K$ must be negative and therefore K must be between 0 and 1. When $\Delta G°$ is 0, since neither R nor T can be 0, $\log_e K$ must be 0 and therefore K must equal 1. Finally, when $\Delta G°$ is negative, the same argument leads to the conclusion that K must be greater than 1 (see Figure 3-1).

Many of the reactions of life are unfavorably situated with regard to their equilibria, so that if they are to proceed, they must be linked to spontaneous reactions. This sort of linkage may be given, in the simplest case, as one reaction pulling another by removing products. If the reaction

$$A \rightleftharpoons B$$

is not spontaneous (if its equilibrium is far to the left), any spontaneous reaction resulting in the removal of B will tend to pull the first. For example,

$$B \rightleftharpoons C$$

may proceed well and be far removed from equilibrium. If this is so, the net reaction

$$A \rightleftharpoons B \rightleftharpoons C$$

will go nicely. It should be intuitively clear that the progress of the whole sequence to the right will depend on whether the equilibrium of $B \rightleftharpoons C$ is farther to the right than that of $A \rightleftharpoons B$ is to the left; the reaction farthest removed from its equilibrium will win out.

It should be clear that the same result can be given in terms of the $\Delta G°$'s. We can express it by saying that a reaction with a negative free-energy change may drive a reaction with a positive one. This may be put differently by writing the net reaction as the sum of the two parts:

$$
\begin{array}{lll}
A & \rightleftharpoons B & \Delta G_1° \\
B & \rightleftharpoons C & \Delta G_2° \\
\hline
A & \rightleftharpoons C & \Delta G_1° + \Delta G_2°
\end{array}
$$

Thus, if there is to be a net reaction $A \rightarrow C$, $\Delta G_1° + \Delta G_2°$ must be less than zero. A biologist might put this still differently and say that energy-producing reactions drive energy-requiring reactions. In fact, in the biological world, discounting recourse to magic, the only way in which one reaction may drive another, and hence the only way that free energy may be transferred between one set of reactants and another is by a sequence of reactions such as

$$A \rightleftharpoons B \rightleftharpoons C$$

where one reaction pulls another by virtue of the relative positions of their equilibria. It is clear that much of the complex array of metabolic reactions is nothing more than a mechanism for the linkage of unfavorable (and energy-requiring) reactions, such as the syntheses of cell constituents, with those that are farther from their equilibria (energy-producing), such as the metabolic breakdown of sugars.

An electrochemical version of free energy

The reader is aware from a course in introductory chemistry that many reactions involving the movement of electrons (oxidation or reduction) may be followed electrically by the movement of those electrons. Such a situation is called an *electrochemical cell* and, without considering the experimental details, it is worth noting that there is a direct relationship between the electrical work done by such a reaction and its free-energy change. In other words, the electrical work is done totally at the expense of free energy, and the relation between the two is given as

$$\Delta G° = -nF \Delta E°$$

where $\Delta G°$ is the standard free-energy change, $\Delta E°$ the difference in standard potential (in volts), n the number of electrons transferred per molecule of reactant, and F, the Faraday, is a constant representing the electrical capacity to do work and equals 96,500 coulombs. Since free-energy changes are generally given in terms of calories, a more useful figure for F is 23,063 calories per volt.

The electrochemical representation of free energy is of interest to biologists mostly in the study of such oxidation-reductions as the electron-transferring reactions of the cytochrome chain (see Chapter 6). In many such cases, measurements of equilibrium constant are well-nigh impossible, but electrical measurements are quite accessible and hence provide a means of obtaining thermodynamic information. Inasmuch as these electron-transferring reactions are central to cellular energy production, knowledge about their free-energy changes is especially desirable.

The influence of pH on free-energy changes

The importance of specifying conditions under which free-energy changes are measured has been mentioned. The standard free-energy changes given thus far have referred to standard conditions, where reactants are at 1 M concentrations. This means that, when H^+ is involved in a reaction, standard state implies an H^+ concentration of 1 M, which is pH 0 (see the definition of pH below). This is hardly a physiological pH, and, in treatments related to biological chemistry, standard conditions are often redefined as at pH 7. It is obviously important to be clear as to what pH is specified in a given discussion. It is also clearly useful to have in hand an expression relating $\Delta G°$ to a change in pH. Such an equation is easy to

Figure 3-2 Dependency of free-energy change upon pH. The example given is for the free energy of hydrolysis of adenosine triphosphate (ATP).

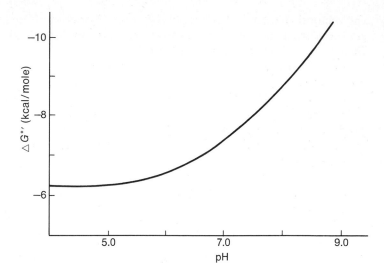

derive[2] and, for the case where hydrogen ion is a product of the reaction, may be written

$$\Delta G^{\circ\prime} = \Delta G^{\circ} - 2.303RT(\text{pH})$$

where $\Delta G^{\circ\prime}$ is the standard free-energy change at the given pH, ΔG° that at pH 0, R the gas constant, and T the absolute temperature; "pH" refers to the pH at which the reaction is actually measured. This equation serves as evidence that the conditions under which a free energy change takes place may affect the magnitude of the change and must always be specified in any real situation.

The steady state

Most of the preceding discussion of energy is strictly applicable only to completely reversible processes, which is to say those that are at equilibrium. Thus, calculations of free energy related to biological reactions are valid only when the reactions are at equilibrium. Such situations seldom (or never) obtain in living cells. In general, cellular reactions tend to be very far removed from their equilibria, so that thermodynamic calculations represent limiting cases from which it is often quite impossible to derive useful information. In other words, the essentially "closed" systems studied at equilibrium do not give results directly applicable to the cell, which is often described as "open," with a significant flow of nutrient, waste product, and heat across its boundaries. Furthermore, ordinary thermodynamic arguments involve scalar quantities, which fail to reflect

---

[2] The interested reader may follow the derivation by considering the relation between ΔG° and K and including the hydrogen ion concentration in the equilibrium. This strategy, together with the definition of pH as $1/\log(H^{+})$, should suffice.

the vectoral character of many cellular events. For example, many chemical transformations of the cell involve the transport of molecules across a membrane or are directed with respect to an enzyme surface, and it is important that the vectoral aspects of such matters be included in energetic considerations.

Although the area of nonequilibrium thermodynamics involves somewhat formidable mathematics and is largely outside the scope of this book, it is worth considering one central feature of it, the idea of the steady state, which replaces the central role of the equilibrium in ordinary thermodynamics.

If one considers systems that do not come to equilibrium, it is useful to think in terms of flow, whether it be flow of molecules through a membrane, flow of heat between regions of different temperature, or the flow of matter through a biochemical pathway. Furthermore, in extending the analogy with liquid flow, one might say that a flow is proportional to some sort of driving force. This might be written, in the simplest case,

$$\mathbf{J} = L\mathbf{X}$$

where $\mathbf{J}$ is a vector representing a flow rate, $\mathbf{X}$ a vector representing the driving force, and L a proportionality constant (scalar). The driving force might be a pressure (driving a liquid flow), the difference between concentrations of some compound on two sides of a membrane (driving transport across the membrane), or the free-energy difference in a series of reactions.

The mathematical description of flow can in many instances be extremely complex, but there is one situation, the *steady state*, that represents both relative simplicity and applicability to the processes of life. A process is in the steady state when its properties are not changing in time. An example of the steady state is found in a river that is flowing at a uniform rate, so that, for each arbitrary segment, as much water enters as leaves (see Figure 3-3). Examples of the steady state on the cellular scale include a portion of a metabolic pathway where an intermediate is formed and utilized at the same rate, so that its concentration remains constant with time. Other examples will be discussed in later chapters.

Clearly, in connection with a system at the steady state, thermodynamic functions such as energy and entropy take on a different aspect. For example, differences in free energy or electrochemical potential may be regarded as elements of the driving force instead of parameters of a stationary equilibrium situation. Similarly, we found that entropy does not change in a system at equilibrium, although it is continuously changed in flowing systems. Indeed, it is possible to define equilibrium as that condition where

$$\frac{dS}{dt} = 0$$

and the steady state as that where

$$\frac{dS}{dt} = \text{a constant}$$

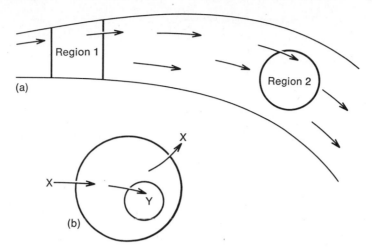

Figure 3-3 Concept of the *steady state*. (a) In any arbitrary portion of the river, exactly as much water enters as leaves, so that the amount of water in that portion remains constant. This condition is true of any portion, no matter what its geometry: Thus it applies to region 1 as well as to region 2. (b) In a highly diagrammatic cell, compound X (the magic ingredient) enters the cell and can either leave the cell by the same route (across the plasma membrane) or can become converted to Y in the interior of an equally diagrammatic organelle. In the steady-state condition, the amount of X in the cell remains constant in time, so that its entry must exactly be offset by the sum of its exit and its conversion to Y. Notice that the size of the "pool" of X within the cell would be constant and would not be related to the flow rates of X.

The constant production of entropy in steady-state systems is a central feature of such systems, with important consequences for the cell. In addition, the rate of entropy production approaches a minimum value when a system nears the steady state, a property of the state also significant in its description. Growing cells, which, by their nature, are collections of flow systems, must decrease entropy at the expense of free energy. It is significant that dynamic cellular processes approach the steady state wherever possible, a tendency that has the effect of minimizing the production of entropy and thus conserving free energy for other purposes.

Power versus control in machines

Many of the important results of thermodynamics have come from the study of energy transformations involved in the action of man-made machines, such as steam or hot-air engines. This portion of classical thermodynamics remains somewhat inaccessible to the biologist, owing to superficial dissimilarities between energy as applied to an engine and energy as involved in a reaction within the cell. Nonetheless, the study of machines has provided interesting analogies for those involved in the living world, and the recent development of information theory and the

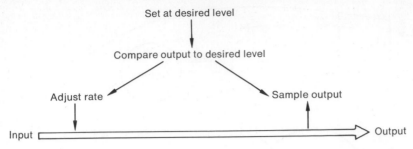

Figure 3-4 Feedback system. The general pattern applies to a wide variety of situations from industrial processes to the synthesis of many intracellular components.

study of servomechanisms suggests that the end of such analogies is not in sight.

For our purposes, a machine is something that does work, and doing work means quite literally using energy. In other words, a machine might lift something and take in energy to do so. It should be obvious that the ability of the machine to work is limited by the design of the machine and related matters such as efficiency but, most fundamentally, by the energy at hand. A machine cannot expend in work any more energy than it has available, a truth known, in various guises, as the *first law of thermo-dynamics*. It may also be expressed by saying that energy may be neither created nor destroyed, or by asserting that "you can't get something for nothing."[3] In any case, the relation between power input and work output by machines is well known and valuable to us in our general understanding of energy (see Figure 3-4).

However, contemporary machines are noted for doing more than con-verting energy into brute work. Most of the advances in machine design during the present century have been related, not to the problem of how to get more work out of a mechanical system, but, more, how to control that work to a higher degree. Machines have evolved to the extent where the control is impressive indeed, culminating (for the present) in automated machines of fantastic subtlety.

These advances suggest that the design of machines, from an energy point of view, involves two separate but related systems—the power part, by which we mean the portion that directly converts input (such as energy) into output (such as mechanical work), and the control part. Both require energy, and both must be taken into consideration when the energy input and output of a machine is considered, but it appears uni-versal that the power system uses the lion's share of the total energy avail-able. The dramatic automation of a mechanical process is added to the uncontrolled process at a surprisingly low expense of additional energy. Control of the utilization of massive amounts of energy is effected by machinery using almost no energy at all but operating with great delicacy and precision. An example might be the home furnace, which transmits

---

[3] The *second law of thermodynamics*, which is based on the nonreversibility of real processes, adds that "you can't even break even."

the chemical energy of fuel into heat on a large scale but is under the control of a thermostat, whose power requirements are indeed minimal.

The distinction between power and control is equally valid in consideration of energy transformations by the cell. We have said that cells require energy in respectable amounts to perform their many activities, chemical and mechanical. The matter of driving thermodynamically unfavorable reactions by linking them to those that go spontaneously is essentially in the realm of providing power. Respiration and photosynthesis supply power for the performance of work in a manner analogous to the power input of a large machine. But anyone who has studied any part of the living world must be impressed not only by the availability of energy to do things, but by the finely tuned control of what is done. Cells frequently have a way of not making more of anything than they need and of not doing more mechanical or electrical work than appropriate. Just as an automated factory controls the various processes so that all are consistent with the final rate of output, the cell is able to mesh the myriad chemical events so that none exceeds the need of the whole and nothing is wasted. As in the case of a machine, the control mechanisms of the cell differ from the processes that they control in requiring only small amounts of energy and in essentially representing modes of communication between different events.

Control through feedback

In discussing the exact form that control mechanisms in machines or cells may take, it is useful to return to the example of the home furnace and its control by a thermostat. To begin with, the role of a furnace is to provide heat: a question of power. But a furnace that does nothing but provide heat is thought by many to be a rather barbaric form of machine. A more acceptable one would be a furnace that could provide just the right amount of heat, that would maintain a set and civilized temperature within a house. Excluding difficulties in deciding what represents a civilized temperature, it must be said that the task of the furnace is then a difficult one and must suddenly add several functions to the one of transferring heat.

In the first place, the system must be able to measure the temperature to see how it is doing. There must be a provision for comparing the actual temperature with the temperature at which the thermostat is set. The system must convert any difference between the two into a signal that will communicate to the power end of things (the furnace) the need for more or less heat. To make matters more difficult, furnaces generally exist in two states, on and off, and human beings (especially Americans, it is claimed) tend to be rather particular about the temperature's not fluctuating wildly. Clearly, the message from the thermostat must be of a yes-or-no character, but it must be at the same time capable of avoiding excessively uneven output. Thus, the control mechanism must have the ability to measure the difference between the desired output and the actual situation, and to encode the difference in a form that is "understandable" to the machine producing the output. The requirements for control on the cellular scale will be seen to be the same.

For example, a basic and typical control problem on the cellular scale is that of making enough, but not too much, of some cell product. An amino acid should be synthesized at a rate sufficient to allow its use in the manufacture of protein, but there is little benefit—and possibly considerable harm—in making an excess. The synthetic machinery of the cell faces a problem identical to that of the thermostat in most details. The amount of product of a set of reactions must be somehow monitored and compared to the required amount. If an excess is synthesized, somehow the synthetic route must be shut off until the proper level is attained. Such a situation, where excess output leads to a shutting off, is called *negative feedback* and is extremely common in the biological world. In the case of amino acid synthesis, feedback control is exerted either by the inhibition by the final product of an enzyme involved in the synthesis or by the suppression of the synthesis of an enzyme, again by the final product. Thus, in both mechanisms, the formation of an excess of the amino acid (or whatever) leads to a reduction in its rate of synthesis, thereby maintaining a suitable level. The reverse sort of feedback, not surprisingly termed *positive feedback*, is less prevalent (or less obvious) but examples do exist in the economy of the cell, perhaps most clearly in the case of glycolysis, where ATP stimulates a process leading to its synthesis.

Information transfer

It should be clear that feedback control of some event, whether it be in a machine or a cell, involves the problem of communication between the input and output. The information about output must be compared (an algebraic operation) with a set value and then communicated to another part of the whole process, whereby the rate is regulated. This sort of information transfer is perhaps the least complicated of the many encountered in the cell. In fact, any form of regulation in cellular processes must involve some form of information transfer. As the fundamental basis for the regulation of the activities of living things resides in the genetic systems of the organism, it is not at all odd that the terminology of modern genetics is full of references to information storage and transfer. Such phrases as the "genetic code," "genetic information," and "nonsense" suggest a debt to information theory that is not accidental.

To be more explicit about the role of information in cellular affairs, we should first be a bit more exact about what we mean by information. The difficulty in talking about information is that everyone has a general idea as to what it is and therefore it resists exact description. Let us then be formal and unbending and look at information in fairly explicit terms. Figure 3-5 shows a generalized system for information transfer, whether in the cell or at both ends of a telegraph line. Information is encoded at

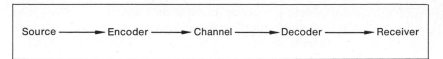

Figure 3-5 Information channel.

the source and decoded at the receiver. The idea of information as a code means only that if one sends information about a thing (such as the heat content of a living room in our example above), one does not need to send the thing itself. If one wishes to send the information that it is raining in Moose Jaw, it is not required to send a bucket of rainwater; it would be adequate (and even preferable) to send a message, and in this case the code would probably be something approximating the English language.

In our diagram, the information is said to be carried on a channel, which is nothing more than the physical mechanism for communication. An information channel might be a telegraph wire, an electromagnetic vibration of such-and-such a frequency, or a series of linked reactions in the chemistry of the cell. The channel is distinguished by having a limit on its ability to carry information and by the fact that it is susceptible to random interference, called *noise*, of which radio static is an example. Noise tends to limit information transfer and can be countered by such devices as repeating portions of the message (redundancy), which, however, obviously decreases efficiency.

Noise is a very useful concept in this connection and has been applied to the cell in the context of genetic mutations as random interference in genetic information transfer. Indeed, a consideration of what noise is, and specifically why it is not information, can give insight into the character of information itself. Consider (in Figure 3-6) an information source emitting a signal. The reader is asked to decide whether this is, in any sense, information, but without knowing anything about what goes on inside the source. The signal might be information or, conversely, it might be nonsense (a signal lacking information), perhaps due to an infinitely noisy channel. How can one decide? One might be tempted to note that real information should be orderly and the order contained in the remark "emgfitheeikkkmrughf," while perhaps eloquent if delivered with feeling, is not immediately identifiable. But that is the problem: Order, like beauty, is at least partly in the eye of the beholder. The signal emitted by our source might well be completely random (without order), made up from random spins of a roulette wheel. But it is also possible that it could be meaningful and even beautiful to a Basque shepherd. Finally, it might be a prearranged signal meaning something like "help," or "it is raining in Moose Jaw," or perhaps both. Just looking at the signal, we are hard pressed to decide if there is any information at all, and this inability leads

Figure 3-6 Information source generating a signal.

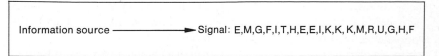

to a fundamental conclusion about information transfer. It is clear that any decisions about information require the presence of a source *and* a receiver able to receive the decoded measage. Information without a receiver just does not exist.

But there is yet another restriction on what information is. Not only must a recipient exist, but the communication must say something to it that is really information, in the sense that it did not "know" it already. This very important restriction means that even if the code is recognizable, the message must say something not totally irrelevant to the receiver. If I am in Moose Jaw, and I receive the glad tidings that "it is raining in Moose Jaw" while the rain drips down my nose, the information content of that message is rather low, in both a colloquial and strict sense. Since I was completely certain about it to begin with, I gained nothing from the message. Had I been only about 50 percent sure to begin with, the message would have increased the probability for me from 0.5 to 1, and the increase in probability would give a measure of the information transferred. This is true of all information transfer, and one can thus define information in terms of its ability to increase the probability of something being true. This may be stated

Information $= -\log P$

where P is the probability of the message being known to begin with and the log is taken to any base that is convenient. Note that if a message is totally certain to begin with, $P = 1$, and the information content of the message equals the log of 1, or 0.

The intimate relation between information and probability may be seen in another light by examining a message generated by some completely random process (such as a throw of dice). We saw earlier that the most probable state was a random one, and therefore a completely random message is completely probable, that is, $P = 1$. If the message is totally probable, it is totally predictable and therefore completely unable to decrease our uncertainty about anything. Therefore, the information content is zero.

Entropy and information

We said earlier in the chapter that real events tended to slip in the direction of disorder, which was more probable than order. This slippage was represented as entropy in some of our equations, and entropy was thus defined as a measure of probability. We have now discovered that *information* is related to probability in a similar fashion. Obviously there is some connection between the two, and it is interesting that information theorists have a habit of using the terms information and entropy interchangeably (although when they use the term *entropy*, it is with the opposite sign from the *entropy* of thermodynamics).

Since a change in entropy is also a form of a change in energy, it follows that information is also a species of energy. Whenever information is transferred, energy is consumed. One does not obtain order out of disorder without putting in energy. Writers about biological matters are fond of pointing out that the orderly character of a cell represents an island in a sea of disorder and that cells (and organisms) must obtain energy to combat the increase in disorder (entropy) that accompanies any process. Some have spoken of organisms feeding on negative entropy (which one might wish to call information) in order to obtain order. Although this does not perhaps help very much in the sense of suggesting concrete experimental approaches to the living world, it is certainly true that the maintenance of order through the specific control mechanisms of the cell does require some energy. It is likewise clear that energy "used up" in information and control is very like entropy, and that the only source of it in the cellular world is in the free-energy changes accompanying spontaneous reactions.

Some energy calculations relevant to the cell

The reader, having endured a theoretical discussion of the nature of energy, may wish to return to more biological affairs and consider the application of these notions to the cell. We should first consider a general question, namely the worth of such calculations to one trying to understand the nature of cellular processes. The answer seems to be that some thermodynamic computations appear to be highly instructive whereas others are not. For example, we have seen that calculations about the free-energy change of reactions are really also calculations about the equilibrium constants of the same reactions. This, in turn, is a quantitative way of saying which reactions take place spontaneously and which do not. Since cellular chemistry is a fabric of spontaneous reactions in the process of being linked to those that are not, such considerations are extremely important. We shall illustrate calculations of this kind in a moment.

On the other hand, biophysicists have been (for some years) fond of calculating the entropy of certain cells and other biological systems. These calculations are based on the statistical definition of entropy and, necessarily, involve many assumptions and approximations, owing to the dissimilarity between cells and the ensembles of uniform molecules or atoms that statistical mechanics normally treats. Indeed, one can argue that the results of such computations are more a property of the assumptions than of the cells. Even if these computations were rigorous (which they obviously are not), they would give rise to figures of questionable usefulness; one can discover only that cells are highly "improbable" in the sense that they are unlikely to occur as the result of random processes. One might reach the same conclusion from the examination of the ultrastructure of cells. The conclusion would be admittedly subjective, but it may be as useful as a number expressed in entropy units up to six decimal places, and based

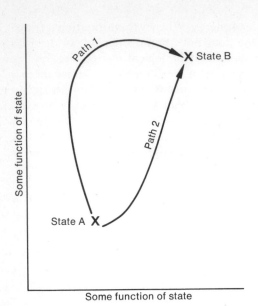

Figure 3-7 Transitions involving functions of state. The initial and final states are independent of the pathways connecting them.

on an arbitrary input. The conclusion that we are belaboring is that computations, like measurements, are interesting only when they inform one about nature by providing information that was not already at hand.

Before proceeding to some calculations that are of interest and utility, it is first necessary to make one final definition that simplifies many calculations. We define as a *function of state* any thermodynamic quantity whose calculated value is independent of a way in which the system, such as a cube of gas molecules or a cell or you, goes from state A to state B; then, the route by which it goes will not make any difference in the values of any state function that are computed.[4] In a sense, the distance between two locations on the surface of the earth provides an analogy. The distance, as the crow flies, between Ithaca, New York, and the North Pole is the same no matter how one actually makes the trip. Making a side trip to, say, the Australian outback alters *the path taken but not the net distance*. It is just so with state functions. States A and B in Figure 3-7 could indicate, for example, the condition of a muscle fiber before and after contraction. Paths 1 and 2 might indicate two different ways of carrying out the contraction (perhaps in one step or in two stages), but it is significant that the free-energy difference, ΔG, between the two states is the same no matter what path is taken: *Free energy is thus a state function.*

Other examples of state functions are enthalpy and entropy; heat and temperature, on the other hand, are not state functions. When one deals with state functions, one can ignore routes and simply consider the algebraic addition of the quantities. This is most convenient, as will now be shown.

---

[4] It is possible to offer a much more explicit definition of *state function* based on the fact that a certain integral equation describing it vanishes. Experience with the author's students suggests that this approach is not necessarily useful, since, when we take it, the student sometimes vanishes as well.

Computation 1 *Change in free energy obtained from free energies of formation.* Consider the reaction

$$A \rightleftharpoons B \rightleftharpoons C$$

The free-energy change, ΔG, associated with the transformation of A into C is exactly equal to the sum of the two free energies associated with

$$A \rightleftharpoons B$$

and

$$B \rightleftharpoons C$$

We have already seen that it is exactly this additive nature of free energy that enables a spontaneous reaction to "pull" one that is not spontaneous. As long as the sum of the free energies is negative, the net reaction will occur. There is, however, another benefit stemming from the additive nature of state functions. If one is interested in the free energy of a given reaction, one can obtain it from the *free energies of formation* of the reactants and products. There is a wealth of measurements available on the free energy of formation of compounds from their constituent elements. These can be summarized by

$$\text{Constituent elements}_A \xrightarrow{\Delta G_a^\circ} A$$
$$\text{Constituent elements}_B \xrightarrow{\Delta G_b^\circ} B$$

Then, if one wishes to find out the free-energy change associated with the reaction

$$A \rightleftharpoons B$$

it is clear that it may be obtained simply by finding the difference between the free energies of formation of A and B, $\Delta G_a^{\circ\prime}$ and $\Delta G_b^{\circ\prime}$. For example, an important reaction in aerobic cells is the conversion of fumarate to malate upon the addition of water across a double bond:

$$
\begin{array}{c}
COO^- \\
| \\
CH \\
\| \\
HC \\
| \\
{}^-OOC
\end{array}
+ H_2O \rightleftharpoons
\begin{array}{c}
COO^- \\
| \\
CH_2 \\
| \\
HCOH \\
| \\
COO^-
\end{array}
$$

To obtain the free-energy change associated with this reaction it is only necessary to subtract the free energy of formation of the reactants from that of the product. The requisite values are those shown in Table 3-2, from which one obtains

$$\Delta G^{\circ\prime} = (-201.98) - (-56.69 - 144.41)$$
$$= -201.98 - (-201.10)$$
$$= -201.98 + 201.10$$
$$= -0.88 \text{ kcal}$$

The reader will note that this method has the obvious advantage of requiring only simple addition and the obvious disadvantage of yielding the answer as a very small difference between two very large numbers. Thus, accurate determinations of the large numbers are very important and, fortunately, such determinations have been made with great precision. As an added example of the same approach (and, also, as added evidence that the individual steps in a reaction do not influence state calculations), consider the overall biological reaction whereby glucose is broken down aerobically to yield two molecules of pyruvate:

Glucose + O_2 $\rightleftharpoons$ 2 Pyruvate + 2 H^+ + 2 H_2O

In this case, one must include the stoichiometry of the reaction. Thus, the free energy of formation of pyruvate is -113.32 kcal, so that $\Delta G^{\circ\prime}$ for two molecules of pyruvate would be 226.64 kcal, that for two waters would be -113.38 kcal, that for H^+ is zero (see Table 3-2), that for glucose is -219.38 kcal, and that for molecular oxygen is also zero. The free-energy change asociated with the net reaction written above would be

$$\Delta G^{\circ\prime} = (-226.64 - 113.38) - (-219.38)$$
$$= -120.64 \text{ kcal}$$

Clearly the net reaction is spontaneous and hence energy-yielding. We shall see in Chapter 5 that the way in which the energy is conserved from this net reaction is by linking it to the nonspontaneous synthesis of adenosine triphosphate (ATP). Since the formation and hydrolysis of ATP will be shown to be central to all (or almost all) energy transfer in cells, we turn now to the important question of the free energy associated with these reactions.

*Table 3-2   Some free energies of formation**

Compound	$\Delta G^{\circ\prime}_{formation}$ (kcal)
Water	-56.69
Malate	-201.98
Fumarate	-144.41

* Values based on measurements with 1-M solutions at pH 7 and 25°C. As a reference, the free energy of formation of H^+ under these conditions is taken as zero.

Computation 2 *Free-energy change on hydrolysis of ATP.* Throughout this book, we shall be encountering the importance of ATP as the energy currency of the cell. Given the central role of this compound, it is perhaps strange that the exact value for the free energy of its formation or hydrolysis has been determined with accuracy only quite recently. The reason for this difficulty is instructive. It develops that the most suitable way of determining such a quantity is based on the relationship between the standard free-energy change and the equilibrium constant as given in the equation

$$\Delta G^{\circ\prime} = -RT \log_e K$$

Thus, a determination of the reaction to be studied requires carrying it out only until equilibrium is attained and then measuring the concentration of the reactants and products in order to compute K. This tactic works best, however, when K is in the region of unity as one is not then forced to measure very low concentrations of any of the compounds involved. It will become clear that it was this restriction that most severely limited measurements of ATP hydrolysis and formation. The reaction of most basic interest with regard to the function of ATP in cellular-energy transfer is

$$ATP + H_2O \rightleftharpoons ADP + phosphate$$

When this reaction proceeds to the "right," that is, in the hydrolytic direction, it is energy-yielding and is usually coupled to energy-requiring processes. The reaction in the reverse direction is the energy-conserving synthesis of ATP occurring in oxidative and photosynthetic ATP synthesis. In either case, the reaction represents addition or removal of a single phosphate group.

   To estimate the equilibrium constant (and therefore the free-energy change), it is only necessary to allow the hydrolysis to proceed to equilibrium and to measure the concentrations of the products. Unfortunately, the reaction is massively in the direction of the hydrolytic products, and it is technically difficult to determine the concentrations with sufficient accuracy. This is clearly a problem in any measurements that involve reactions with large negative free energies. For this reason, it is necessary to base one's strategy on the additive nature of free energies. Thus, one measures the equilibrium positions of two reactions that, added together, give the reaction in question. This results in splitting the free-energy change into two components, each of which yields equilibrium constants of a more measurable nature. The hydrolysis of ATP can be considered to be composed of two steps—(1) the phosphorylation of glucose by ATP and (2) the hydrolysis of glucose phosphate:

(1)   Glucose + ATP $\rightarrow$ Glucose-6-phosphate + ADP

(2) Glucose-6-phosphate + H_2O $\rightarrow$ Glucose + phosphate

The net reaction is then

$$ATP + H_2O \rightarrow ADP + phosphate$$

since the other compounds "cancel out" in the net reaction equation. It must be added that the two reactions require the addition of enzymes for their catalysis so that they will be complete in a reasonable time span. The first is catalyzed by the important enzyme *hexokinase*, while the second is catalyzed by a *phosphatase*. It will be discovered in Chapter 4 that the addition of a catalyst does not alter any state function associated with the reaction but only the rate, which it does by altering the reaction path to include an enzyme–reactant complex.

The measured equilibrium constant for the first reaction is 661. The associated free-energy change at 25°C is then given by

$$\Delta G^{\circ\prime} = -RT \log_e (661)$$

$$= -(1.98)(298)(6.7)$$

$$= -4.0 \, \text{kcal}$$

By a similar method, the equilibrium constant of reaction 2 is 171 and the free-energy change is -3.3 kcal. Therefore, one can obtain the free-energy change for the net reaction by addition and

$$\Delta G^{\circ\prime} = (-4.0) + (-3.3)$$

$$= -7.3 \, \text{kcal}$$

It is comforting that the free energy for ATP hydrolysis obtained in this indirect manner is consistent with determinations made by other (indirect) means, so that the value is almost certainly correct. It is also necessary to add that such determinations depend on the choice of temperature, concentrations, and pH. In this case, the calculation also includes the assumption that the ATP is in the form of its magnesium salt.

Table 3-3 illustrates some other free energies of hydrolysis of phosphate compounds and is included to give some idea of the variation of values obtained with different phosphate esters. It is seen that ATP, in this connection, is intermediate in the free-energy change associated with the hydrolysis of the terminal phosphate. This should dispel some of the magic

Table 3-3   *Free energy of hydrolysis of some biological phosphates*

Compound	$\Delta G^{\circ\prime}$ (kcal)
Phosphoenolpyruvate	-14.8
1,3-Diphosphoglycerate	-11.8
Phosphoarginine	-7.7
Adenosine triphosphate*	-7.3
Glucose-1-phosphate	-5.0
Glucose-6-phosphate	-3.3
Glycerol-1-phosphate	-2.2

* Value is for the hydrolysis of the terminal phosphate.

that is often attributed to the "high-energy bonds" of this compound. The magic, if any, resides not in an unusual value for $\Delta G^{\circ\prime}$ but rather in the wealth of metabolic reactions that connect ATP synthesis and hydrolysis in the cell. This wealth will become increasingly apparent on reading the following several chapters.

Computation 3 *Free energies obtained from electrochemical potential.* Finally, let us consider the computation of free-energy changes from measured potential. There is one area of cell study where this is particularly useful and that is the process of cellular respiration. Briefly, respiration includes the passage of electrons along a chain of electron carriers, the *cytochromes*. The cytochromes operate in a chain where a reduced cytochrome passes an electron to the next (oxidized) one, which, in turn, passes it to the next. Eventually, the electron is passed to the final acceptor, molecular oxygen. Without worrying about how it might occur, it is possible to state that electrons start in a high-energy location and they end up in a relatively low-energy one. The difference in energy between the two states represents that available for biological function and it is conserved in the form of ATP. Now, one might wish to know where in the chain of carriers ATP synthesis actually occurs; let us look first at the electrochemical potentials of the carriers themselves. Consider the following portion of the chain:

Cytochrone *b* → Cytochrome *c* → Cytochrome *a*

The potentials of these carriers have been determined, either by direct electrode measurement, or by way of equilibrium measurements (which may be described as, technically, quite ghastly). Recent values for cytochrome *c* are about 0.26 volt, for cytochrome *a* 0.28 volt, and for cytochrome *b* about 0 volt. Thus, between cytochrome *c* and *d* there is a difference of about 0.02 volt (or 20 millivolts) and between *b* and *c*, about 0.26. We now ask what sort of a potential difference between subsequent carriers is required for there to be sufficient energy to synthesize a molecule of ATP on the passage of a single electron. The free-energy change required was seen above to be −7.3 kilocalories. We found earlier that free-energy changes could be obtained from potentials by the following equation:

$$\Delta G^{\circ\prime} = -nF\,\Delta E^{\circ\prime}$$

This may be solved for $\Delta E^{\circ\prime}$ to give

$$\Delta E^{\circ\prime} = \frac{\Delta G^{\circ\prime}}{-nF}$$

From this, the potential difference corresponding to −7.3 kilocalories is

$$\Delta E^{\circ\prime} = \frac{(-7.3)}{-(2)(23.1)}$$

$$= 0.16 \text{ volt}$$

Note that this calculation is based on the passage of two electrons ($n = 2$) and refers to the *standard* free energy and potential. Thus, the calculation refers to the situation where the concentration of, say, phosphate is $1\ M$ and there is an equal amount of ADP and ATP present. These conditions are obviously not necessarily met in the cell (and, indeed, it is unlikely that they could be met there). For this reason the calculation is not precise and only gives an estimate of the real situation *in vivo*. On the other hand, it does provide a useful guide. For instance, it is clear from the preceding discussion that there is unlikely to be enough of a potential difference between cytochromes *c* and *a* to accommodate synthesis of ATP, while there is probably a sufficient difference between cytochromes *b* and *c*; this impression is confirmed by independent evidence, which will be presented in Chapter 6.

This last calculation is typical of many that are made about cellular-energy matters, in which even approximate figures may be useful provided they are considered in light of experimental evidence. In cell study, approximate figures are often the only ones attainable, and they can still be important if their approximate nature is recognized and if the calculations are made in an effort to answer real questions and not for their own sake alone.

Suggested Reading

Books

Bray, H. G., and K. White, *Kinetics and Theormodynamics in Biochemistry*, 2nd ed., Academic Press, New York, 1966.

Florkin, M., and E. H. Stotz (eds.), *Bioenergetics*, vol. 22 of *Comprehensive Biochemistry*, American Elsevier Publishing Co., New York, 1967.

George, F. H., *Cybernetics and Biology*, W. H. Freeman and Co., San Francisco, 1965.

Kalekar, H. M., *Biological Phosphorylations: Development of Concepts*, Prentice-Hall, Englewood Cliffs, N.J., 1969.

Kaplan, N. O., and E. P. Kennedy (eds.), *Current Aspects of Biochemical Energetics*, Academic Press, New York, 1966.

Katchalsky, A., and P. F. Curran, *Non-equilibrium Thermodynamics in Biophysics*, Harvard University Press, Cambridge, Mass., 1965.

Klotz, I., *Energy Changes in Biochemical Reactions*, Academic Press, New York, 1967.

Klotz, I., *Introduction to Chemical Thermodynamics*, W. A. Benjamin, New York, 1964.

Lehninger, A. L., *Biochemistry*, Worth Publishers, New York, 1970, Ch. 14.

Lehninger, A. L., *Bioenergetics*, W. A. Benjamin, New York, 1965, Chs. 2, 12.

Morowitz, H. J., *Energy Flow in Biology*, Academic Press, New York, 1968.

Morowitz, H. J., *Entropy for Biologists*, Academic Press, New York, 1970.

Wiener, N., *Cybernetics*, 2nd ed., MIT Press, Cambridge, Mass., 1961.

Articles

Alberty, R. A., "Effect of pH and metal ion concentration on the equilibrium hydrolysis of adenosine triphosphate to adenosine diphosphate," *J. Biol. Chem.*, **243**, 1337 (1968).

George, P., and R. J. Rutman, "The high energy phosphate bond concept," *Progr. Biophys. Biophys. Chem.*, **10**, 1 (1960).

Ingraham, L. L., and A. B. Pardee, "Free energy and entropy in metabolism," in *Metabolic Pathways*, 3rd ed., D. M. Greenberg (ed.), vol. 1, Academic Press, New York, 1967, p. 2.

Lipmann, F., "Metabolic generation and utilization of phosphate bond energy," in *Advances in Enzymology*, F. F. Nord (ed.), vol. 1, John Wiley & Sons (Interscience Division), New York, 1941, p. 99.

4 Proteins and enzymes

The class of compounds called *proteins* comprises about 15 percent of the total mass of cells. They represent important structural elements of the cell, membranes being about one-half protein, and it is obvious that changes in the cell environment affecting the properties of protein will produce important effects. Moreover, nearly all proteins that have been studied exhibit enzymic activity—that is, they are able to serve as catalysts in one or more biochemical reactions—and, again, it is clear that small changes in environment that influence proteins will produce dramatic alterations in the metabolic processes of the cell.

Amino acids

Proteins are polymers composed of *amino acids*, and many of the properties of the whole protein reflect those of the subunit. Proteins differ from such polymers as glycogen, which are made up of many identical units, in incorporating various amino acids in a single molecule, thus permitting enormous variety. Most amino acids may be described by the general formula

$$H_2N-\overset{\displaystyle R}{\underset{\displaystyle H}{\overset{|}{\underset{|}{C}}}}-COOH$$

where R represents a chemical group that might be H (glycine), CH_3 (alanine), $-CH_2SH$ (cysteine), or any one of 16 or so others. Ignoring the nature of the possible R groups (which are summarized in Figure 3-1), the amino acid is seen to have two functional groups, a basic amino group and an acidic carboxyl group. These account for two most important properties of amino acids—their interaction with hydrogen ion and their coupling together to form protein. In the first case, it is important to note that these compounds have, discounting the R group, two sites where ionization can take place and therefore where hydrogen ion concentration can have an effect. Above pH 2, the carboxyl group becomes ionized,

$$COOH \rightleftharpoons COO^- + H^+$$

Figure 4-1 Structures of amino acids commonly found in proteins.

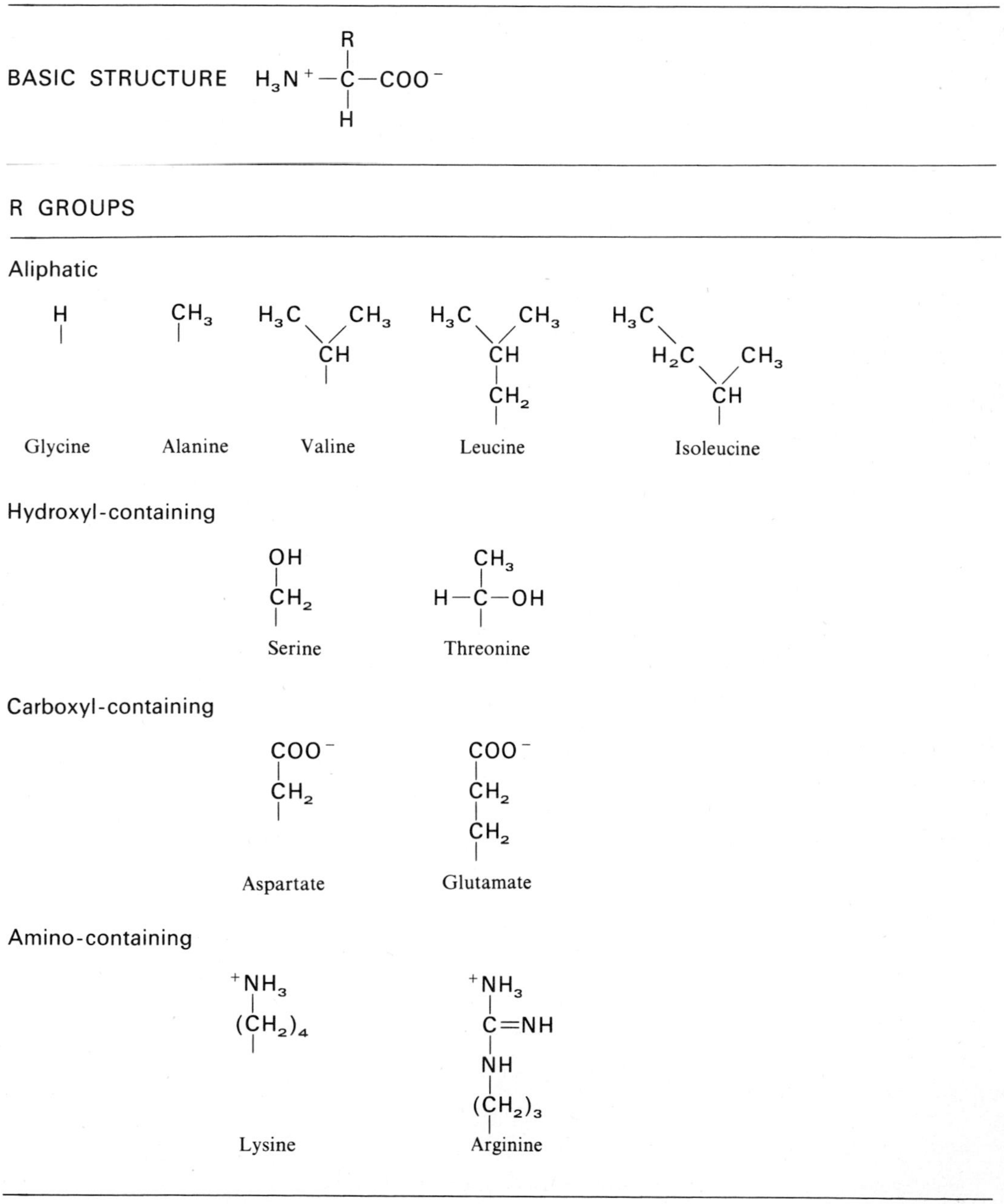

BASIC STRUCTURE

$$H_3N^+ - \underset{\underset{H}{|}}{\overset{\overset{R}{|}}{C}} - COO^-$$

R GROUPS

Aliphatic

H	CH_3	H_3C CH_3 $\diagdown$ $\diagup$ CH	H_3C CH_3 $\diagdown$ $\diagup$ CH CH_2	H_3C H_2C CH_3 $\diagdown$ $\diagup$ CH
Glycine	Alanine	Valine	Leucine	Isoleucine

Hydroxyl-containing

OH
|
CH_2
|

Serine

CH_3
|
H $-$ C $-$ OH
|

Threonine

Carboxyl-containing

COO^-
|
CH_2
|

Aspartate

COO^-
|
CH_2
|
CH_2
|

Glutamate

Amino-containing

$^+NH_3$
|
$(CH_2)_4$
|

Lysine

$^+NH_3$
|
C $=$ NH
|
NH
|
$(CH_2)_3$
|

Arginine

continued

Figure 4-1 (continued)

Sulfur-containing

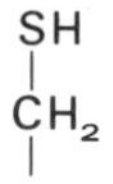

Cysteine (and cystine*) 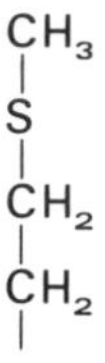   Methionine

Aromatic

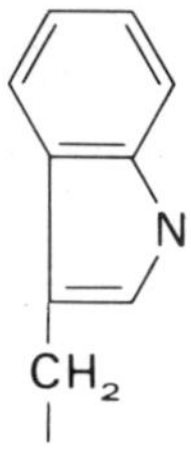

Tryptophan 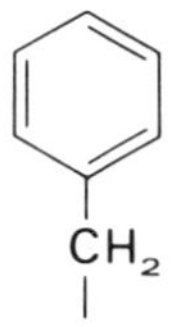   Phenylalanine 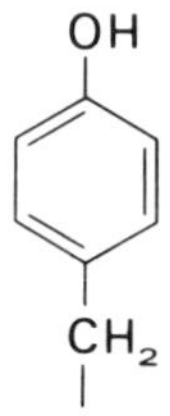   Tyrosine

Heterocyclic

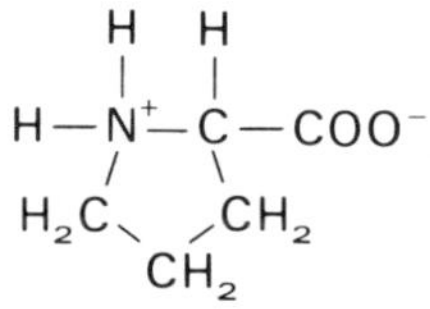

Proline 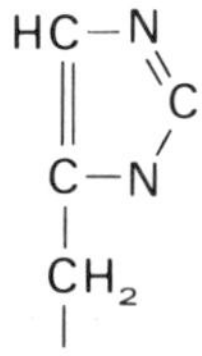   Histidine

Amide-containing

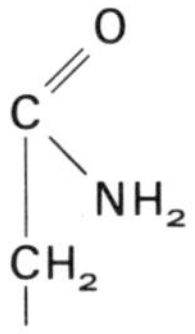

Asparagine 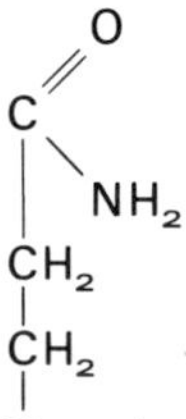   Glutamine

---

* Cystine should be considered to be two molecules of cysteine connected together at their sulfur atoms. The fusion is an oxidation leading to loss of the two hydrogens of the SH groups so that a disulfide bridge (—S—S—) remains.

whereas below pH 8 to 9, the amino group becomes positively charged,

$$NH_2 + H^+ \rightleftharpoons NH_3^+$$

Thus, the electrostatic state of the amino acid depends sharply on the pH of its environment, and properties change accordingly. At neutral pH, two major contributions to the structure will be the double (dipolar) ion,

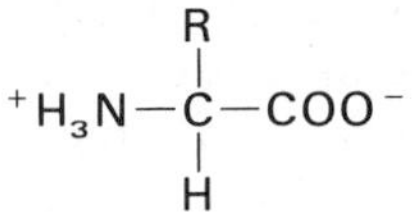

$$^+H_3N-\underset{\underset{H}{|}}{\overset{\overset{R}{|}}{C}}-COO^-$$

In no case does the uncharged species (the way one normally writes the structure) predominate in aqueous solution. In fact, with many amino acids (see Figure 4-1) additional reactions with H^+ are possible because

Figure 4-2 Examples of nonprotein amino acids. It should be mentioned that these represent only a few of the many amino acids that are not primarily subunits of the proteins. For instance, plants contain a wide variety of such compounds, many of which are toxic to animal cells. Likewise, bacterial cell walls contain a variety of amino acids, including some in the D (or "unnatural") configuration.

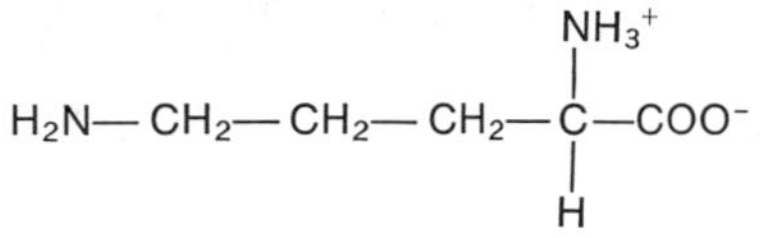

$$H_2N-CH_2-CH_2-CH_2-\underset{\underset{H}{|}}{\overset{\overset{NH_3^+}{|}}{C}}-COO^-$$

Ornithine

Found in liver: important in nitrogen metabolism

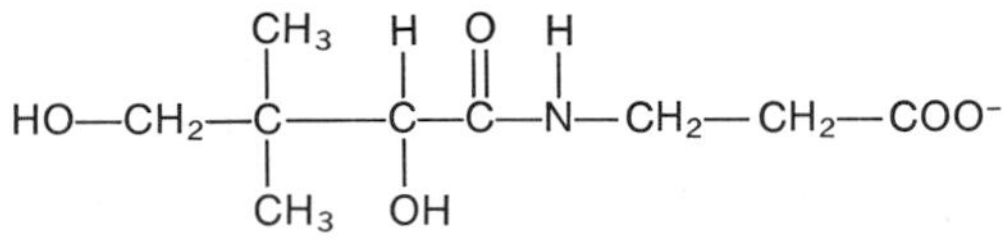

$$HO-CH_2-\underset{\underset{CH_3}{|}}{\overset{\overset{CH_3}{|}}{C}}-\underset{\underset{OH}{|}}{\overset{\overset{H}{|}}{C}}-\overset{\overset{O}{||}}{C}-\overset{\overset{H}{|}}{N}-CH_2-CH_2-COO^-$$

Pantothenic acid

Precursor of coenzyme A

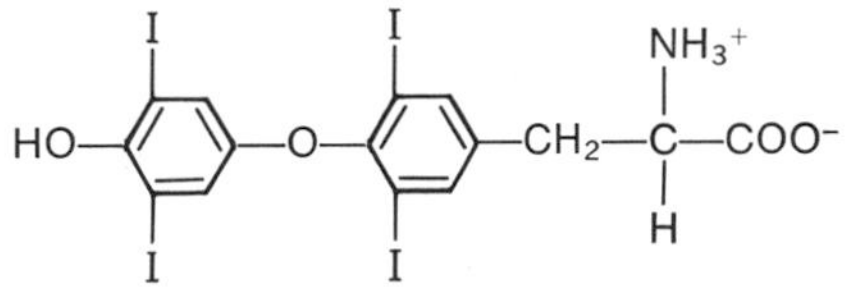

$$HO-\bigcirc-O-\bigcirc-CH_2-\underset{\underset{H}{|}}{\overset{\overset{NH_3^+}{|}}{C}}-COO^-$$

Thyroxine

Product of thyroid gland

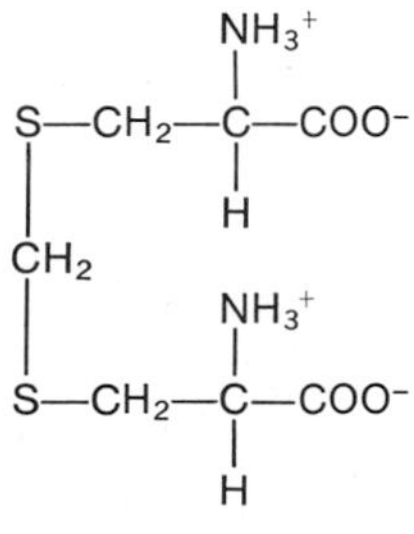

$$S-CH_2-\underset{\underset{H}{|}}{\overset{\overset{NH_3^+}{|}}{C}}-COO^-$$
$$|$$
$$CH_2$$
$$|$$
$$S-CH_2-\underset{\underset{H}{|}}{\overset{\overset{NH_3^+}{|}}{C}}-COO^-$$

Djenkolic acid

Found in certain plant seeds

97

of the presence of ionizable R groups, and the sum of possible interactions may become quite complex. However, since most biological functions of amino acids occur in the vicinity of neutral pH, the more complicated interactions may be ignored.

All of this is to say that the hydrogen ion concentration in the environment of an amino acid or protein may influence its state in a significant way. On the other hand, amino acids, either alone or linked in a protein, influence the environment in the sense that they can serve as a buffer. The functional R groups maintain a stable pH by taking up or expelling $H^{\cdot}$ as the H^+ concentration fluctuates, thereby counteracting these fluctuations. The action of amino acids as both weak acids and bases spreads the stabilizing influence over a considerable range of pH, and it is precisely this wide range that makes amino acids effective buffers. The side chains of the amino acids provide the proteins with large numbers of free ionizable groups, giving them impressive buffering capability. Since the free amino acid content in the cell tends to be quite low relative to the content in proteins, it is likely that protein plays a major role in keeping the interior of cells at the stable pH required for stability of function. The role of amino acids as pH stabilizers extends to a practical level, as they are often included in reaction mixtures in enzyme studies as buffers, with glycine or histidine frequently used.

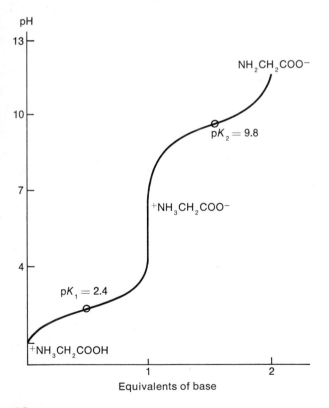

Figure 4-3 Amino acids as buffers: a titration curve of glycine. Note that there are two pK's corresponding to ionization of the carboxyl group (pK_1) and of the amino group (pK_2). Since there are two pK's, there are two regions where buffering capacity is high—that is, where addition of acid or base has relatively little effect on the pH.

The peptide bond

Although amino acids possess properties and undergo many reactions of interest to the organic chemist, it is safe to say that in the biological world one reaction eclipses all others in importance. That reaction is the formation of the *peptide bond*, which enables amino acids to serve as the subunits of peptides and proteins, giving them a place in practically all activities of the cell. Since the only two chemical groups common to all amino acids are the amino and carboxyl groups, polymerization of amino acids may be expected to involve these structures. Amino acids polymerize to form proteins by the condensation of the amino group of one molecule with the carboxyl group of another, thus forming a peptide bond. As an example, we may consider the synthesis of a dimer of glycine, glycylglycine, but the argument holds for the formation of any polypeptide. Note, in the first place, that the condensation of an amino and a carboxyl group is nothing more than formation of an amide linkage with elimination of water, a fact not altered by the presence of both groups on one molecular species. An amide condensation of two molecules of glycine might then be written

$$
\begin{array}{ccc}
\overset{+}{N}H_3 & \overset{+}{N}H_3 & \overset{+}{N}H_3 \\
| & | & | \\
CH_2 & + CH_2 & \rightleftharpoons \quad CH_2 + H_2O \\
| & | & | \\
COO^- & COO^- & CO \\
& & | \\
& & NH \\
& & | \\
& & CH_2 \\
& & | \\
& & COO^- \\
\end{array}
$$

and the peptide (amide) linkage could be written more explicitly as

$$
\begin{array}{c}
\ \ \ O\ \ \ \ H \\
\ \ \ \|\ \ \ \ | \\
-C-N- \\
\end{array}
$$

In addition, the opposite reaction, hydrolysis of the peptide bond, occurs under a variety of conditions, including enzyme catalysis and extremes of

Figure 4.4   Geometry of the peptide linkage. A segment of a polypeptide chain consisting of three amino acids (and part of another) is shown.

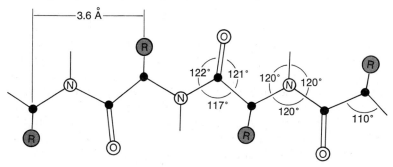

pH. The action of proteolytic (protein-hydrolyzing) enzymes involved in digestion amounts to the addition of water across the peptide bond and leads to the liberation of free amino acids.

Levels of protein structure

Since it is clear that what works in the synthesis of a dipeptide also works in the synthesis of a protein, although on a larger scale, one might suppose that protein structure (and therefore enzyme structure) would be simply a question of amino acid content. This would be the case if proteins were simple polymers of one amino acid, but the actual situation is more complex. Nearly all proteins, whether they have several hundred subunits or only 50, contain from 15 to 20 different kinds of amino acid. Thus, knowledge of the structure must include, as a minimum, not only information about amino acid content but also their sequence in the protein chain. Although many proteins have been purified and examined carefully for amino acid composition, knowledge about sequence is slow and difficult, and the amino acid sequence of only a few proteins has been obtained to date.

In those few cases where the unique sequence of amino acids in a protein (primary structure) is known, one might regard the structure as fully elucidated, but, in fact, even the heroic efforts of protein chemists in this area have failed to give us the complete picture. Although our knowledge of the sequence of subunits may be complete, the most interesting aspect of structure (especially with regard to enzyme action) has to do with the three-dimensional configuration of the protein in space—the way that the chain might be coiled or wrapped around itself.

Studies of a number of proteins by X-ray diffraction have led to the conclusion that there is a repeating three-dimensional unit found in a large number of different proteins. A considerable portion of the total chain length of many proteins appears to consist of a helical configuration, with an average of 3.6 amino acids per turn of the helix. This arrangement of the protein chain in space is called the *secondary* structure and is consistent with what is known about the bond angles and hydrogen-bonding capability of amino acids. Now, if one imagines a protein to be composed of a chain (primary structure) where large portions of the chain are in the form of a regular coil, then the actual disposition of that coil in space represents the *tertiary structure*. The chain, helical regions and all, is seen to be wrapped about itself in an exceedingly complex but consistent fashion, often forming a rather compact, globular molecule. Owing to advances in the technique of X-ray diffraction and the development of electronic computational methods for analyzing data, considerably more extensive and detailed information is available about the tertiary structure of proteins than about the primary structure, which is not at all inappropriate in view of the fact that it is the tertiary structure that appears to be most relevant in the analysis of enzyme action.

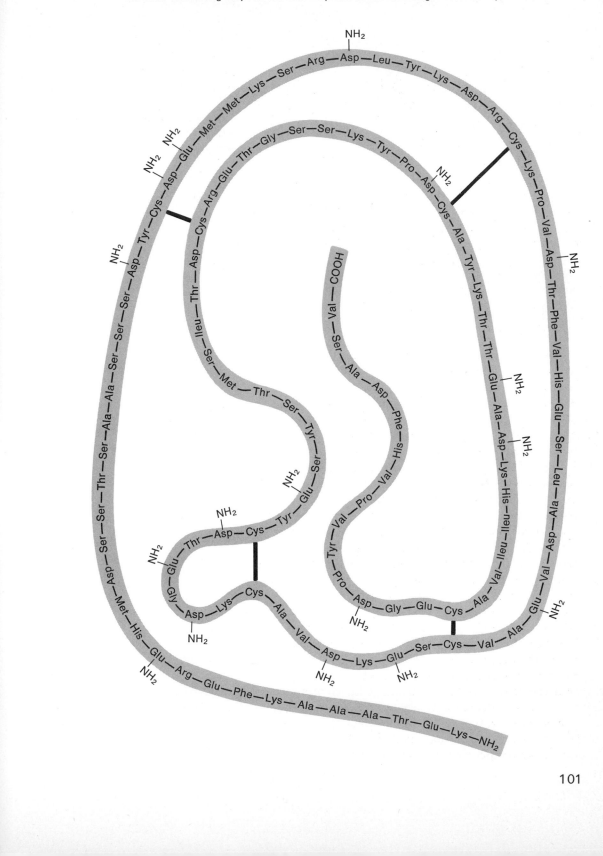

Figure 4-5 Primary structure of the enzyme ribonuclease isolated from mammalian pancreas. The lines connecting loops of the chain represent disulfide bridges between cysteine units.

Carboxyl-terminal end

Heme ring system

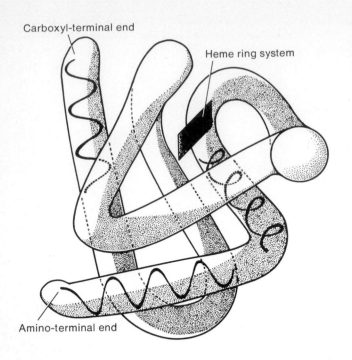

Amino-terminal end

Figure 4-6 Tertiary structure of myoglobin, a heme-containing, oxygen-binding protein from muscle. Some regions of the protein that contain alpha helical structure are indicated. This view of the protein is obtained from high-resolution X-ray diffraction measurements. Such studies have reached the point where the individual amino acids may be defined as to their position in the protein molecule.

One might well wonder what the mechanism could be for winding and coiling in so complex and specific a manner. In fact, it appears that there need be no special mechanism at all, for the secondary and tertiary structures follow inevitably from the amino acid sequence. In other words, the final total structure is the most favored one in a thermodynamic sense and, given the proper sequence, the protein approaches the most stable configuration spontaneously. The structure is determined by intramolecular electrostatic and hydrogen bonds as well as by a limited number of covalent interactions between the individual amino acids of the chain. The spontaneity of tertiary structure may be seen in the instance where it may be disarranged by heat treatment and then reconstituted, a process usually called *reversible denaturation*.

It might be imagined that the limitations imposed by the existence of proteins in three-dimensional space would prevent consideration of the structure of proteins beyond the *tertiary* level. This is certainly true for any single polypeptide chain, however large or intricately coiled. On the other hand, many proteins exist as aggregates of more than one polypeptide unit and the nature of their association is termed *quaternary structure*. In the case of an enzyme, catalytic activity is frequently associated with a particular degree of polymerization—a tetramer perhaps being active, while monomeric units as well as octomers and larger aggregates are relatively inactive. Thus, if changes in the environment of an enzyme lead to changes in degree of association of subunits, they may also produce significant alterations of the rate at which the reaction is catalyzed. It will be seen later in the chapter that such phenomena provide a basis for a very general type of enzyme regulation.

The measurement of tertiary and quaternary structure

Not only do higher levels of protein structure play a role in the regulation of enzyme activity, but there is evidence that the very process of catalysis involves changes in the shape of enzyme molecules. Crystallographic technique for examining tertiary structure (or conformation) applies, not surprisingly, to the molecule in the crystalline state instead of in solution, where catalysis is generally studied. Moreover, since the analysis of the structure of a single protein may require months of measurement and computer time, such approaches are hardly suited to rapid changes in conformation accompanying processes of regulation or catalysis. Fortunately, several techniques exist that enable one to monitor alterations in the shape of proteins and one of these even appears suited to the study of proteins still located in cell organelles.

These methods for studying structural changes are all optical in nature and include measurement of light scattering by proteins in solution as well as optical rotation of light passing through such solutions. Light-scattering measurements can give information about the overall shape of a protein, while optical rotatory measurements also provide information as to the percentage of alpha helix in a given polypeptide chain. In both instances, there are technical problems in measurement, and the theoretical treatment of such results is rather complex. Recently, an added technique has become available that appears to be applicable to a wide variety of proteins in varying states of purity, including the total impurity of proteins in their physiological locus in cell organelles. This approach depends on the sensitivity of the fluorescence of certain organic molecules to the medium that surrounds them. One binds such a molecule (such as 1-anilino-8-naphthalenesulfonic acid) to the protein in question. When a conformational change occurs, the immediate environment of the bound molecule is alerted and the degree of fluorescence is changed. This

Table 4-1 Quaternary subunits of some proteins*

Protein	Number of subunits	Molecular weight of subunits	Total molecular weight
Insulin	2	5,733	11,466
Hemoglobin	4	16,000	64,000
Alcohol dehydrogenase (liver)	4	20,000	80,000
Enolase	2	41,000	82,000
Aldolase (muscle)	4	40,000	160,000
ATPase (mitochondrial)	10	26,000	280,000
RNA polymerase	2	440,000	880,000
Turnip yellow mosaic virus	150	21,000	5,000,000
Tobacco mosaic virus	2,130	17,500	40,000,000

* These figures are given as examples only. Many of the determinations are quite approximate, but the ranges of size are instructive.

approach requires rather simple instrumentation and has even been used to follow apparent conformational changes associated with the action of respiratory enzymes situated within intact mitochondria.

Enzymes as proteins

All enzymes are proteins, and many of their properties reflect those of proteins in general. For example, enzymes are sensitive to extremes of heat and hydrogen ion concentration, so the isolation of enzymes from tissue must be carried out at carefully controlled pH and usually at a low temperature. Furthermore, the action of enzymes often exhibits a sharp temperature and pH optimum, the reasons for which may be attributed to the properties of proteins and even amino acids.

Figure 4-7 shows the influence of pH on an enzyme-catalyzed reaction and indicates that the measurement of enzyme reactions must always take it into consideration. The exact position of the pH optimum differs for different enzymes, but it is often in the region of neutrality. From Figure 4-8 one can see that an enzyme exhibits an optimum for temperature as well.

While the pH effect probably represents an influence on a large number of ionizable groups of the enzyme, temperature dependency is much simpler to interpret. The left side of the peak in Figure 4-7 (rising activity with rising temperature) is an expression of the fact that chemical reactions of all sorts are temperature-sensitive and go faster at elevated temperatures. The reason for this may be summarized briefly by saying that, in most reactions, there is an activated intermediate with a higher energy than the reactants (see Figure 4-9). An increase in temperature increases the total energy available for formation of the activated intermediate and thus increases the amount of the intermediate. Since the rate of the reaction as a whole is proportional to the concentration of the intermediate, the increase in temperature promotes the reaction. In the case of enzymes, however, there is a temperature optimum beyond which additional

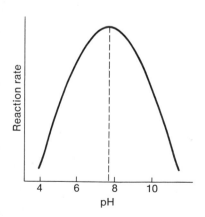

Figure 4-7　Influence of pH on the rate of an enzymic reaction. The optimum pH is indicated by the dashed line.

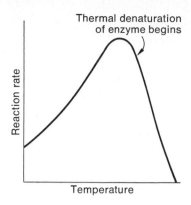

Figure 4-8 Influence of temperature on the rate of an enzyme-catalyzed reaction. In the cases of many cellular enzymes, the temperature at which thermal inactivation begins to destroy the enzyme is in the range 45 to 55°C.

increases slow down the reaction. This is a separate effect, superimposed on the general increase of reaction rate with temperature. The explanation is simply that, since enzymes are proteins, they are inactivated by too much heat and finally destroyed altogether, a process called *irreversible denaturation* (and exemplified by the boiling of an egg). The important point is that the increase in temperature beyond the optimum interferes not with the reaction catalyzed by the enzyme but with the integrity of the enzyme itself.

Figure 4-9 Course of a chemical reaction: here (a) denotes the energy level of the transition state of a reaction that is not catalyzed by an enzyme. Thus, an amount of energy corresponding to the height of (a) above the initial state (and termed the *energy of activation*) must be applied to the system in order for the reaction to proceed to completion. When the temperature of a reaction mixture is raised, thermal energy is increasingly available to provide activation. The role of an enzyme might be said to be a lowering of this energy barrier. Thus (b) denotes the energy level of the transition state of an enzyme-catalyzed reaction. Comparison of the two courses of the reaction [via (a) and (b)] indicates that, at the same temperature, the presence of enzyme greatly increases the probability that some molecules will have sufficient energy to pass the transition-state barrier and undergo reaction. The rate of the reaction accurately reflects this probability.

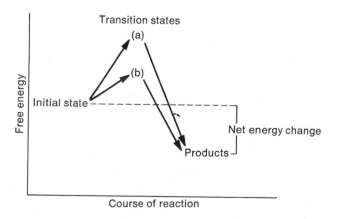

The rate of enzyme reactions

To say anything intelligible about the role and regulation of enzymes within the intact cell, it is first necessary to examine the exact mechanism of enzyme catalysis in a little detail. We may begin by noting that the role of enzymes is to promote reactions (make them go faster). For example, consider the reaction

$$A \rightleftharpoons B$$

Let us imagine that this reaction is thermodynamically possible. Now, even though the reaction may be spontaneous, it may be slow. It might require several years to be detectable. The cell cannot usually afford to wait, and it is not surprising that the addition of the specific enzyme that catalyzes the conversion of A to B might well accelerate the rate several million-fold.

Another property of the reaction $A \rightleftharpoons B$ is the position of equilibrium (see Chapter 2). The position of equilibrium is given by the equilibrium constant

$$K = \frac{(B)}{(A)}$$

and is a thermodynamic property of the reaction. Now, a very important feature of enzymes is that, although they greatly accelerate the reaction, they do not shift the equilibrium. K is K no matter what; with an enzyme, you just get there quicker. A more general way of saying the same thing is that enzymes do not alter thermodynamic properties of reactions, only the kinetic ones (those having to do with rates).

The specificity of enzymes

From the point of view of regulation, one of the most important aspects of enzyme action is its high degree of specificity. An enzyme that reacts with, say, glucose will not be likely to react at a comparable rate with any other sugar, however similar to glucose in structure. Although some enzymes are less particular about their substrate than this, reacting for example with all D-amino acids, they are exceptions. This specificity of enzymes, taken together with the very large number of chemical reactions occurring in a cell, suggests that each cell must contain an extensive selection of enzymes in order to manage the totality of metabolism.

Furthermore, the fact that single enzymes do not promote a number of different reactions enables one to speak of metabolic pathways where a sequence of enzymes directs a flow of compounds along a well-defined route. It is very important to point out that enzyme specificity leads directly to the possibility of tightly controlled metabolism, since it is quite possible for a single enzyme to be crucial in the synthesis of a given end product. Since this enzyme is generally doing nothing else, it is possible to turn the pathway on or off by controlling the synthesis or reactivity of the individual enzyme, which becomes a control point for the pathway.

The control of enzymic reaction rates by reactant concentration

The small molecule that reacts in the presence of an enzyme is called the *substrate*. It is possible to alter the rate of an enzymic reaction by changing the concentration of the substrate, a reflection of the more general fact that chemical reactions tend to depend on concentration. In the simplest case of a nonenzymic reaction, the velocity of conversion of A to B may be described as

$$\frac{dA}{dt} = k(A) \tag{4-1}$$

where dA/dt is the velocity of the reaction, (A) the concentration of A, and k the rate constant of the reaction. In this simple case, where velocity is dependent on the concentration of a single reactant, the reaction is called a *first-order reaction*, and k is known as the first-order rate constant. Were there two reactants, such as in the reaction

$$A + B \rightleftharpoons C$$

the expression for velocity would become

$$\frac{dA}{dt} = k(A)(B) \tag{4-2}$$

Now, the expression for the rate of an enzymic reaction will be more complicated than this simple case. However, it remains true, as in these equations, that the rate of an enzymic reaction is proportional to concentrations of reactants. This is seen in Figure 4-10, which shows the effect of plotting velocity against the concentration of substrate. In the case of the first-order reaction (Figure 4-11) one finds a straight line of which the slope is the rate constant k. In the enzymic reaction, there is a region where rate is proportional to the substrate concentration, but this passes into a region of the curve where additional substrate makes little difference in the reaction rate. In this portion of the curve—that is, where substrate is in excess—the rate of reaction is limited by the concentration of the enzyme, so that doubling the enzyme concentration would lead to a curve similar to that in Figure 4-11 but with the plateau elevated to twice the level shown. The dependency of the enzyme-catalyzed rate upon the enzyme concentration when the substrate is not limiting is significant because it allows the cell to regulate activity by regulating the synthesis

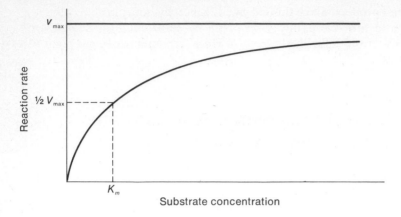

Figure 4-10   Influence of substrate concentration upon the rate of an enzyme-catalyzed reaction. The curve is a rectangular hyperbola and is asymptotic to a value of velocity called V_{max}. An important parameter of enzymes, the Michaelis constant (K_m), which is described in detail later in the text, is seen to be the value of the substrate concentration when the velocity of the reaction is equal to $\frac{1}{2}V_{max}$.

Figure 4-11   Effect of reactant concentration on the rate of (a) a first-order reaction and (b) a zero-order reaction. Note that a zero-order reaction is defined by the independence of rate upon concentration—that is,

$$\frac{dA}{dt} = k$$

as opposed to

$$\frac{dA}{dt} = k(A)$$

in the case of a first-order reaction. The value of k is reflected in the slope of the line in the first-order case and the height of the line in the zero-order case.

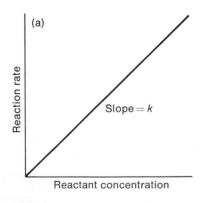

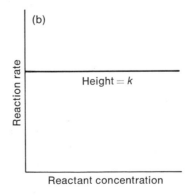

108

Figure 4-12 Influence of enzyme concentration on reaction rate. When other factors are held constant, the rate of an enzymic reaction is proportional to the amount of enzyme present. For this reason, it is possible to equate activity of an enzyme preparation and the amount of enzyme present, provided no other factors interfere.

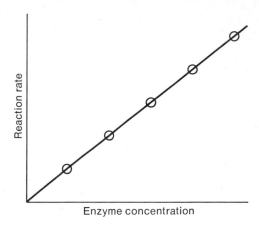

of a particular enzyme (see Figure 4-12). Moreover, the relation between enzyme concentration and rate enables one to use the rate of a particular enzyme-catalyzed reaction as a measure of the amount of enzyme present, a procedure that frequently simplifies analysis of complex cellular systems where any other measure of enzyme concentration would be quite impossible.

The curve shown in Figure 4-11 summarizes results from a large number of experiments performed with many different enzymes during the last half-century. Indeed as long ago as 1913, Michaelis and Menten[1] were able to show that the substrate dependency curve of Figure 4-11 could be explained on the basis of the formation of an enzyme–substrate complex that broke down to give products according to the scheme

$$E + S \ \rightleftharpoons \ ES \ \rightarrow \ \text{Products}$$

Thus the way in which the enzyme reaction rate varies with the substrate concentration is interesting not only from the point of view of cell regulation of chemical processes but also as a way of exploring the intimate character of enzyme catalysis. To gain a clearer basis for understanding enzyme action in general, we shall now turn to a more complete analysis of the effect of substrate concentration on the reaction. In the first place, consider the reaction

$$E + S \ \underset{k_2}{\overset{k_1}{\rightleftharpoons}} \ ES \ \overset{k_3}{\rightarrow} \ X + E$$

where E represents the enzyme, S the substrate, ES the intermediate complex, X the products of the reaction, and k_1, k_2, and k_3 the appropriate rate constants. It is important to mention that the enzyme, being a protein, is normally much larger than the typical substrate, which is commonly a small molecule with a molecular weight of a few hundred or so. For this reason, the substrate is generally present in a concentration

---

[1] L. Michaelis and M. L. Menten, *Biochem. Z.*, **43**, 333 (1913).

in great excess with respect to the enzyme. The image of a few large enzyme molecules in the presence of many small substrate ones is central to the discussion that follows.

To begin, let e be the total concentration of enzyme, p the concentration of the enzyme–substrate complex, and s that of the substrate. At any given moment during the reaction, some of the enzyme will actually be part of the complex, so that the concentration of *free* enzyme is given by $e - p$. Since the substrate is present in huge excess when compared to the enzyme–substrate complex, we can ignore the vanishingly small amount of substrate taking part in the complex. In other words, s and $s - p$ are, for all practical purposes, equal.

Finally, one other assumption is necessary before beginning. Since there are no branch points in our reaction scheme, it is obvious that the rate of the whole reaction $S \rightarrow X$ will be the same as any partial reaction such as $ES \rightarrow X + E$. In other words, we can consider the rate of the total reaction, v, to be equal to the rate of the breakdown of ES in the direction of products. But in the fashion of Equation 4-1, this is given by

$$v = \frac{dx}{dt} = k_3 p \tag{4-3}$$

Note that this reaction is given as irreversible. This simplifies things and is valid, since we shall consider the reaction only in fairly early stages, before much X has been produced and therefore before any back reaction is significant.

It is also extremely important (and interesting) to note that when one actually measures an enzyme reaction, it usually appears that the reaction proceeds at a constant rate for some time until the reactant begins to approach some sort of equilibrium. If this is so, it must be the case that the rate of the reaction $ES \rightarrow X$ must remain constant for a considerable period. This could not be so if the concentration of ES, p, were changing, since k_3 is a constant in any case. Thus, p must also remain constant throughout much of the reaction, existing in a steady state, where the rate of formation of ES exactly equals its breakdown. This seemingly innocuous point turns out to have exciting consequences and leads directly to much of what we know about the general features of enzyme action.

For example, we see that, if p is constant, the rate of the one reaction leading to its formation must exactly equal the sum of the two leading to its breakdown. Following the method in Equations 4-1 and 4-2, we can then write the following equation:

Formation of ES = Breakdown of ES

$$k_1(e - p)s = k_2 p + k_3 p$$

and this can be written

$$k_1(es - ps) = (k_2 + k_3)p$$

Then, dividing both sides by k_1,

$$es - ps = \frac{k_2 + k_3}{k_1} p$$

and solving for *es*,

$$es = ps + \frac{k_2 + k_3}{k_1}p$$

$$= p\left(s + \frac{k_2 + k_3}{k_2}\right)$$

so that

$$p = \frac{es}{\dfrac{k_2 + k_3}{k_1} + s} \qquad (4\text{-}4)$$

Now, we have already seen that the rate of the whole reaction, v, is the same as that of any segment and, most particularly, identical to that of the breakdown of the enzyme–substrate complex to form products. This was given in Equation 4-3 by

$$v = k_3 p$$

If we substitute the value of p from Equation 4-4, this becomes

$$v = \frac{k_3 es}{\dfrac{k_2 + k_3}{k_1} + s} \qquad (4\text{-}5)$$

This is a useful equation, giving rate as a function of a set of the constants k_1, k_2, and k_3, as well as the concentration of the enzyme and the substrate. However, it may be simplified by an additional consideration. Consider the case where the enzyme is saturated with substrate (far to the right in Figure 4-10). Under such conditions, the rate, v, might be said to approach a maximum value, which we call V_{max}. The reason for this is that in the presence of excess substrate practically all the enzyme is in the form of the complex. In other words, as $s \to \infty$,

$$v \longrightarrow V_{max}$$

because

$$p \longrightarrow e$$

From this, it is clear that, substituting in Equation 4-3,

$$V_{max} = k_3 e$$

which, in turn, may be substituted in Equation 4-5 to give

$$v = \frac{V_{max}s}{\dfrac{k_2 + k_3}{k_1} + s} \qquad (4\text{-}6)$$

One further simplification in form may be gained by noting that the ratio of rate constants has passed through the derivation unaltered, in a way that enables us to collect them together as a single constant, which we call the *Michaelis constant*, and write K_m, where

$$K_m = \frac{k_2 + k_3}{k_1}$$

Thus, the basic enzyme equation, Equation 4-6, may be written[2]

$$v = \frac{V_{max}s}{K_m + s} \qquad (4\text{-}7)$$

In addition, it turns out that K_m represents not only a minor simplification of the equation but has something of a life of its own. Equation 4-7 was derived by Michaelis and Menten from rather different postulates, among which was the definition of K_m as a dissociation constant for the enzyme–substrate complex to free enzyme and substrate. Although this view is generally invalid, they correctly noted that K_m provided a measure of "affinity" between the enzyme and the substrate molecule. The modern reader may wish to describe affinity in somewhat more chemical terms, but the fact is that K_m does provide a fair measure of the tightness of binding of the ES complex. Thus, together with V_{max}, it provides a useful parameter for describing the action of a given enzyme, and a considerable part of the description of an enzyme involves the measurement of the two.

It might be thought that the evaluation of K_m requires the estimation of the individual rate constants. Fortunately, this is untrue, as rate constants for enzyme reactions are notoriously hard to get at. Likewise, the measurement of V_{max} would seem to involve direct measurement of rate at infinite substrate concentration, an approach that is often impossible for a variety of technical reasons. It should be added that when it is possible to obtain values for V_{max}, one may immediately obtain K_m as well since it has the units of concentration and, in fact, is the value for s when $v = \frac{1}{2}V_{max}$. To see why this is so, try substituting $\frac{1}{2}V_{max}$ for v in Equation 4-7.

Happily, a technique is available that circumvents the need for measurements of rate at infinite values of s and which yields both V_{max} and K_m from a single graphical presentation. First, by simple algebra (the reader should check this), one may obtain a reciprocal form of Equation 4-7, which becomes

$$\frac{1}{v} = \frac{K_m}{V_{max}}\frac{1}{s} + \frac{1}{V_{max}} \qquad (4\text{-}8)$$

One may then plot data in the form of $1/v$ versus $1/s$, producing the graph shown in Figure 4-13. On this graph it is clear that at the $1/v$ axis

---

[2] This equation is frequently written differently after dividing both numerator and denominator by s as

$$v = \frac{V_{max}}{1 + \dfrac{K_m}{s}} \qquad (4\text{-}7a)$$

Figure 4-13 Determination of kinetic constants by a reciprocal graph. This is commonly called a *Lineweaver–Burke* plot and yields K_m and V_{max} from the intercepts.

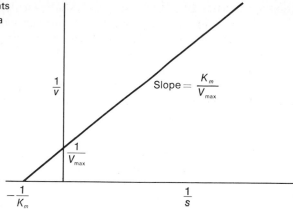

intercept, $1/s$ is 0, so that s must approach infinity. Thus $1/v$ becomes $1/V_{max}$. Furthermore, by elementary manipulation of Equation 4-8, it is seen that, when $1/v$ is 0 (the intercept with the $1/s$ axis), $1/s$ becomes $1/K_m$. Thus, a set of relatively accessible measurements and a single graphical representation of data give the two most important kinetic variables for an enzymic reaction.

Now, it is one thing to obtain kinetic measurements describing the activity of enzymes in a test tube and another to apply such results to the myriad enzymes of the living cell. It should be said here that the results obtained *in vitro* represent limiting cases for enzymes in their natural habitat—the cell. Thus, although the activity of an enzyme *in vivo* may be complicated by many things, it is unlikely to be inconsistent with

Figure 4-14 Determination of K_m and V_{max} by an alternative method. This graph is named for *Eadie* and has the virtue that experimental points tend to be evenly distributed along the line, instead of clustered near the $1/v$ axis, as in the case of the Lineweaver–Burke graph. Additional ways of plotting enzyme data are available. For example, if one graphs s/v versus s, the graph looks like that in Figure 4-13. The intercept on the s axis is $-K_m$ while that on the s/v axis is K_m/V_{max}. The reader is invited to demonstrate the feasibility of these methods by showing to his own satisfaction that the intercepts are as stated.

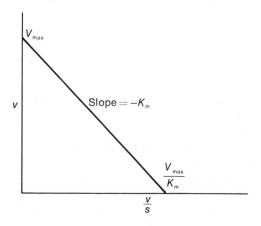

those properties observed when one studies a purified, intact enzyme. Furthermore, the kinetic properties of purified enzymes are interesting from the point of view of understanding their cellular role. For example, K_m has been said to represent a sort of affinity between enzyme and substrate. If this be so, it is useful to know K_m's when one is trying to unravel such matters as the competition of several enzymes for one substrate, since, all else being equal, the enzyme with the highest affinity will come out ahead.

The inhibition of enzymic reactions

There are a large number of ways to inhibit an enzymic reaction, including many that amount to destroying the enzyme. For example, the addition of strong acid will generally have an adverse effect, either by placing the protein in an unfriendly pH environment for activity or, if the acid is at a high concentration, even by hydrolyzing the protein. Furthermore, a number of reagents, including those that bind to the sulfhydryl groups of the protein, inhibit by blocking portions of the enzyme or by producing drastic changes in the tertiary structure of the enzyme. Such reagents frequently lack specificity, inhibiting a wide selection of enzymes and telling very little of interest about any one. They tend to represent rather brutal treatment of the enzyme molecule and are scarcely able to provide answers about the subtle question of enzyme catalysis. In contrast to these inhibitors, a second class, the *competitive inhibitor*, represents an important probe in the study of a given enzyme mechanism while also playing an important role in the regulation of enzyme activity within the cell.

This extremely interesting set of enzyme inhibitors appears to act by virtue of an ability to compete with the substrate for the active site of the enzyme. In other words, the inhibitor is able to bind to the enzyme site (and has the equivalent of a K_m describing this binding) but is unchanged by the enzyme. The reaction sequence for the normal action of the enzyme on its substrate

$$E + S \rightleftharpoons ES \rightarrow P$$

becomes, in the case of the inhibitor I,

$$E + I \rightleftharpoons EI$$

Since the two reactions of E (with S and I) are in competition, they will be affected both by the affinities between the enzyme and S and I, and by the relative concentrations of the two compounds. If one adds some inhibitor and obtains perhaps 50 percent inhibition, it is possible to add more substrate and reduce the inhibition to a lower value. In this case, adding more substrate enables it to compete more effectively for a place on the enzyme surface, a possibility reflected in the common term for such effects as *competitive inhibition*.

114

There is an interesting corollary to the ability to reverse competitive inhibition with excess substrate. Since by adding enough substrate one can, in principle, reverse inhibition completely, a competitive inhibitor cannot be said to alter the maximum velocity V_{max} of a reaction. For this reason, a plot of $1/v$ versus $1/s$ gives characteristic results in the case of a competitive inhibitor and can be used to identify such inhibition. Figure 4-15 shows a family of lines representing various inhibitor concentrations, all of which pass through a common value for $1/v_{max}$, all with different apparent Michaelis constants, as shown by $1/K_m$. This is entirely in keeping with our remarks about K_m representing a measure of affinity between enzyme and substrate, since a second molecule competing at the active site would certainly alter the affinity of that site from the point of view of the substrate (it would tend to gum it up).

We can be a bit more explicit about competitive inhibition by considering the reaction between enzyme and inhibitor in somewhat greater detail. As we have seen, the reaction between enzyme and substrate may be written

$$E + S \underset{k_2}{\overset{k_1}{\rightleftharpoons}} ES \overset{k_3}{\rightarrow} P$$

and, by analogy, the reaction between enzyme and inhibitor becomes

$$E + I \underset{k_6}{\overset{k_5}{\rightleftharpoons}} EI$$

where the reaction does not proceed to the formation of products. If we again let p represent the concentration of ES, and e that of the total enzyme, then i can be the concentration of inhibitor and q that of the EI complex. We then make use of the steady-state assumption exactly as before and can write

$$k_1 s(e - p - q) = (k_2 + k_3)p$$

Figure 4-15   Lineweaver–Burke plots of two classes of enzyme inhibition. These differ in competitive inhibition (a), which reflects competition between substrate and the inhibitory molecule at the same enzyme site. In the case of noncompetitive inhibition (b), the two molecules act at different loci. In the latter case, the affinity of the site for the substrate molecule is unaltered, so that K_m is likewise unaltered. A different V_{max} is noted in the presence of inhibitor.

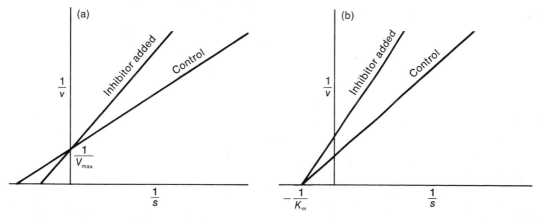

and

$$k_5 i(e - p - q) = k_6 q$$

Moreover, we saw in Equation 4-3 that

$$v = k_3 p$$

If we then solve this set of equations for v, we see that

$$v = \frac{k_3 e}{1 + \dfrac{k_2 + K_3}{k_1 s}\left(1 + \dfrac{k_5 i}{k_6}\right)}$$

This may be simplified by defining an inhibitor constant, K_i, by analogy to the Michaelis constant, as k_6/k_5. We can also substitute V_{max} for $k_3 e$ as before. Then,

$$v = \frac{V_{max}}{1 + \dfrac{K_m}{s}\left(1 + \dfrac{i}{K_i}\right)} \tag{4-9}$$

It is clear that when the concentration of inhibitor i becomes 0, the right side of the denominator becomes 1 and the equation becomes identical to the situation when no inhibitor is present, as in Equation 4-7a. The inhibitor constant, K_i, is of importance in characterizing a given inhibitor with respect to a given enzyme and can be easily evaluated from a reciprocal plot of the sort shown in Figure 4-13. On such a graph, the intercept when a competitive inhibitor is present is given by

$$\frac{\dfrac{1}{-K_m}}{1 + \dfrac{i}{K_i}}$$

and it is clear that, when the inhibitor is omitted (that is, when $i = 0$), the intercept becomes $-1/K_m$, as seen before.

Since one of the most significant features of enzymes is their enormous specificity, the existence of molecules other than the substrate that may be bound to the site of activity suggests the possibility of probing the geometry of the site. Indeed, a survey of competitive inhibitors shows that, in nearly all cases, the inhibitor bears a very close structural similarity to the substrate molecule. For example, the enzyme succinate dehydrogenase acts by withdrawing two electrons from carbons 2 and 3 of succinate and passing them to the respiratory chain (see Chapter 6). Succinate is converted by this action to fumarate as follows:

$$\begin{array}{ccc}
\text{COOH} & & \text{COOH} \\
| & & | \\
\text{CH}_2 & \longrightarrow & \text{CH} \\
| & & \| \\
\text{CH}_2 & & \text{HC} \qquad + 2\,e + 2\,H^+ \\
| & & | \\
\text{HOOC} & & \text{HOOC} \\
\text{Succinate} & & \text{Fumarate}
\end{array}$$

116

Now, it is interesting that a potent competitive inhibitor of this reaction is the dicarboxylic acid malonate, which is seen to bear a striking resemblance to the substrate. On examination of the structure of malonate,

$$
\begin{array}{l}
\text{COOH} \\
|\\
\text{CH}_2 \\
|\\
\text{COOH}
\end{array}
$$

Malonate

one is tempted to say that its symmetrical geometry with the two carboxyl groups makes it look very like succinate to the enzyme, but the lack of a second CH_2 makes withdrawal of the two electrons impossible, and therefore blocks any possible reaction leading to a reaction product. In other words, the formation of half a double bond makes very little sense in this instance. Although malonate is a potent and exhaustively studied inhibitor of the reaction, it is interesting that a number of other inhibitors exist, including, among others, the following:

$$
\begin{array}{llll}
\text{COOH} & \text{COOH} & \text{COOH} & \text{COOH} \\
| & | & | & | \\
\text{C}{=}\text{O} & \text{CH} & \text{CH} & \text{CH}_2 \\
| & || & || & | \\
\text{CH}_2 & \text{HC} & \text{CH} & \text{C}{=}\text{O} \\
| & | & | & | \\
\text{COOH} & \text{HOOC} & \text{COOH} & \text{CH}_3
\end{array}
$$

Oxaloacetate   Fumarate   Maleate   Acetoacetate

Of these, the most effective is oxaloacetate, which has a greater binding capability than even malonate, whereas acetoacetate is less effective than any of the others by a factor of about 10. This might be taken to indicate that two carboxyl groups are required for effective binding to the active region of succinate dehydrogenase, a conclusion for which there is considerable evidence.

It is also very interesting that two of the competitive inhibitors of the enzyme are the product, fumarate, and its isomer, maleate. Indeed, the observation that the product of an enzyme reaction inhibits the reaction is a general one. A large number of such cases have been demonstrated and described as *product inhibition*. The discovery of competitive inhibition by a product should not be very surprising, since there are very few enzymes that produce such a radical structural change in their substrate that the product looks too drastically different from the substrate. In other words, products tend to look somewhat like the substrates and so might be expected to inhibit competitively.

In addition, it should be emphasized that product inhibition, as in all competitive inhibition, is the result of the binding of the inhibitor to the enzyme surface. It is not identical to the much more general "inhibition" that occurs in any reaction where the product builds up. The latter case represents nothing more than an approach to equilibrium with consequent decrease in reaction velocity.

Table 4-2   *Inhibitors of some enzymes*

Compound	Enzyme inhibited
Iodoacetate	Pyrophosphatase, peptidase, glyceraldehyde phosphate dehydrogenase
Cyanide	Cytochrome oxidase, peptidase, succinate dehydrogenase
Malonate	Succinate dehydrogenase
N-Ethylmaleimide	Enzymes with active SH groups
Fluoride	Enolase, pyrophosphatase

The widespread occurrence of product inhibition is also interesting from the point of view of cellular control of chemical reactions. The economy of the cell is such that reactions are usually allowed to proceed only as fast as the product can be utilized in still other reactions. Product inhibition provides a partial means whereby reactions do not exceed the requirements placed on them by the cell, in that, when the product builds up to an excessive level, the reaction tends to be inhibited. This example of negative feedback has an added feature of competitive inhibition—inhibition is minimal, as long as the substrate is in excess compared to the product. Thus, the reaction is able to approach the correct level very quickly before regulation by product inhibition takes place, a feature that is of considerable importance in maintaining responsive, but highly controlled, metabolic pathways in the cell.

End-product inhibition

Not only may control of an individual reaction be exerted by means of product inhibition, a form of competitive inhibition, but mechanisms exist where the final product of a metabolic sequence inhibits a reaction near the beginning of that sequence. Consider a synthetic pathway proceeding from compound A via several intermediates, B, C, and D, to the finished product:

$$A \rightarrow B \rightarrow C \rightarrow D \rightarrow \text{Product}$$

A number of instances are known where such a pathway is inhibited by the product, but rather than the inhibition's taking place between D and the product, as in the case of ordinary product inhibition, it occurs at an earlier site, often between A and B. Thus, if the product is available to the cell (say in the diet), its synthesis is inhibited, as is that of intermediates leading to it. However, there are many cases where intermediates are shared by several pathways. In such cases, inhibition of the individual reactions by the final product occurs only back to the branch point, a modification that prevents inhibition of synthesis of other end products when only one is present in excess.

It seems clear that the end-product inhibition does not occur by means of simple product (or competitive) inhibition. For one thing, the structure of the end product is less likely to resemble that of an early precursor. Likewise, the specificity of inhibition in branched pathways suggests a different sort of mechanism. In fact, it appears that enzymes inhibited in this manner possess two sites on the enzyme surface, one concerned with the activity of the enzyme, the other a site of attachment for the end-product inhibitor. The binding of the end product to the inhibitory site produces changes in the tertiary or quaternary structure of the enzyme, which, in turn, alter the reactivity of the catalytic portion of the enzyme molecule. The noncatalytic site of inhibitor binding has come to be called the *allosteric site* of the enzyme, and enzymes with such an arrangement—and consequently subject to end-product inhibition—are known as *allosteric enzymes*.

A good example of such an allosteric enzyme is aspartate transcarbamylase, which is on the pathway leading to the formation of cytidine triphosphate (CTP). Activity of the enzyme is inhibited by the pathway's end product, CTP, which is structurally quite unrelated to the substrate of the enzyme, aspartate. Apparently there are two sites on the enzyme surface, one carrying out catalysis and the other involved in regulation by CTP. This is suggested by the ability of the sulfhydryl-binding inhibitor, *p*-chloromercuribenzoate, to eliminate control by CTP, even though it does not inhibit the activity of the enzyme (when CTP is absent). It appears that *p*-chloromercuribenzoate inhibits the binding of CTP to the allosteric site but does not influence directly the reactivity of the site where catalysis occurs.

Recent studies reveal, furthermore, that the catalytic and allosteric sites of the enzyme occur on separate protein subunits. The subunit bearing the catalytic site is possessed of enzymic activity but is not subject to regulation by CTP. The subunit bearing the CTP-binding site is devoid of activity but, when mixed in correct proportions with the catalytic subunit, confers on it the property of allosteric regulation by CTP.

Mechanisms of enzyme catalysis

The impressive ability of enzymes to effect large increases in the rates of specific reactions demands explanation. The search for such information has been occupying biochemists for several decades, and a great deal of information is presently available about general enzyme catalysis and about specific enzyme reaction mechanisms. The latter are beyond the scope of this book and we shall offer only some general comments, together with the suggestion that the interested reader may pursue the matter further in a number of the books listed in the Suggested Reading at the end of the chapter.

We have seen that an essential feature of enzyme action is the formation of a complex between enzyme and reactant, and the central issue in enzyme chemistry is the following: How does the character of that complex enhance the reaction rate in question? In other words, one may ask what properties of the complex lead to a lowering of the activation energy (see Figure 4-8). It is likely that the answer to this question involves several distinct mechanisms for a given enzyme. For example, one role of enzymes often appears to be the enforcement of proximity between two reactants bound on the enzyme surface. Thus, reactants are bound at the active site in such a configuration that the reaction between them is favored. A second contributing effect in some cases appears to be the exclusion of water from the site where the reaction occurs, an exclusion that can profoundly affect the reactivity of certain compounds. This effect would be expected to be of most significance in those (frequent) cases of enzymes where the active site is located within a deep cleft in the three-dimensional structure of the protein. That cleft provides an environment quite unlike that existing in the free solvent away from the enzyme, and reactions that are favored, for example, by a hydrophobic medium would have enhanced rates there. It has also been suggested that the tight binding of substrate to enzyme produces a sufficient distortion of chemical bonds within the substrate to significantly increase its reactivity and, hence, the rate of the reaction. This suggestion is sometimes called the "strain model" of enzyme catalysis. Another theory is that the binding of the substrate to the enzyme followed by a major change in conformation may lead to added distortion of the substrate molecule, as a sort of bending after it is firmly fixed to the enzyme surface. This effect has been called the "rack" model of catalysis, and it is supposed that the distortion of bonds in the substrate molecule may be sufficient to render it more highly reactive.

Finally, it should be added that many enzymes appear to work in a manner similar to inorganic catalysts but with much greater specificity. An important form of general catalysis is that produced by acids and bases, and the mechanisms of a number of enzymes appear to represent special cases of this. For example, transient donation and acceptance of hydrogen ion (which is the essence of acid-base catalysis) appears to be central in the

Table 4-3 Some examples of coenzymes*

Coenzyme	Type of reaction
Coenzyme A	Transfer of acyl groups (e.g., acetyl, malonyl, acetoacetyl, etc.)
NAD^+	Oxidation-reduction reactions leading to energy conversion (see Chapter 6)
$NADP^+$	Oxidation-reduction reactions, mostly leading to reductive syntheses (see Chapter 8)
Pyridoxal phosphate	Reactions involving amino acids
Thiamine pyrophosphate	Reactions involving decarboxylations
Tetrahydrofolic acid	Reactions involving transfer of a single carbon in metabolism
Flavin coenzymes	Oxidation-reduction reactions

* Additional information is to be found in the appropriate parts of this text.

action of the enzyme ribonuclease, and in this case, the donor-acceptor appears to be the functional group of the amino acid histidine.

Linked-enzyme systems

It is all too clear that much of the discussion in this chapter applies, in a strict fashion, only to isolated, *single* enzymes under certain conditions. The situation in the cell is much more complex and, whereas some properties of purified enzymes (for example, product inhibition) are obviously applicable to the intact cell, others are less clearly so. Although a discussion of the role of enzymes in metabolism in subsequent chapters will put considerable flesh on the bones of pure enzyme chemistry, at this stage it is worth making several observations about enzyme-mediated pathways in general.

Consider a general pathway

$$A \xrightarrow{1} B \xrightarrow{2} C \xrightarrow{3} C \xrightarrow{4} \text{ etc.}$$

where it is clear that the rate of each reaction will be, in part, determined by the others. The substrate for reaction 2 is the product of reaction 1, so the rate of 2 may be directly related to the progress of 1. Furthermore, if reaction 2 is inhibited by its product, its rate will also be regulated by the third reaction, which will be engaged in removing that product. Real metabolic pathways are still more involved, with branches and sundry other complications, and the complete description of a pathway in terms of steady-state enzyme kinetics, where every velocity is a function of all the other velocities, gives rise to exceedingly complex mathematics. Fortunately, nature sometimes simplifies things a bit, such as in cases where one reaction is so much slower than the others that it can be regarded as rate-limiting. However, there is usually no easy way out, and the sum of a number of enzymic reactions, which can be treated rigorously when taken singly, together must be examined in a relatively crude fashion, often being regarded not so much in terms of molecular kinetics as a sort of analogy to fluid flow.

Properties of enzymes in the intact cell

Most of the comments that we have made concerning enzymes and the rates at which they catalyze reactions apply, rigorously, as we said, to enzymes that have been purified to some extent and are in solution. They are frequently studied in the presence of concentrations of substrates far in excess of those experienced in the cell and of other compounds, such as organic buffers, which occur not at all in nature. The reader may well wonder to what extent the results (and algebra) associated with such artificial situations apply to living cells. The answer is two-fold. In the first place, enzymes within cells frequently behave exactly as they do in a test tube and it is often possible to perform kinetic experiments and measure maximum velocity and inhibition constants *as if* the cell were merely an enzyme container. In the case of the enzyme *cytochrome oxidase* oxygen and carbon monoxide compete for a site on the enzyme, oxygen being its natural substrate. The competition may be studied with highly purified cytochrome oxidase and the results thus obtained may be applied directly to the intact aerobic cell. The results of the *in vitro* experiment may be highly instructive in understanding *in vivo* phenomena, such as why cells are rapidly killed by carbon monoxide.

On the other hand, the leap from the scale of enzyme molecule to that of the cell may not be taken with too much carelessness. In the case of intracellular enzymes it is often difficult to know exactly how much substrate there is to react with a given enzyme and to what extent products are being removed by other reactions. It is not always even clear what substrate reacts with a given enzyme in the cell, and there are cases of infamous oxidation enzymes, known among the cognoscenti as *nothing dehydrogenase*, for which no physiological substrate has been discovered and which are only known by their ability to react with synthetic organic chemicals that do not occur in the cell at all.

It is likewise frequently the case that enzymes occur in the cell bound in a specific fashion to cellular membranes, and their activity within the cell may reflect their immediate environment, the phospholipid of the membrane. Their stability may be greatly influenced by binding to membranes and, as a general rule, bound enzymes become extremely labile when released from their membrane location. For example, the enzyme succinate dehydrogenase, mentioned above, is stable in a mammalian cell for a long time at body temperature ($37°C$), but when it is removed from its locus on the inner mitochondrial membrane, it becomes so fragile that it loses most of its activity in a few hours, even though kept at zero degrees in the absence of oxygen. In a similar manner, another protein of the mitochondrial inner membrane, cytochrome b, not only becomes unstable on separation from that membrane (the separation requiring strong detergents or other harsh treatment) but undergoes significant changes in reactivity. The native enzyme reacts with another cytochrome, c_1, but not with carbon monoxide, while the isolated enzyme, if sufficiently pure, reacts not with cytochrome c_1 but with carbon monoxide. In these cases, and in many others, study of the enzyme *in vitro*, that is, after it has been separated from the cell, can tell one a great deal about its role and properties in the cell, provided only that considerable caution be exercised.

The location of enzymes in the cell

To continue our consideration of the role of enzymes in cells, we must point out that they are not distributed in cells in a random fashion but are located in specific sites. Indeed, the locations of particular enzymes are significant in determining their role in the cell, and a full understanding of metabolism requires that such morphological considerations be kept in mind. In some instances, the mere presence of a particular enzyme in a certain locus provides a clue as to what its function may be. For example, the presence of ion-stimulated ATPase activity in plasma membranes (as well as in certain other interior membranes) has led to the suggestion that these may be involved in the transport of ions across those membranes. Likewise, the existence of certain nonheme iron proteins in the mitochondrial inner membrane led to the suggestion that those proteins were probably involved in cell respiration, since many other proteins of that particular membrane play such a role. Subsequent studies of nonheme iron activity have confirmed this supposition.

During a large part of the remainder of this book we shall encounter enzymes and enzyme systems and shall discuss their intracellular location. For our present purposes, we shall only mention the importance of location and shall summarize (in Table 4-4) the parts of cells in which certain functions occur. The reader should note that this table is presented only to give a general idea and that many exceptions exist. Often a given enzyme occurs in more than one location or in different locations in different cell types. In a number of instances, a particular enzyme activity will be measured in two locations, but with different properties in each case, indicating that two different enzyme molecules exist. For example, both the inner and outer mitochondrial membranes contain NADH–cytochrome c reductase activity. That of the inner membrane is inhibited by the metabolic poison rotenone, while that of the outer membrane is not. Finally, the activity of the outer one appears to be that of a single enzyme, while that of the inner membrane reflects the activity of a chain of at least four molecular components.

The reader will also notice that the table refers only to the loci of enzymes in eukaryotic cells. Similar information exists in the case of

Table 4-4   *Locations of some enzymes and metabolic processes in the organelles of eukaryotic cells*

Region	Metabolic activities
1. Plasma membrane	Transport proteins
2. Nucleus	
a. Interior	Enzymes of DNA, RNA metabolism
b. Envelope	Transport proteins, respiratory enzymes
3. Mitochondria	
a. Inner membrane	Enzymes of respiration, ATP synthesis, transport proteins
b. Outer membrane	Kynurenine hydroxylase, monoamine oxidase
c. Interior (matrix)	Enzymes of the Krebs cycle, fatty acid oxidation
4. Ribosomes	Protein synthesis
5. Endoplasmic reticulum	Enzymes of lipid metabolism
6. Lysosomes	Lytic (breakdown) enzymes
7. Chloroplasts	Enzymes connected with photosynthesis
8. Nonmembranous cytoplasm	Numerous enzymes, including those associated with glycolysis (sugar breakdown) and many biosynthetic processes

prokaryotic cells, although the matter appears somewhat simpler, since many fewer types of intracellular organelles exist in these. In general, one can say that, in prokaryotic cells, the plasma membrane contains those enzymes that would be found in either the plasma membrane, mitochondrial inner membrane, or chloroplast thylakoid membranes of higher forms. Present knowledge suggests that other enzymes are likely to be found in the remaining, nonmembranous, cytoplasm, with little or no apparent specificity.

Multienzyme systems

We saw earlier in the chapter that many proteins are functional only when aggregated to form a specific quaternary structure. Thus, an enzyme may be active only in the form of a *tetramer*, a collection of four subunits, called *protomers*. The protomers fit together in such a way that their binding is spontaneous. In other words, a negative free-energy change is associated with the reaction. Spontaneity of this kind is often called *the principle of self-assembly* (about which more later).

There are additional systems of protein subunits in which the active form is a combination of subunits but where each subunit represents a different reaction. A whole metabolic sequence of reactions can be carried out by one such complex and the study of such *multienzyme complexes* can be of importance in linking the study of purified enzymes with the study of enzymes under cellular conditions.

The great advantage of multienzyme complexes is that the close proximity of the several enzymes enables the product of one reaction to react with the next enzyme without diffusion to a distant site. Many enzymes in the cell exist in the nonmembranous ("soluble") portion, where they are essentially free in solution in the cytoplasm. In such cases, the product of a reaction must diffuse through the cytoplasm until it meets the surface of the next enzyme to which it must bind. Since the rate of diffusion can, then, limit the rate of the whole sequence of reactions in general, the cell must maintain a relatively high concentration of the enzymes in question. In other words, in the case of a multienzyme complex, the cell can get on with a lower enzyme concentration—a situation that leads to somewhat greater efficiency.

Multienzyme complexes exist in two basic geometrical arrangements. In some cases, including tryptophan synthetase and fatty acid synthetase discussed below, the enzymes of the complex are bound together, so that the complex represents simply a protein–protein interaction. On the other hand, in the case of the respiratory chain enzymes of the mitochondrion, the proteins are bound not only to each other but to the structure of the mitochondrial membrane itself, so that complexing between protein and phospholipids appears to be very important. In these cases, it appears that the proteins should be thought of as forming part of the membrane itself and not simply stuck on like flies on flypaper. These two basic arrangements of multienzyme systems are illustrated in Figure 4-16.

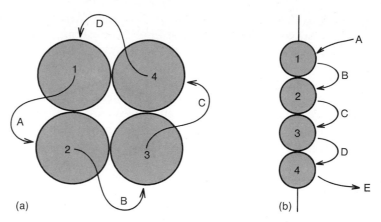

Figure 4-16 Two configurations of multienzyme systems. Part (a) represents a collection of four enzymes that catalyze four sequential reactions. In some such instances, a sequence of reactions involves a number of circuits around the set of enzymes. For example, when fatty acids are synthesized (see Chapter 8), each cycle adds two carbons to a growing acyl chain, so that formation of a 16-carbon acid requires eight cycles. Part (b) represents a collection of enzymes located in a specific fashion on a cellular membrane. Here bonds between the phospholipids of the membrane and the enzymes are of great importance in holding the complex together.

A well-characterized example of a multienzyme system is the tryptophan synthetase that has been studied in the bacterium *Escherichia coli*. This is a relatively simple complex, being formed from only two enzyme subunits. The complex has been highly purified to the extent of being crystallized and has a molecular weight of about 135,000 daltons. There are two subunits, each associated with a separate reaction, and the amino acid sequence of these subunits has been determined. The complex is made up of two *alpha* and two *beta* chains, so that it can be designated alpha$_2$beta$_2$. The net reaction catalyzed by the complex is the synthesis of the amino acid tryptophan,

$$CH_2CHNH_2COOH$$

from indole-3-glycerol phosphate and the amino acid serine. This reaction occurs in two steps. The first is catalyzed by the alpha subunit, although, when the individual units are not incorporated in the complex, the reaction that they carry out proceeds at a low rate. This reaction is

Indole-3-glycerol phosphate $\rightleftharpoons$ Indole + glyceraldehyde phosphate

The second reaction requires the dimer of the beta subunit (beta$_2$) and likewise proceeds at a low rate when the beta units are employed in the absence of alpha. The reaction is

Indole + serine $\rightleftharpoons$ Tryptophan

125

and requires also the coenzyme pyridoxal phosphate. Low activities that are obtained when the complex is less than complete suggest that binding between the subunits produces changes in the conformation of the individual proteins that, in turn, lead to changes in the catalytic activity. Finally, we should add that tryptophan synthetase, one of the most simple multienzyme complexes as well as an important enzyme in amino acid metabolism, has been studied intensively in numerous ways. Not only is it one of the relatively few enzymes whose amino acid sequence is established, but it has been studied for a long time with regard to its genetic regulation as well.

A considerably more complicated multienzyme complex that has also been extensively characterized is the *fatty acid synthetase* system of yeast and animals. In this case, the net reaction consists of no less than six individual reactions, each catalyzed by a separate protein subunit. These reactions will be considered in detail in Chapter 8; our purpose now is only to point out something of what is known about the arrangement of enzymes in this aggregate. In the first place, there is a great deal of indirect and direct (electron-microscopic) evidence that the individual proteins are arranged in a sort of cartwheel configuration (Figure 4-17). The reader will notice that there is an extra subunit in this figure (beyond the six required to catalyze the six reactions). The seventh protein is called the *carrier protein* and serves to bind the fatty acid precursors, which then react with the various enzymes in sequence, being released from the carrier only when the series of reactions is complete. It is supposed that the central molecule in the complex is the carrier and that a flexible side chain attaches to the fatty acid precursor by means of a sulfhydryl group. The carrier protein is small, with a molecular weight of about 16,000 daltons, and its amino acid sequence is established. The advantage of a carrier protein is clearly to remove ordinary diffusion as a potential rate-limiting step in the set of reactions. The image that one envisions of the action of the complex is that the bound precursor pivots from one enzyme to the next, reacting at each active site in turn.

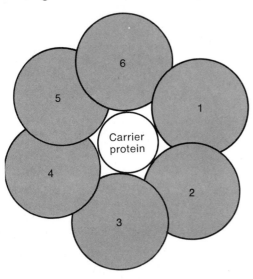

Figure 4-17 Cartwheel configuration of fatty acid synthetase.

Isoenzymes

A final aspect of complexes of enzyme subunits is the interesting topic of *isoenzymes*. There has grown up a body of evidence that many enzymes exist in the cell in a number of different forms, which are identifiable by physical techniques such as electrophoresis, or by differences in the catalytic activity. For example, rat lactate dehydrogenase, which catalyzes the reaction

Lactate + NAD$^+$ $\rightleftharpoons$ Pyruvate + NADH + H$^+$

and which plays an important role in sugar metabolism, exists in rat tissue in five distinct forms, or *isoenzymes*. These all exhibit the same molecular weight (about 130,000 daltons) and are made up of four, poly-peptide subunits. The different isoenzymes can be separated by electro-phoresis, which indicates that they have differing charges and, although catalyzing the same reaction, differ decisively in their kinetic properties, including the K_m.

These features of lactate dehydrogenase turn out to result from its sub-unit composition. There are two basic subunits—the M polypeptide, which predominates in the enzyme isolated from skeletal muscle, and the H polypeptide, which is prevalent in the enzyme isolated from the heart. Thus, the five common forms of the enzyme can be construed as different mixtures of these two units. Thus, the heart muscle form of the enzyme is made up of only the H subunits, while the skeletal muscle form is made up of four M subunits. The combinations of the enzyme found in rat tissue, then, are

HHHH   (heart)
HHHM
HHMM
HMMM
MMMM   (skeletal muscle)

The tetrameric form of the enzyme can be dissociated into the individual subunits, and the subunits are themselves devoid of activity. When the sub-units are mixed *in vitro*, they assemble into the tetramers and regain activity. Since the various isoenzymes of lactate dehydrogenase differ in important ways with respect to their activity, it is clear that different patterns of assembly from subunits can play an important regulatory role in the cell.

Thus, lactate dehydrogenase exists in different cells with somewhat different functions, although the same reaction is obviously involved. It appears that the different molecular forms are correlated with the differing roles that the reaction plays in different cell types, and this is borne out in situations where the function can be shown to change during embryonic development. In such cases, the subunit composition of the enzyme can also be shown to change at the same time. For instance, the forms of the

127

enzyme with more M subunits are associated with cells that obtain a large part of their energy without the use of oxygen, while those with more H subunits prevail in cells with an aerobic pattern of energy metabolism. Embryos initially depend on less aerobic mechanisms and become more dependent on oxidations as they develop. While this happens the number of MMMM and HMMM forms diminishes and they are replaced with more HHHH and HHHM, providing direct evidence of the connection between the nature of energy metabolism and the isoenzyme content of the cell.

The evolution of enzyme structure

Often the study of cell physiology and biochemistry proceeds in the absence of any consideration of the evolution of the systems being studied. Indeed, at the level of fundamental cellular processes, evolutionary aspects are frequently difficult to uncover. This may be due, in some cases, to the identity or near-identity of a process no matter where one encounters it. For instance, the genetic code (Chapter 16) appears to be universally the same from one end of the living spectrum to the other, a constancy that renders its evolution inaccessible of study (although many interesting speculations have been made). In addition, however, less fundamental processes are sometimes difficult to place in an evolutionary context, owing to the great variation that can occur even in closely related organisms. Certain fermentative pathways in bacteria are in this category.

On the other hand, recent advances in the study of protein and enzyme structure have provided unique opportunities to study evolution on the molecular level. The detailed knowledge of primary structure of proteins now available yields a kind of information that can readily be used in a comparative fashion, and this comparative biochemistry informs our views of evolutionary pathways in much the same way that comparative anatomy does. Moreover, a comparative study of, say, amino acid sequences of enzymes provides opportunities to be quantitive that are of considerable help. For example, similarities between structures (such as a part of the inner ear in certain species of fishes) may be obvious to the informed observer, but there is often an uncertainty in deciding just how similar things really are. In contrast, if one is considering protein primary structures, there is a very satisfactory precision to any comparison. For example, one may say that of the 142 amino acid residues in enzyme X, only 14 differ in the enzyme isolated from the unicorn when compared to the griffon. If 34 differ when one compares the unicorn and the kraken, it may be concluded that the unicorn and griffon are rather closely related evolutionarily, while the unicorn and kraken are much less so. (This conclusion conforms to one's everyday experience, which is so much the better.)

In addition, we possess considerable information about proteins from more prosaic beasts. If one compares the amino acid sequence of the blood oxygen carrier, hemoglobin, and the muscle oxygen-binding protein, myoglobin, one discovers considerable similarity. For example, of the 153 amino acid positions in whale myoglobin, 38 are filled with the same amino acids as the corresponding positions in the alpha chain of human hemoglobin. This is far too much similarity to be accounted for by pure chance, and the inescapable conclusion is that both proteins have evolved from a common precursor. Even more striking is the fact that the alpha chains from humans and horses have 123 amino acids in common, so they must be regarded as evolutionarily quite close. In this manner, it becomes possible to construct a measure of evolutionary relationships, and it is significant that such results bear a strict similarity to evolutionary trees based on comparative anatomy.

A similar argument may be made in the cases of the protein-hydrolyzing enzymes trypsin and chymotrypsin. These proteins have a very similar active site, containing a serine residue, and function by means of similar catalytic processes, although they differ somewhat in substrate specificity. Almost half the amino acid positions are filled by the same amino acids in the two chains, suggesting, again, that they have evolved from a common precursor. It is also interesting that there are proteolytic enzymes with an active site very similar to that in chymotrypsin and trypsin but with practically no other similarity in primary structure. These enzymes should probably be taken as examples of the parallel evolution of the same site and the same reaction mechanism from quite different precursors. Thus, there would appear to be a selective pressure favoring any really successful enzyme mechanism or active site configuration, so that these would tend to crop up in quite evolutionarily unrelated organisms.

Finally, the electron carrier of the respiratory chain, cytochrome c, is beginning to provide a wealth of very interesting information about the course and rate of evolution. Cytochrome c is a relatively small protein with only 104 amino acids, and its primary structure has been determined in the cases of almost 40 organisms, including the gray whale, the rattle-snake, and the penguin. It is of great significance that, in the case of this important protein, much of the structure is invariant in all organisms studied. What changes there are are often minimal, as in the substitution of a similar amino acid for an existing one (for example, phenylalanine for tyrosine or glutamate for aspartate). The protein carries a heme prosthetic group where oxidation and reduction occurs, and its points of attachment in the peptide chain are invariant as well.

Again, a measure of evolutionary relatedness may be obtained from a tabulation of similarities of primary structure. Such data are given for cytochrome c in Table 4-5. There, it is seen that man is much more closely related to the horse than to wheat, a conclusion in keeping with other available information. As increasing information becomes available about cytochrome c from other organisms, it is likely that more interesting conclusions will be made as well. It will also be interesting to relate changes in primary structure with those in tertiary configuration. Unfortunately, for technical reasons, the study of cytochrome c conformation by X-ray diffraction has not yet advanced as far as one would hope.

Finally, it has been possible to relate the changes in protein primary

129

Table 4-5   *Differences in amino acid sequences in cytochrome c from various sources**

For example, the cytochrome isolated from turkey differs in 11 amino acid positions from that obtained from horse tissue.

	Man	Rhesus monkey	Horse	Dog	Turkey	Penguin	Tuna	Moth	Wheat	Yeast
Man	0	1	12	11	13	13	21	31	43	51
Rhesus monkey	1	0	11	10	12	12	21	30	43	51
Horse	12	11	0	6	11	12	19	29	46	51
Dog	11	10	6	0	10	10	18	25	44	49
Turkey	13	12	11	10	0	2	17	28	46	51
Penguin	13	12	12	10	2	0	18	27	46	51
Tuna	21	21	19	18	17	18	0	32	49	48
Moth	31	30	29	25	28	27	32	0	45	47
Wheat	43	43	46	44	46	46	49	45	0	50
Yeast	51	51	51	49	51	51	48	47	50	0

* Data are adapted from R. E. Dickerson and I. Geiss, *The Structure and Action of Proteins*, Harper & Row, New York, 1969, p. 65.

structure with time on the evolutionary scale. Given knowledge (obtained from more traditional data) as to when various evolutionary lines branched apart, one can estimate the approximate rate at which amino acids are substituted in protein chains. For example, hemoglobin and myoglobin appear to be changing at a rate of about 6×10^6 years for a 1 percent change in primary structure, while the comparable figure for cytochrome c is about three times as long. Thus, cytochrome c is evolutionarily stable, a property that probably results from a high proportion of possible changes in its structure having lethal effects. This, in turn, may reflect great selectivity as to the conformations that enable it to react with the other members of the respiratory chain, where the fit is probably very important (see Chapter 6). In any event, one may predict an increasing importance in this sort of analysis as more proteins from a wide variety of sources are examined for primary structure, conformation, and other properties.

Suggested Reading

Books

Anfinsen, C. B., *The Molecular Basis of Evolution*, John Wiley & Sons, New York, 1959.

Bailey, J. L., *Techniques of Protein Chemistry*, 2nd ed., American Elsevier Publishing Co., New York, 1967.

Bernhard, S., *The Structure and Function of Enzymes*, W. A. Benjamin, New York, 1968.

Dixon, M., and E. C. Webb, *Enzymes*, 2nd ed., Academic Press, New York, 1964.

Edsall, J. T., and J. Wyman, *Biophysical Chemistry*, vol. 1, Academic Press, New York, 1958, Ch. 3.

Fruton, J. S., and S. Simmonds, *General Biochemistry*, 2nd ed., John Wiley & Sons, New York, 1958, Chs. 2–5.

Hirs, C. H. W. (ed.), *Enzyme Structure*, vol. 11 of *Methods in Enzymology*, Academic Press, New York, 1967.

International Union of Biochemistry, *Enzyme Nomenclature*, American Elsevier Publishing Co., New York, 1965.

Jencks, W. P., *Catalysis in Chemistry and Enzymology*, McGraw-Hill Book Co., New York, 1969. (A definitive, up-to-date treatment of catalytic mechanisms.)

Meister, A., *Biochemistry of the Amino Acids*, vols. I and II, 2nd ed., Academic Press, New York, 1965.

Neilands, J. B., and P. K. Stumpf, *Outlines of Enzyme Chemistry*, 2nd ed., John Wiley & Sons, New York, 1958.

Perutz, M. F., *Proteins and Nucleic Acids*, American Elsevier Publishing Co., New York, 1962.

Schröder, E., and L. Lübke, *The Peptides* (trans. by E. Gross), 2 vols., Academic Press, New York, 1966.

Articles

Atkinson, D. E., "Regulation of enzyme activity," *Ann. Rev. Biochem.*, **35**, 85 (1966).

Cleland, W. W., "The statistical analysis of enzyme kinetics," in *Advances in Enzymology*, F. F. Nord (ed.), vol. 29, John Wiley & Sons (Interscience Division), New York, 1967, p. 1. (An important review and classification of enzyme reaction pathways on the basis of kinetic behavior.)

Cleland, W. W., "Enzyme kinetics," *Ann. Rev. Biochem.*, **36**, 77–112 (1967).

Faller, L., "Relaxation methods in chemistry," *Sci. Am.*, **220**, 30 (May, 1969).

Ginsberg, A., and E. R. Stadtman, "Multienzyme systems," *Ann. Rev. Biochem.*, **39**, 429 (1970).

Koshland, D. E., Jr., and K. E. Neet, "The catalytic and regulatory properties of enzymes," *Ann. Rev. Biochem.*, **37**, 359–410 (1968).

Monod, J., J. P. Changeux, and F. Jacob, "Allosteric proteins and cellular control systems," *J. Mol. Biol.*, **6**, 306 (1963).

Phillips, D. C., "The three-dimensional structure of an enzyme molecule," *Sci. Am.*, **215**, 78 (Nov., 1968).

Shannon, L. M., "Plant isoenzymes," *Ann. Rev. Plant. Physiol.*, **19**, 187 (1968).

Smith, E. L., "The evolution of proteins," *Harvey Lectures*, **62**, 231 (1966/1967).

Stadtman, E. R., "Allosteric regulation of enzyme activity," in *Advances in Enzymology*, F. F. Nord (ed.), vol. 28, John Wiley & Sons (Interscience Division), New York, 1966, p. 41.

Stryer, L., "Implications of X-ray crystallographic studies of protein structure," *Ann. Rev. Biochem.*, **37**, 25 (1968).

Part II

Cell metabolism

5 Enzyme systems

The chemical transformations that take place within cells are known, collectively, as *metabolism*. Most of what a cell does is closely related to one or more chemical reactions, so metabolism is at the heart of much of cell physiology. Likewise, the techniques that cells have evolved for the control of metabolism are the basis for control of almost all other cellular events as well, and their study is of obvious interest.

A complete discussion of metabolism should include an exposition of all the chemical pathways of the cell, as well as their control and inter-relations. Since our knowledge of their control and interrelations is somewhat incomplete in many instances, and since a complete catalog of metabolic reactions or pathways would fill a number of volumes, we shall follow a more modest course, the examination of a few pathways that seem to be quite pivotal. For example, a pathway (known as *glycolysis*) involving the breakdown of glucose is employed by many cells in energy conversion and also in the synthesis of many components, since a number of pathways branch off from it at various points. We will thus follow glucose breakdown, partly because it illustrates a number of features of metabolic pathways in general, and partly because it is itself at the heart of many metabolic sequences in many cells.

Techniques of metabolic studies

To discuss any metabolic pathway in a rational manner, it is necessary to consider how such a pathway can be examined experimentally, since the nature of our knowledge about such matters depends very much on the route by which we gained it. Furthermore, the student of cell biology often deals with metabolic pathways only as rather confusing sequences of organic formulas on paper, a vision of cell life that must be quite dull and which obscures the sense of game and strategy that permeates such areas of cell study.

The examination of metabolism may be conveniently divided into

135

two basic strategies, both of which are generally required for a complete understanding of what is going on. They are (1) examination of the flow of compounds through intact pathways, and (2) the resolution of the pathways into individual enzymic reactions. The ideal confirmation of the validity of a biochemical pathway is the reconstitution of the pathway from the individual enzymes, although unfortunately this is not always possible (or even attempted).

The study of intact pathways

Techniques used in studying the flow of material through a pathway are dependent on the number of ways in which intermediates of the pathway may be measured. Ideally, a method should be used which allows continuous estimation of the concentration of reactants, although this is not always feasible.

Probably the most useful type of determination in the study of metabolism is spectrophotometric. For example, one may follow the progress of many biochemical reactions by watching changes in the absorption of light, either in the visible, ultraviolet, or infrared region of the spectrum. Consider the reaction

$$\text{Substrate} + \text{NAD}^+ \xrightarrow{\textit{dehydrogenase}} \text{Oxidized substrate} + \text{NADH} + \text{H}^+$$

In this case, NADH (the reduced form of nicotinamide adenine dinucleotide, an important coenzyme) has a strong absorption peak in

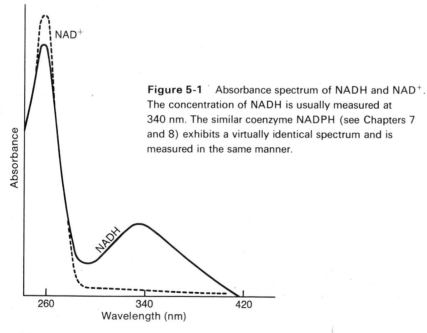

Figure 5-1 Absorbance spectrum of NADH and NAD$^+$. The concentration of NADH is usually measured at 340 nm. The similar coenzyme NADPH (see Chapters 7 and 8) exhibits a virtually identical spectrum and is measured in the same manner.

the ultraviolet region at 340 nm,[1] while NAD^+ has none. Therefore, the progress of the reaction may be followed continuously by watching the increase in absorbance at 340 nm, using a standard visible-ultraviolet spectrophotometer. Such a technique obviously depends on the absence of changes at 340 nm due to other substances, a restriction that usually presents little difficulty. Other forms of spectroscopic measurement used in the study of metabolism include electron paramagnetic resonance spectroscopy (of value in the study of reactions involving free radicals), and spectrophotometry, which utilizes rapid-flow techniques, making it possible to follow reactions over periods as short as a few thousandths of a second.

A valuable nonspectrophotometric method is manometry, in which changes in gas pressure (or volume) are followed. Manometric techniques are especially important in the reactions of respiration (which involve oxygen uptake) or photosynthesis (which involve oxygen evolution). In addition, reactions are often measured by means of polarography (which can also be applied to estimation of oxygen-consuming or -producing reactions) and changes in hydrogen ion, which are often part of biological reactions and are easily followed by means of a glass-electrode pH meter.

Any of the methods described above may be used under appropriate conditions to follow the course of a pathway by examining the disappearance of the initial substrate or the increase in the final product. It is more difficult to dissect the sequence into its parts, but such tricks as the use of inhibitors may often be successful. A metabolic inhibitor that disrupts a pathway may often be useful, as one may detect the buildup of the compound just on the input side of the site of inhibition as the inhibitor dams up the flow. Examples of the importance of inhibition studies will be given later in the context of cell respiration.

Radioisotopes provide an important technique for the elucidation of pathways, and several strategies have been especially successful. In the simplest case, it is often possible to obtain radioactive samples of the starting compound in a metabolic series and, after adding them to cells or cell fractions, to isolate labeled intermediate compounds by chromatographic techniques, after which they may be measured by means of Geiger counting. The rate of appearance of radioactivity in these compounds is especially important, since the high degree of interconnection between metabolic systems makes almost anything radioactive from almost any labeled precursor, if one waits long enough. It is thus important that the compounds thought to be in a metabolic sequence be rapidly labeled, and the rate of labeling should be consistent with the rate of flow through the pathway (if it is known).

A second and very valuable experimental approach to the study of pathways by tracers is that of isotopic competition. This method involves the addition of labeled precursor and measurement of the radioactivity of the final product. If one suspects that a certain compound is an intermediate in the series, the addition of cold (unlabeled) compound should

---

[1] 1 nm = 1 nanometer = 10^{-9} meter. Since 1 Ångström unit (Å) is 10^{-10} meter, 1 nm = 10 Å.

depress the incorporation of label into the final product. This technique is summarized in Figure 5-2.

Finally, a technique of the greatest importance in metabolic study has been the use of biochemically blocked mutants of microorganisms. This technique, which has led to the description of a number of biosynthetic pathways, involves the isolation of a number of different mutants, all unable to synthesize the compound whose synthetic route is to be studied. If one examines mutants unable to synthesize compound E according to the pathway

$$A \rightarrow B \rightarrow C \rightarrow D \rightarrow E$$

an organism blocked between B and C will probably make an abnormally large amount of B and no E. In addition, since the pathway between C and E is intact, the addition of C will enable the organism to make E in a normal fashion. The examination of a number of such mutants blocked in different places leads to knowledge of the complete series, which may then be checked by other means. It is important to note that in all such investigations the worker is greatly assisted by a knowledge of the organic chemical mechanisms that are likely to lead to the final product, so that he does not have to examine an infinite number of possible pathways. Cells, in other words, may be expected to obey the normal rules of organic chemistry.

Figure 5-2 Isotopic approaches to the study of metabolic pathways. (a) Direct-incorporation experiment. One determines whether addition of labeled hypothetical precursor leads to labeling of the product and, if so, at what rate. (b) Isotopic competition. In order to see if B is on the pathway between A and D, measure the rate of labeling of D with and without added cold (unlabeled) B. If B is an intermediate, then cold B will suppress labeling of D.

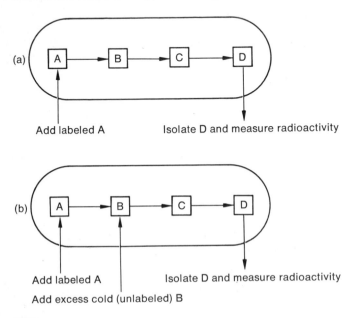

The resolution of a metabolic sequence into individual reactions

The first stage in the fractionation of the machinery by which the cell makes something is essentially a problem in protein purification. It is very important that the individual enzymes be as pure as possible, inasmuch as impure enzymes often have a number of different activities that lead to wild and unreliable results in the study of metabolism. (The reader is urged to refer to any textbook of biochemistry for the techniques of enzyme purification.) Suffice it to say here that useful techniques include salt fractionation, where proteins are separated by their differing solubility in solutions of such a salt as ammonium sulfate. Various forms of chromatography are of central importance in protein and therefore in enzyme fractionation, including ion-exchange column chromatography as well as electrophoretic separation on paper and thin-layer plates. All such approaches depend on varying affinities between different proteins and the material in the column or paper or the surface of the thin-layer plate. Devices are available that collect samples of material as they come from a column and record the amount of protein in each sample, a refinement of great benefit, as a single chromatographic separation may take a number of hours.

Other means of purification include electrophoresis, which is the placing of a sample in a field of high voltage. Proteins may thus be separated on the basis of their charge, and additional resolution may

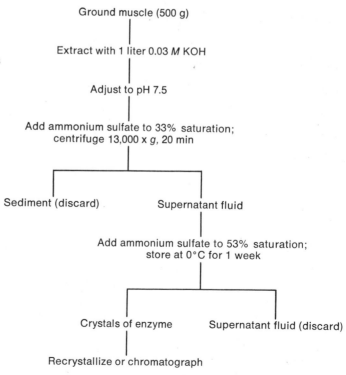

Figure 5-3 Outline of an enzyme-purification procedure. The enzyme used in this example is *aldolase* (discussed later in this chapter) and is uncommonly easy to crystallize. All operations are carried out in a cold room, close to 0°C.

Ground muscle (500 g)

Extract with 1 liter 0.03 M KOH

Adjust to pH 7.5

Add ammonium sulfate to 33% saturation; centrifuge 13,000 x g, 20 min

Sediment (discard) Supernatant fluid

Add ammonium sulfate to 53% saturation; store at 0°C for 1 week

Crystals of enzyme Supernatant fluid (discard)

Recrystallize or chromatograph

often be obtained by changing the pH under which electrophoresis is carried out, thereby altering the charge due to the free ionizable groups of the protein. Finally, the technique of centrifugation is important in enzyme separations, inasmuch as a number of enzymes are found as aggregates large enough to be sedimented in a high centrifugal field.

Figure 5-3 summarizes enzyme purification techniques by showing how a representative enzyme might be isolated from cell material. At each stage of the purification, the activity of the enzyme is determined as a measure of how much enzyme there is, and recorded as total and specific activity, the latter expressed as activity per weight, giving a measure of purity.

The real test of a proposed metabolic pathway is the reconstitution of the pathway from known enzymes, each isolated by such techniques as described above. This is a difficult and time-consuming affair and has been done in only a limited number of cases. Figure 5-4 is a somewhat simplified version of a reconstituted system which served as confirmation of a number of the reactions of glycolysis (the degradation of glucose) and also as a useful analytic method for the determination of an important compound, adenosine triphosphate (ATP).

The technique consists in mixing a sample containing ATP (which is, itself, quite difficult to measure directly) with excess glucose and the

Figure 5-4 Measurement of ATP by an enzymic technique. The method consists in allowing the ATP to phosphorylate glucose to form glucose phosphate. In the presence of excess glucose, each ATP molecule will give rise to one of glucose phosphate. The hexokinase is then precipitated by adding trichloroacetic acid. Analysis of glucose phosphate is accomplished by the addition of the enzymes shown, as well as excess ATP and NADH. Oxidation of NADH to NAD^+ is followed with a spectrophotometer at 340 nm, where NADH (but not NAD^+) absorbs strongly. Note that for every molecule of glucose phosphate, two of NADH are oxidized.

(a) Add excess glucose and the enzyme hexokinase:

$$\text{ATP + Glucose} \xrightarrow{\text{hexokinase}} \text{Glucose phosphate}$$

(b) Precipitate the hexokinase and measure the glucose phosphate formed by means of the following reactions:

$$\text{Glucose phosphate} \xrightarrow{\text{isomerase}} \text{Fructose phosphate}$$

$$\text{Fructose phosphate + ATP} \xrightarrow{\text{phosphohexokinase}} \text{Fructose diphosphate + ADP}$$

$$\text{Fructose diphosphate} \xrightarrow{\text{aldolase}} \text{2 Triose phosphate}$$

$$\text{2 Triose phosphate + 2 NADH} \xrightarrow{\substack{\text{glycerol phosphate} \\ \text{dehydrogenase}}} \text{2 Glycerol phosphate + 2 NAD}^+$$

Overall reaction

$$\text{Glucose phosphate + ATP + 2 NADH} \longrightarrow \text{2 Glycerol phosphate + ATP + 2 NAD}^+$$

enzyme hexokinase, which phosphorylates the glucose to form glucose phosphate, after which the reaction is terminated by precipitating the hexokinase with acid. Then the mixture is neutralized and the enzymes shown in the figure in parentheses are added to allow the stepwise conversion of glucose phosphate to two molecules of triose phosphate, which then oxidizes the coenzyme NADH to form NAD^+. The significance of these reactions to the cell will be discovered later in the chapter. As stated earlier, NADH absorbs strongly at 340 nm, whereas NAD^+ does not. Thus each molecule of ATP ultimately leads to the oxidation of two molecules of NADH, leading to a decline in absorbance at 340 nm. The above represents a good example of the use of enzymes and enzyme systems as analytical reagents for the determination of their substrates, an approach that is of great value to the cell physiologist.

Cell metabolism: carbon sources and energy sources

Aerobic cells are generally able to break down sugars, such as glucose, in order to support life. It is easy to show that administration of radioactive (^{14}C) glucose to cells leads to a rapid labeling of a number of cell compounds; the inference is that glucose is serving as a source of cell carbon. Similarly, it may be shown that cells also use glucose as an energy source; that is, cells are able to link thermodynamically unfavorable reactions to the degradation of glucose. Since one process appears to fulfill two necessities of the cell, it is very important to see how the two are related and to remember at all times that both are operative.

For example, it is known that administration of compounds that block energy transfer reactions prevent the incorporation of ^{14}C from glucose into cell material, indicating that mechanisms for carbon assimilation and energy conservation are closely interconnected. Other examples of the interconnection will be discussed later.

The utilization of glucose: glycolysis

Since it is impossible to give a complete description of the metabolic activities of the cell in a short space, we shall content ourselves with citing some pathways that both serve to illustrate important general aspects of metabolism and are of central importance in themselves. The assimilation of carbohydrates (such as the sugar glucose) is a good case in point, as it not only leads to energy conservation in the cell, but also provides the starting point for the synthesis of a large number of cell components. We shall therefore adopt the tactic of imagining a cell floating in a medium (or in an intact organism), and we shall feed it some glucose and follow the process whereby it is taken up and used.

141

Moreover, we shall consider but one of the many routes of glucose degradation found in biology. That route, glycolysis, is a primary pathway in most higher organisms as well as in many bacteria, while many other bacteria have evolved quite different paths.

The first event is the obvious one of taking the glucose into the cell interior, where all subsequent reactions must take place. This is not necessarily a simple matter, as many cells possess permeability barriers to carbohydrates, and those that are able to admit them often do so by means of an energy-requiring "active transport" mechanism. Since such matters are discussed in detail in Chapter 12, we content ourselves here with pointing out that the barrier is likely to exist and that the uptake of the molecule, itself an energy source, may well require the expenditure of energy.

In understanding the subsequent chemical transformations of the glucose molecule, it is helpful to examine the structure of the molecule itself (see below). It has been known for a long time that glucose exists in solution largely as a six-membered-ring configuration. The reader should become familiar with both structures, as both are useful in visualizing the geometry of some of the reactions of interest. The reader is also reminded that, as a six-carbon sugar, glucose is one of a class of molecules called the *hexoses*, which, in turn, form a subclass of the monosaccharides.

After glucose has penetrated the cell interior, it is able to come into contact with the glycolytic enzymes found in the cytoplasm, apparently not associated with any organelle. The first reaction that takes place is the hexokinase-catalyzed phosphorylation of glucose to form glucose-6-phosphate, the phosphate coming from an adenosinetriphosphate (ATP) molecule. Since energy conservation through ATP synthesis will be seen to be one of the net results of glucose assimilation, we see a situation in the economy of the cell where a certain amount of energy must be expended to produce more, an occurrence reminiscent of "pump-priming" in economics.

Glucose-6-phosphate is then converted to its isomer, fructose-6-phosphate, which has a five-membered ring:

Glucose-6-phosphate[2]   Fructose-6-phosphate

The importance of the isomerization becomes clear at once when it is seen that the following reaction involves a second phosphorylation, that of fructose-6-phosphate by ATP, to yield fructose 1,6-diphosphate:

---

[2] We adopt the convention of writing the phosphate group as $\circledP$.

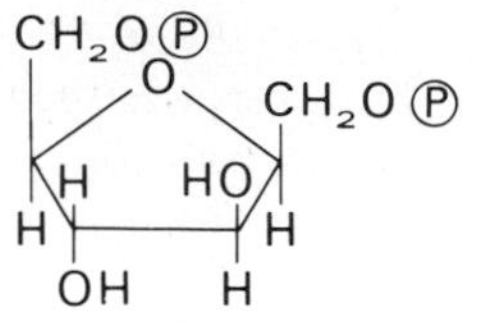

Fructose-1,6-diphosphate

The enzyme that catalyzes this phosphorylation is known as *phosphofructokinase* and has been extensively studied from the point of view of the regulation of the glycolytic pathway. This enzyme appears to be an example of an allosteric enzyme, and its activity is inhibited by relatively low concentrations of ATP. Since ATP is an end product of the glycolytic sequence of reactions, it is likely that its inhibition of phosphofructokinase represents an important regulatory loop, wherein excess ATP production leads to diminished activity in reactions leading to its synthesis. In aerobic cells, another product of glycolysis is citrate (as we shall see), and it is interesting that citrate has also been shown to inhibit the phosphofructokinase reaction in a similar fashion, providing additional feedback inhibition.

It is clear from the structure of fructose-1,6-diphosphate that the isomerization and subsequent phosphorylation have left an essentially symmetrical molecule, with a phosphate projecting from each end. Since one of the primary events in glycolysis is the cleavage of a six-carbon sugar to form two similar three-carbon units, the symmetry is of interest.

The cleavage, catalyzed by the enzyme aldolase, results in the formation of two similar, but not identical, triose phosphates, which are themselves readily interconverted by an isomerase:

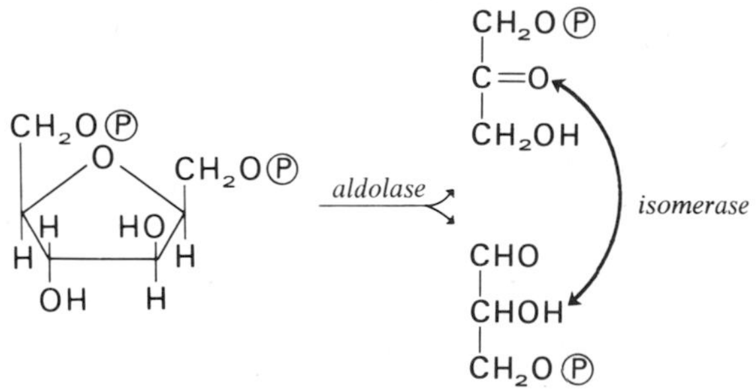

Thus the hexose unit has been twice phosphorylated and then cleaved to yield two identical triose phosphate molecules. Note that two molecules of ATP are required to bring things to this state, not including a possible requirement for ATP in connection with the transport of glucose into the cell. Since ATP is in a real sense the energy currency of the cell, it is plain that glycolysis must yield at least two molecules of ATP just to break even.

In any case, since the two triose phosphate molecules are interconvertible, and since the cell has evolved a major pathway for the removal of one of them, glyceraldehyde phosphate, we may consider it to represent the important product of hexose diphosphate cleavage. In this instance

143

the removal of a reaction product determines not only which of the triose phosphates is on the main stream of glycolysis but even whether the main stream can move forward at all. It turns out that the equilibrium for the aldolase reaction is about 90 percent in the direction of fructose diphosphate, so that if that reaction is not coupled to the removal of triose phosphate by subsequent enzymes, the pathway would not function. The fact that a major enzyme in an important pathway is found to have an equilibrium far in the opposite direction of its normal function illustrates the great importance of sequential reactions in metabolism. One might say that the aldolase reaction is thermodynamically unfavorable and therefore energy-requiring, and the energy is provided by coupling it to spontaneous reactions, in this case those involved in the removal of glyceraldehyde phosphate.

The next stage in glycolysis is the oxidation of glyceraldehyde phosphate. Oxidation is another way of saying that one or more electrons are removed from the molecule, and, the world being what it is, if an electron is removed, it must be put somewhere. In this case the electron is transferred to a coenzyme molecule—specifically, a coenzyme that exists to transfer electrons from one molecule to another, nicotinamide adenine dinucleotide (NAD^+) (see Figure 5-5). NAD^+ is reduced by picking up

Adenosine Triphosphate (ATP)

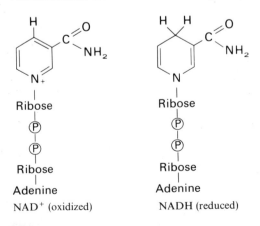

Nicotinamide Adenine Dinucleotide (NAD)

Figure 5-5 Structure of ATP (adenosine triphosphate) and NAD (nicotinamide adenine dinucleotide).

144

two electrons together with a hydrogen atom; it is then denoted NADH. Since there is a net positive charge on the oxidized molecule, we shall write it NAD^+. NAD^+ participates in a good number of cell reactions as an electron donor or acceptor (as a reducing or an oxidizing agent). We call it a *coenzyme* instead of a substrate for historical reasons and because it reacts with a number of different enzymes and might be said to assist them in their role of passing electrons about. In addition, NAD^+ exists in equilibrium between two interconvertible forms within the cell (oxidized and reduced), also a characteristic of coenzymes and not of substrates. Glyceraldehyde phosphate is oxidized in the presence of inorganic phosphate and NAD^+ to give diphosphoglycerate[3]:

$$Glyceraldehyde\ phosphate + H_3PO_4 + NAD^+ \rightleftharpoons$$

$$Diphosphoglycerate + NADH + H^+$$

(see also Figure 5-6). In this case the equilibrium is far to the right (towards products), so that this reaction might be said to pull the previous one. In addition, certain other gains have been made in the matter of extracting energy during the breakdown of glucose, but they are not obvious without consideration of the structure of diphosphoglyceric acid. The reader has already noted that compounds involved in energy matters often have phosphate groups, as in ATP. A *high-energy compound* is generally one that is able to come apart very spontaneously, but in such a fashion as to allow coupling to other energy-requiring reactions. Hydrolysis of the terminal phosphate of ATP is a good case in point and will be seen in a number of contexts. Now, the fact is that all phosphorylated compounds are not necessarily "high-energy" compounds nor do all high-energy compounds contain phosphates, and it is worthwhile considering what constitutes a high-energy bond in a biological context. It will be recalled that a reaction (such as the hydrolysis of a bond) yielding energy is a spontaneous reaction, so a high-energy bond is generally one at which reactions occur readily. Most energy-yielding reactions in biology involve hydrolysis (the addition of water across a bond with subsequent cleavage). The spontaneity of cleavage of the last phosphate bond of ATP implies a large free-energy change, and the bond is often written as a wiggly line. Indeed, other high-energy bonds of the cell have a somewhat similar configuration around them and in many cases may be written

$$
\begin{array}{ccc}
Y & & OH \\
\parallel & & | \\
-X-O- & P & -OH \\
& | & \\
& O^- &
\end{array}
$$

where Y represents carbon or oxygen and X is either carbon or phosphorus. In other words, a phosphate group attached to an atom that

---

[3] The salt of diphosphoglyceric acid. Since biochemical reactions take place around neutral pH, the salts are the major species and we shall denote them this way throughout.

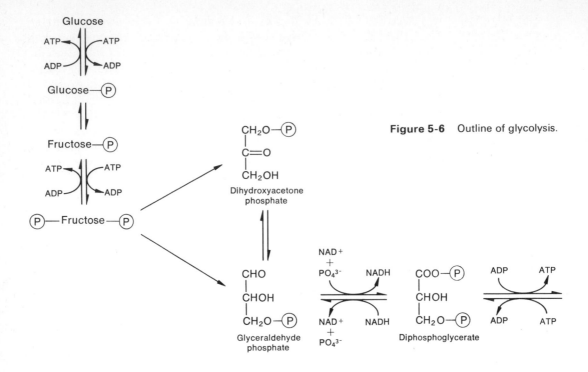

Figure 5-6 Outline of glycolysis.

also bears a double bond may often be a high-energy phosphate compound and commonly written

$$-\overset{\overset{\displaystyle Y}{\|}}{X}\sim PO_4$$

The relatively high free energy of hydrolysis of a $\sim$ bond is related to the electronic structure of the molecule. For example, an important (but not total) reason for the high-energy character of the terminal phosphate bond of ATP is the fact that, at neutral pH, hydrolysis of the bond yields two negatively charged species, phosphate$^-$ and ADP$^-$. Since these similarly charged molecules tend to repel each other, the back reaction is minimized, and equilibrium is rather far in the direction of the products. A similar argument may be made for the other $\sim$ bonds, since the excess electrons on X due to the double bond impart a relative negative charge, which again tends to repel the phosphate group.

In any case, ATP conforms to this general structure with phosphorus as X and oxygen as Y. Diphosphoglycerate has a similar structure,

$$
\begin{array}{c}
\overset{\displaystyle O}{\underset{\displaystyle \|}{}}\quad\quad\overset{\displaystyle O^-}{\underset{\displaystyle |}{}} \\
C\sim O\!-\!\!-\!\!-P\!\!-\!\!-\!\!-OH \\
| \qquad\qquad | \\
H\!-\!C\!-\!OH \quad O^- \\
| \\
CH_2O\!-\!\textcircled{P}
\end{array}
$$

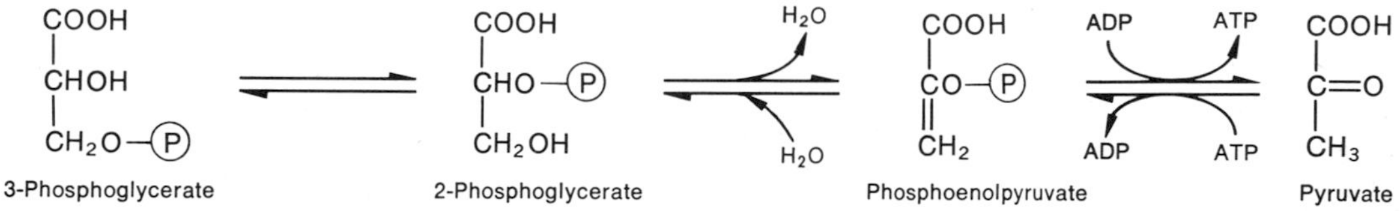

COOH   COOH H_2O COOH   ADP   ATP   COOH

| CHOH | $\rightleftharpoons$ | CHO—(P) | | CO—(P) | | C=O |

CHOH   CHO—(P)   CO—(P)   C=O

CH_2O—(P) CH_2OH H_2O CH_2 ADP   ATP CH_3

3-Phosphoglycerate   2-Phosphoglycerate   Phosphoenolpyruvate   Pyruvate

this time with carbon as the X atom and oxygen as Y. Thus, the reactions of glycolysis thus far should be regarded as a means of forming the characteristic high-energy phosphate arrangement. Once this is accomplished, all that remains is the transfer of the phosphate group from the high-energy "end" of diphosphoglyceric acid to adenosine diphosphate (ADP),

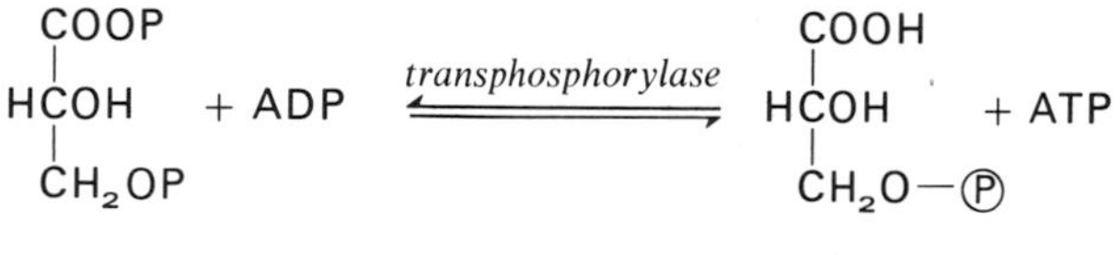

COOP   COOH

HCOH   + ADP $\xrightarrow{\textit{transphosphorylase}}$ HCOH   + ATP

CH_2OP CH_2O—(P)

leading to the synthesis of two molecules of ATP for each one of glucose originally taken up.

Having made some ATP, the glycolytic pathway next involves structural rearrangements, preparatory to achieving the high-energy configuration again to make some more. First, phosphate is transferred from the 3 to the 2 position of glycerate:

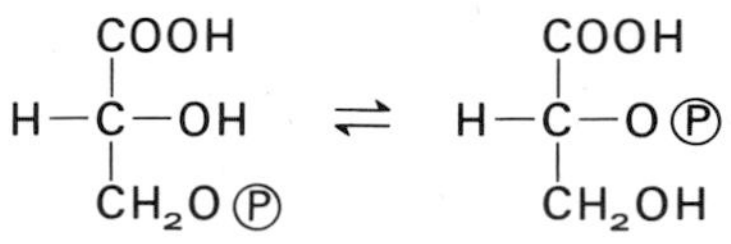

COOH   COOH

H—C—OH $\rightleftharpoons$ H—C—O(P)

CH_2O(P) CH_2OH

Then, the enzyme enolase, which requires magnesium for activity, catalyzes a dehydration, thereby inserting a double bond in the molecule:

$$
\begin{array}{c}
\text{COOH} \\
| \\
\text{H—C—O—ⓟ} \\
| \\
\text{CH}_2\text{OH}
\end{array}
\rightleftharpoons
\begin{array}{c}
\text{COOH} \\
| \\
\text{C}\sim\text{O—ⓟ} \\
|| \\
\text{CH}_2
\end{array}
+ \text{H}_2\text{O}
$$

The resulting compound, phosphoenol pyruvate, is seen to exhibit the now familiar high-energy configuration with a double bond adjacent to a phosphate group. It is again possible to transfer phosphate from an energetic configuration in the phosphoenol pyruvate molecule to one at the terminal phosphate of ATP:

$$
\begin{array}{c}
\text{COOH} \\
| \\
\text{C}\sim\text{O—ⓟ} \\
|| \\
\text{CH}_2
\end{array}
\quad
\begin{array}{c}
\text{ADP ATP} \\
\curvearrowright \\
\rightleftharpoons \\
\curvearrowright \\
\text{ADP ATP}
\end{array}
\quad
\begin{array}{c}
\text{COOH} \\
| \\
\text{C}=\text{O} \\
| \\
\text{CH}_3
\end{array}
$$

Four molecules of ATP have now been formed for every one of glucose utilized.

The other product of the reaction, pyruvate, is one of those pivotal compounds that lead in a number of directions, of which three are summarized in Figure 5-7. Note that pyruvate is reduced in the cases of lactate and ethanol formation. Note also that the reducing agent is NADH. It will be recalled that NADH was formed earlier in the pathway

Figure 5-7 Some reactions of pyruvate. Note that the aerobic metabolism of pyruvate leads to the formation of NADH, while anaerobic metabolism leads to that of NAD$^+$. Under aerobic conditions, the NADH is oxidized by the respiratory chain with associated production of ATP (see Chapter 6).

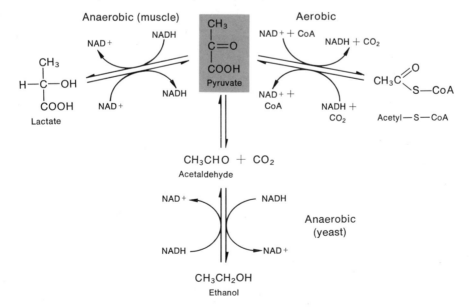

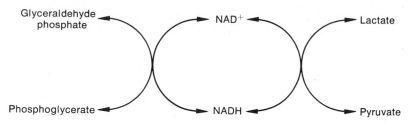

Figure 5-8 Cyclic role of NAD$^+$ in anaerobic metabolism. The function of NAD$^+$ can be summarized by noting that it permits the removal of two electrons in the oxidation of glyceraldehyde phosphate to form phosphoglycerate. These electrons are then used to reduce pyruvate to lactate. Such electron transfer from one part of a metabolic sequence to a subsequent one is characteristic of a fermentation.

in connection with glyceraldehyde phosphate oxidation. The effect of these later reactions leading to NADH oxidation is to regenerate NAD$^+$, which can react again with glyceraldehyde phosphate to keep things going. This cyclic character in the role of NAD$^+$ in glycolysis is summarized in Figure 5-8. Neither the reactions leading to ethanol or lactate are of importance in aerobic cells, and we shall consider only the one major route, decarboxylation and fusion of the 2-carbon units to a coenzyme A residue to form acetyl—S—CoA. Coenzyme A, which bears one of the less informative names in biological chemistry, is a complex molecule,

$$\text{Adenine}-\text{ribose}-\textcircled{P}-\textcircled{P}-\overset{\textcircled{P}}{\underset{\underset{CH_3}{|}}{\overset{|}{C}}}-\overset{H_2}{\overset{|}{C}}-\overset{CH_3}{\overset{|}{C}}-COCO-N\overset{H}{\overset{|}{C}}-\overset{H}{\overset{|}{C}}-CON\overset{H}{\overset{|}{C}}-\overset{H_2}{\overset{|}{C}}-SH$$

of which the business end is a sulfhydryl group. For this reason, we write it CoA—SH, to provide at least a little information about its functional nature. It is a molecule that appears again and again in cell physiology, largely in connection with fat metabolism, and so represents another branch point in the breakdown of glucose.

The precise mechanism of the oxidative decarboxylation of pyruvic acid is somewhat more complex than as described in Figure 5-7. The process may be broken down into several steps, the first being the decarboxylation itself. The reaction requires the presence of a derivative of thiamine (vitamin B$_1$), *thiamine pyrophosphate*,

which turns out to be required for a number of other decarboxylations as well. This is a good example of the importance of vitamins as cofactors

149

in essential reactions of the cell, an observation that we shall make again in other contexts. The reaction also involves a second cofactor, a derivative of lipoic acid lipoamide, which is likewise associated with decarboxylations of other compounds.

The energy yield of glycolysis

It is clear from the preceding discussion that an important result of the glycolytic breakdown of sugars is the synthesis of ATP and that a certain amount of ATP is used in the process. Thus we find a total of four moles of ATP synthesized per mole of glucose broken down, two each by phosphate transfer from diphosphoglycerate and phosphoenol pyruvate. On the other hand, the initial steps of glycolysis require the use of two molecules of ATP, which serve as phosphate donors. In other words, there is a net synthesis of two moles of ATP per glucose molecule. This represents the yield of that part of glucose breakdown that can occur in the absence of oxygen. We shall see in Chapter 6 that aerobic processes involved in the further degradation of pyruvate to CO_2 and H_2O can lead to much greater energy yields (see Table 6-1).

Of course, glucose is by no means the only carbohydrate used by cells, and energy yields depend on the exact situation. For example, glycogen, an extremely large and highly branched polymer of glucose, is an important intracellular carbohydrate storage form in many animal cells. The breakdown of glycogen requires inorganic phosphate and hence leads to the production of glucose phosphate without a requirement for ATP. For this reason, the formation of fructose diphosphate requires one, rather than two, molecules of ATP, and the net yield of ATP from glycogen is three, rather than two, as in the case of glucose. On the other hand, glycogen synthesis requires energy in the form of ATP, so that when its synthesis is taken into consideration, it is a no more efficient energy source than is glucose.

The Krebs cycle

Further breakdown of glucose beyond acetyl —S—CoA requires the presence of oxygen and, from an energetic point of view, is by far the most important route for glucose metabolism. The first clues as to the mechanism by which it takes place were obtained prior to 1940, when it was learned that oxidation of pyruvate was stimulated by several dicarboxylic acids, of which one example was malic acid:

COOH
|
CHOH
|
CH₂
|
COOH

It developed that oxidation of acetyl—S—CoA proceeds by a pathway involving four-carbon acids, a series of reactions commonly named after Krebs, who first described its cyclic nature.

The "primary" event of the Krebs (or citric acid) cycle is the fusion of a molecule of acetyl—S—CoA to a four-carbon dicarboxylic acid, oxaloacetate, to form citrate:

$$CH_3-C{\overset{O}{\underset{S-CoA}{\diagup\diagdown}}} + {\overset{COOH}{\underset{\underset{COOH}{CH_2}}{\underset{|}{C=O}}}} \rightleftharpoons HO-{\overset{CH_2COOH}{\underset{CH_2COOH}{CHCOOH}}} + CoA-SH$$

The rest of the cycle is concerned with the conversion of citrate (six carbons) back to oxaloacetate (four) by a series of decarboxylations and oxidations, which will be seen to be associated with energy conversion through ATP synthesis. The reactions by which this is accomplished are summarized in Figure 5-9 and represent a series of structural rearrangements leading to oxidations. For example, citrate is converted to isocitrate, which is then oxidized.

Some remarks about the individual reactions may be of interest. There are four oxidations in the cycle, of which two are linked to NAD^+ reduction. Of the others, isocitrate oxidation is catalyzed by isocitrate dehydrogenase and involves passage of an electron to $NADP^+$, which is similar to NAD^+, but with an additional phosphate in the ribose part (see above). Oxidation of succinate occurs by the passage of electrons to oxidized flavin bound to the enzyme that catalyzes the oxidation, succinate dehydrogenase. Obviously, mechanisms must exist for the re-oxidation of the NADH, NADPH, and reduced flavin molecules formed. The nature of these oxidations, which take place by way of the cytochrome chain, is the subject of Chapter 6.

It is also seen that there are two decarboxylations between citrate and oxaloacetate. Of these, the decarboxylation of α-ketoglutarate is very similar in mechanism to that of pyruvate, both leading to a coenzyme A derivative and both requiring a thiamine derivative and lipoamide

$$\overset{S-S}{\diagup\diagdown}\underset{H}{\diagdown}CH_2CH_2CH_2CH_2CONH_2$$

as intermediates. Both are oxidative decarboxylations and the oxidizing agent (electron acceptor) is NAD^+. The similarity between the two reactions illustrates nature's tendency to operate with economy and to use similar techniques to accomplish similar ends wherever appropriate.

The reader will note that the purpose of the Krebs cycle, if it is only to degrade a six-carbon compound to one with four carbons, is fulfilled when succinyl—S—CoA is reached. The rest of the cycle might be looked on as only structural rearrangements leading to oxaloacetate, which

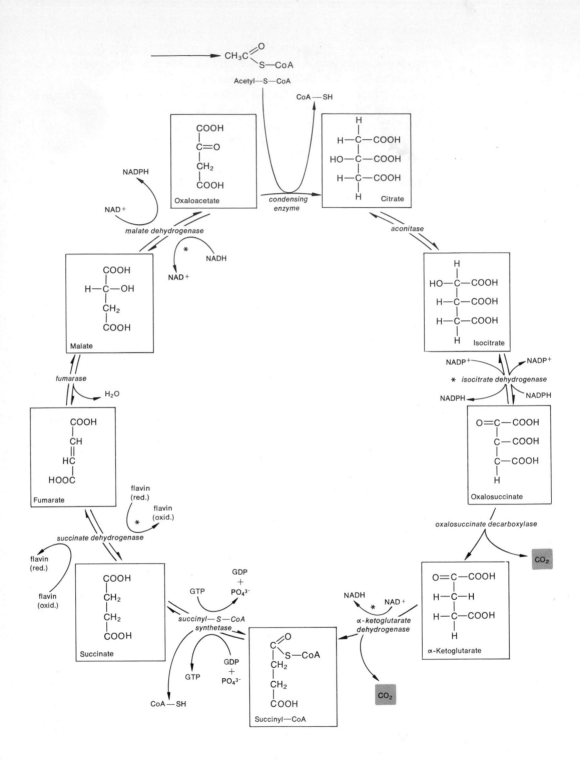

Figure 5-9 Citric acid (Krebs) cycle. Names of enzymes are shown in italics. Dehydrogenase is abbreviated dh. Decarboxylations are emphasized by shading the CO_2 produced. Oxidative steps are denoted with an *.

would then pick up another acetyl—S—CoA molecule. This would be a serious mistake. It is a crude oversimplification to regard the Krebs cycle as only a mechanism for pyruvate removal. In fact, one of the chief functions of the cycle is to lead to ATP production, and the means whereby this is done will be described in detail later. For the time being, it is sufficient to note that ATP production is associated with each oxidative site of the cycle. More exactly, it is associated with the oxidation of NADH, NADPH, and reduced flavoprotein formed at the oxidative sites. Thus, much of the organic chemistry of the Krebs cycle must be seen in the light of arrangements in structure that make oxidation possible; the nonoxidative steps prepare the way for the oxidative ones.

Although most of the ATP synthesis via the Krebs cycle, occurring by way of reoxidation of NADH or another electron donor, will be discussed in Chapter 6, there is one example of ATP production occurring in the cycle itself at the substrate level, so to speak. The formation of succinate from succinyl—S—CoA leads to one molecule of ATP by a two-step mechanism,

$$\text{Succinyl—S—CoA} + \text{GDP} + \text{phosphate} \rightleftharpoons \text{Succinate} + \text{GTP}$$

where GDP and GTP are guanosine di- and triphosphate, closely similar in structure to ADP and ATP. This is followed by a phosphate-group transfer reaction, leading to ATP synthesis:

$$\text{GTP} + \text{ADP} \rightleftharpoons \text{GDP} + \text{ATP}$$

The role of the Krebs cycle in synthesis

A major function of the Krebs cycle is the coupling of hexose break-down to a series of oxidations that lead to ATP formation. However, this is not the whole story, inasmuch as the cycle also plays an important role in the synthesis of a number of compounds required by the cell. We said that pyruvate and acetyl—S—CoA represent branch points in glycolysis, leading, in the case of acetyl—S—CoA, to the synthesis of fats and related compounds. Similarly, the Krebs cycle includes compounds that are important precursors, most notably in the synthesis of amino acids.

For example, α-ketoglutarate leads directly to the synthesis of the amino acid glutamate by the following reaction:

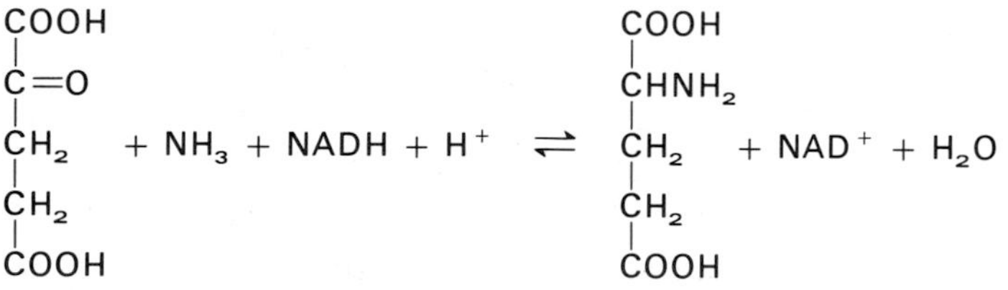

$$
\begin{array}{l}
\text{COOH} \\
\,|\, \\
\text{C}=\text{O} \\
\,|\, \\
\text{CH}_2 \\
\,|\, \\
\text{CH}_2 \\
\,|\, \\
\text{COOH}
\end{array}
+ \text{NH}_3 + \text{NADH} + \text{H}^+ \rightleftharpoons
\begin{array}{l}
\text{COOH} \\
\,|\, \\
\text{CHNH}_2 \\
\,|\, \\
\text{CH}_2 \\
\,|\, \\
\text{CH}_2 \\
\,|\, \\
\text{COOH}
\end{array}
+ \text{NAD}^+ + \text{H}_2\text{O}
$$

153

Glutamate is, in turn, a precursor of a number of other amino acids, including ornithine, proline, and aspartate. The reaction

COOH   COOH   COOH   COOH

$$
\begin{array}{ccccc}
\text{COOH} & \text{COOH} & & \text{COOH} & \text{COOH} \\
| & | & & | & | \\
\text{CHNH}_2 & \text{C}{=}\text{O} & & \text{C}{=}\text{O} & \text{CHNH}_2 \\
| & | & & | & | \\
\text{CH}_2 & +\;\text{CH}_2 & \rightleftharpoons & \text{CH}_2 & +\;\text{CH}_2 \\
| & | & & | & | \\
\text{CH}_2 & \text{COOH} & & \text{CH}_2 & \text{COOH} \\
| & & & | & \\
\text{COOH} & & & \text{COOH} &
\end{array}
$$

Glutamate   Oxaloacetate   α-Ketoglutarate   Aspartate

is an example of the class of reactions known as *transaminations*, which are very important in amino acid metabolism. Fumarate and succinyl— S—CoA both lead to a number of additional synthetic routes, so that Krebs cycle intermediates are constantly siphoned off at a number of locations to manufacture needed cell components. These interrelations of the Krebs cycle are summarized in Figure 5-10 and give a good idea of the central importance of the cycle in synthetic metabolism. The cycle is fundamentally involved in the energy-producing breakdown of hexose as well as many syntheses, most of which require energy. This suggests that the traditional distinction between catabolic (breakdown) and anabolic (synthetic) pathways is quite invalid.

Figure 5-10   Some side reactions of the Krebs cycle. The reactions leading from the keto acids, oxaloacetate, and α-ketoglutarate to amino acids are transaminations. Taken together, these side reactions indicate the importance of the Krebs cycle in cellular synthesis.

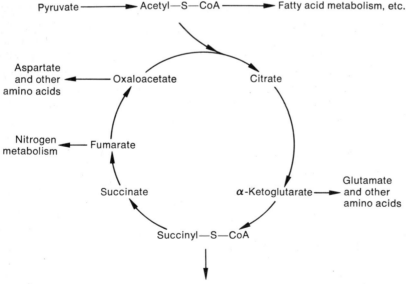

The glyoxylate cycle

The constant siphoning off of intermediates also raises a problem. The continued oxidation of acetyl—S—CoA by way of the cycle requires oxaloacetate to react with it. If there are no losses, each molecule of citrate will lead to one of oxaloacetate and the cycle will be truly catalytic. However, measurements using labeled acetate indicate that in bacteria most of the label is lost in one passage through the cycle—that is, most of the citrate leads to the synthesis of compounds other than oxaloacetate. This poses a serious problem, since if no mechanisms exist for the formation of additional oxaloacetate by routes other than its condensation with acetyl—S—CoA followed by the reactions of the Krebs cycle, the whole affair will grind to a halt.

Figure 5-11 Glyoxylate cycle. The names of the two enzymes of the pathway are given in italics.

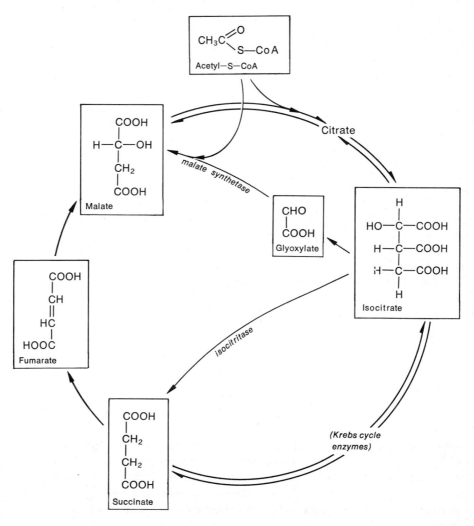

Luckily for us, there are other routes for its synthesis. To cite just one example, in many cells an enzyme, called the *malic enzyme*, catalyzes the formation of malate from pyruvate by the reaction

$$Pyruvate + CO_2 + NADPH + H^+ \rightleftharpoons Malate + NADP^+$$

Note that the NADPH may be formed at the isocitric dehydrogenase step of the Krebs cycle, and that malate can be readily transformed to oxaloacetate by the malic dehydrogenase reaction

$$Malate + NAD^+ \rightleftharpoons Oxaloacetic\ acid + NADH + H^+$$

An additional system for replenishing Krebs cycle intermediates exists in many plant and bacterial cells and is known as the *glyoxylate cycle* (Figure 5-11). It represents a sort of short circuit of the Krebs cycle, providing a way for additional acetyl—S—CoA to add carbon to the cycle intermediates. By it, each molecule of isocitrate leads to the formation of two four-carbon acids, succinate and malate, assisting in the maintenance of the level of oxaloacetate.

The relationship between the Krebs cycle and nitrogen excretion

A final example of the close connections between the citric acid cycle and other metabolic processes is the manner in which the waste products of amino acid breakdown are eliminated. This is largely a problem of what to do with ammonia, since much of amino acid degradation may be looked at as deamination. Once an amino acid is deaminated, the remaining acid may enter metabolism at a number of points. Ammonia is a relatively troublesome waste product, being rather caustic in solution (as NH_4OH). Aquatic vertebrates in many cases are able to wash away the ammonia into their environment rapidly enough to get by, but cells from other animals have found it more advantageous to eliminate nitrogen in the form of urea,

$$
\begin{array}{l}
NH_2 \\
| \\
C{=}O \\
| \\
NH_2
\end{array}
$$

which has the advantage of including two waste products, NH_3 and CO_2 (for example, from the Krebs cycle). An examination of the pathway by which urea is made from ammonia (Figure 5-12) will show interrelations with the Krebs cycle (at fumarate and oxaloacetate) and again illustrates the cycle's central role in the chemical activities of the cell.

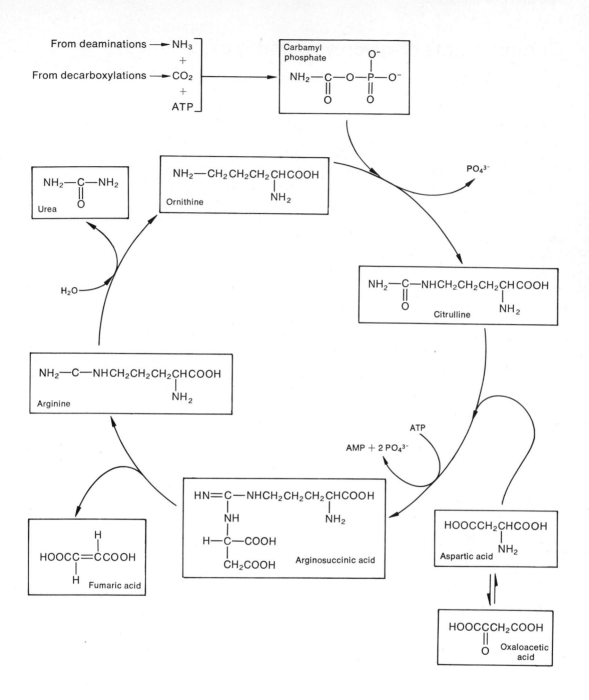

Figure 5-12 Urea cycle. Note the relation of this cycle to the Krebs cycle via oxaloacetate and fumarate.

157

Control points of glycolysis and the Krebs cycle

We have made it clear that the glycolytic and Krebs pathways play a role in the cells where they occur, involving much more than hexose metabolism and forming a sort of main line from which routes to many other syntheses branch off. Thus, it would be hard to imagine a set of reactions more important to the cells that utilize them, and it is inconceivable that they should operate free of tight and sensitive controls. Regulation of these systems should include the possibility of shunting the breakdown of hexose to whatever synthesis is most relevant to the cell at the time and should also provide for adequate synthesis of ATP under whatever conditions prevail in the cell environment.

The most simple control of reactions in any metabolic sequence has to do with the direction in which the pathway may operate. Although all reactions may in principle be reversible, in fact, many are essentially unidirectional under the conditions actually present (see Figure 5-13). Under cellular conditions, for example, many reactions involving the evolution of CO_2 are in effect one way, as mechanisms exist in the cell for the rapid elimination of the CO_2 formed. Since there are several decarboxylations in the breakdown of carbohydrates, these reactions provide a sort of barrier preventing the pathway as a whole from running backward. In contrast, portions of the glycolytic (and Krebs) pathways will run in reverse, and such reversal is of extreme importance, to cite one example, in the light-induced synthesis of carbohydrates in green plants.

A number of comments may be made about the control of the glycolytic pathway. In the first place, control may be affected at the outset by the available hexose within the cell. The entry of hexose in many cases requires the presence of ATP. Thus, it is possible to envision situations in which the available energy level of a cell might become so low that the path for additional ATP synthesis, glycolysis, and consequent relief from the energy depletion might be unavailable. A severe decline in ATP might thus be irreversible. It is likewise possible that the entry of hexose is under some sort of endocrine control in multicellular organisms. There is evidence that insulin affects the rate of hexose transport, providing an additional sort of control at the level of hexose entry.

Once in the cell, hexose must be phosphorylated (twice) using ATP. Since ATP itself is produced farther on in the pathway, this represents a form of positive feedback of "the more you have, the more you get" variety. But a system under the regulation of positive feedback only would be rather explosive. Luckily, the positive influence of ATP is opposed by several important negative controls involving the pyridine nucleotides NAD^+ and NADH, as well as in organic phosphate and ATP itself. NAD^+ plays an especially important role in coordinating the rates of flow through the different parts of the glycolytic pathway. This is because NAD^+ is required at a fairly early stage in glycolysis for the triose phosphate dehydrogenase reaction, leading to NADH formation.

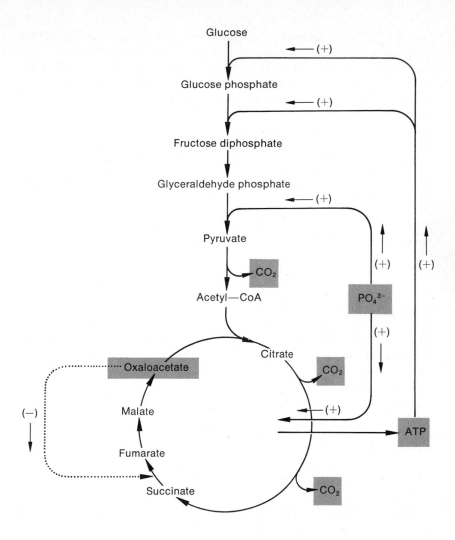

Figure 5-13 Relationship between glycolysis and the Krebs cycle. This figure illustrates negative feedback inhibition of the Krebs cycle by oxaloacetate, competition for phosphate between glycolytic and oxidative phosphorylations, and the requirement for ATP, produced in the oxidative reactions of the Krebs cycle, in the phosphorylation of sugars in glycolysis. This last, positive, effect of ATP should be compared with the inhibitory influence of ATP on phospho-fructokinase, which is outlined in Figure 5-14 and which is of great regulatory significance. Note also the steps in the pathway where decarboxylations occur, rendering those reactions practically unidirectional.

The NADH thus formed is required for the reduction of pyruvate farther along, to form ethanol or lactate. Under anaerobic conditions, where the oxidation of NADH by molecular oxygen does not take place, the linkage between NADH formation and utilization is quite tight and of apparent significance in keeping the two ends of glycolysis going at rates consistent with each other.

The utilization of oxygen leads to a different situation. In the first place, the oxidations of the Krebs cycle, all of which require molecular oxygen, can now take place. For example, NADH oxidation proceeds rapidly by an aerobic mechanism, described in Chapter 6. Thus, the presence of oxygen leads to more NAD^+ and less NADH, which would have the effect of preventing the synthesis of lactate from pyruvate (which requires NADH) and favoring the synthesis of acetyl$-S-$CoA (which requires NAD^+), which then leads to the Krebs cycle.

The Pasteur effect

An oxygen-induced increase in the NAD^+/NADH ratio might lead one to expect stimulation of glycolysis as a whole, owing to an increase in the rate of the triose phosphate dehydrogenase step. Although a reasonable expectation, it is quite untrue. It was discovered about a century ago by Pasteur that oxygen inhibits glycolysis, measured as the rate of carbohydrate breakdown. In other words, it is clear that something other than the NAD^+ oxidation–reduction state is regulating the rate of glycolysis in the presence of oxygen.

This inhibition of glycolysis, which is called the *Pasteur effect*, probably involves a complex mechanism, but it is becoming increasingly clear that an important element of the control is the allosteric inhibition of phosphofructokinase by ATP, mentioned earlier. Thus, oxygen leads to increased ATP synthesis via the oxidations of the Krebs cycle (by a mechanism described in Chapter 6), and the ATP produces inhibition at the phosphofructokinase level. It is interesting that the same enzyme is inhibited also by citrate, which, like ATP, may be regarded as a product of glycolysis. Thus, two feedback loops appear to converge at the same locus of action.

Under anaerobic conditions, the synthesis of ATP occurs to a much lesser extent, and inhibition of glycolysis does not occur. Thus, inability to carry out oxidative ATP synthesis is, in part, counteracted by increased glycolytic ATP formation. Indeed, under such conditions, a fall in intracellular ATP concentration is mirrored by an increase in its hydrolysis products, adenosine di- and monophosphate, ADP and AMP. Significantly, both compounds not only fail to inhibit phosphofructokinase but actually stimulate its activity. Again, such activation appears to be *allosteric*, in reflecting a conformational change in the enzyme molecule.

In addition, a second (and probably secondary) regulatory mechanism contributing to the Pasteur effect resides in the availability of phosphate for the various reactions that require it. Glycolysis requires inorganic phosphate at the level of glyceraldehyde phosphate dehydrogenase,

160

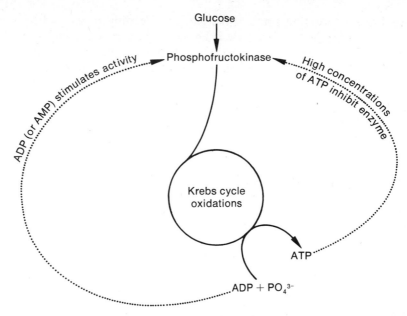

Figure 5-14 Control of glycolysis by allosteric properties of phosphofructokinase. Although ATP is required for the reaction catalyzed by this enzyme,

Fructose-6-phosphate + ATP $\rightleftharpoons$ Fructose-1,6-diphosphate + ADP

high levels inhibit activity. On the other hand, activity is enhanced by ADP or AMP. For this reason, the enzyme serves to "sense" the ATP/ADP ratio, which, in turn, reflects the energy-producing output of the Krebs cycle and associated oxidations.

and in its absence the whole pathway must come to a halt. Phosphate is also required in the various oxidations of the Krebs cycle, and, for reasons discussed in Chapter 6, about three molecules of phosphate are required at each of the four oxidations for the cycle to operate at full rate. This means that, in the presence of oxygen, the Krebs cycle will compete very effectively with glycolysis, and the low availability under aerobic conditions might account in large measure for inhibition of glycolysis, primarily at the glyceraldehyde phosphate dehydrogenase step. It should be added that, because glycolysis is largely cytoplasmic, while Krebs cycle enzymes are located in the mitochondrion, the competition for phosphate is also influenced by the ability of the mitochondrial membrane to effect its transport.

The role of oxaloacetate in regulation

There are also feedback controls operating in the Krebs cycle to maintain its rate between limits. Obviously, oxaloacetate is required at the outset of the cycle to lead to the formation of citrate, so that any process limiting the concentration of oxaloacetate could limit the rate of the whole cycle. Oxaloacetate has, however, an additional controlling effect, as it is known to inhibit some of the reactions leading to its own production. For example, oxaloacetate is known to be a potent competitive inhibitor of succinate dehydrogenase, so that excessive increase in its concentration would lead to the inhibition of its formation until a lower level was attained. Oxaloacetate is also an inhibitor of malate oxidation and appears to influence enough other mitochondrial reactions to suggest that it is a focal point of control in oxidative metabolism as a whole.

Constant proportions of enzymes

We have seen that the concentration of an enzyme in solution directly influences the rate of reaction and that enzyme concentrations are subject to control. Obviously, other forms of regulation will be of little use if enzyme concentrations are either in enormous excess or are insufficient to carry on reactions at a rate commensurate with the rate of the pathway as a whole. In this connection, it is interesting that mitochondria isolated from different tissue contain groups of enzymes that exist in constant proportion to each other, no matter what the source. For example, the enzymes of pyruvate oxidation and the Krebs cycle form a constant pattern in different tissues, with the activity of malate dehydrogenase exceeding that of glutamate transaminase by a factor of about 2. The transaminase, in turn, exceeds succinate dehydrogenase

Figure 5-15 Levels of enzyme activity in mitochondria from different sources. Activities are calculated on the basis of that of cytochrome c. In this figure, "transaminase" refers to glutamate-oxaloacetate transaminase. Data are from an article by M. Klingenberg, in *Methods in Enzymology*, vol. 10, R. Estabrook and M. Pullman (eds.), Academic Press, New York, 1967, p. 3. Additional details may be found there.

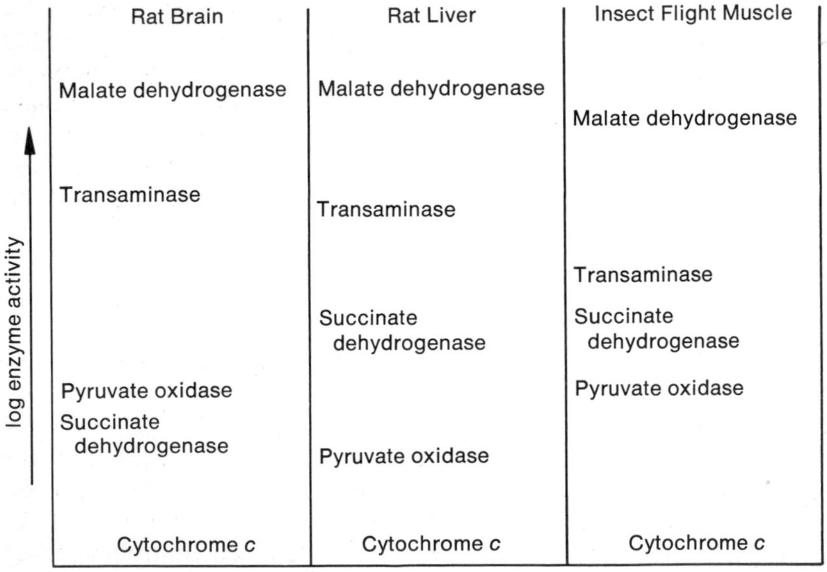

and pyruvate decarboxylase by a factor of about 10. These constant patterns suggest that there are mechanisms by which the production of enzymes is adjusted to levels required for their overall function. The absence or presence of a constant proportion between two enzymes is also useful in determining whether or not the two enzymes are functionally related *in vivo*—that is, whether their connection by a pathway represents an important route of interconversion in the real cellular world.

The preceding paragraphs have given examples of regulatory mechanisms in hexose metabolism, but the examples are by no means inclusive, Additional ones will appear in Chapter 6, and for more detail the reader is directed to the book by Lehninger or the symposium edited by Tager et al., both listed in the Suggested Reading for this chapter. It is useful, however, to point out that there are three general sorts of regulation, which together include the cases that we have examined. In the first place, feedback inhibition of enzyme activity plays an important role, as in oxaloacetate control of the Krebs cycle. A second order of control of the pathways involves regulation of enzyme synthesis, so that enzymes are present in appropriate concentrations to their role. Finally, competition for common intermediates plays a central role in metabolic pathways. Important control points are defined by intermediates used in a number of different reactions and not present in large excess. For example, acetyl$-$S$-$CoA probably plays such a part in that it is at once involved in fatty acid synthesis, the glyoxylate pathway, and the oxidation of pyruvate, so that the processes (among others) are in competition for it and essentially regulate each other by their ability to compete. Other common intermediates that appear to exert control on glycolysis and the Krebs cycle include NAD^+, phosphate, and, as we shall see, ADP. In many cases the regulation depends on a balance between two components, so that glycolysis is really under the control of the NADH/NAD^+ ratio and, for reasons that will shortly become clear, the Krebs cycle is sensitive to regulation by ADP/ATP.

Suggested Reading

Books

CIBA Foundation Symposium, *Regulation of Cell Metabolism*, G. E. W. Wolstenholme and C. M. O'Connor (eds.), Little, Brown and Co., Boston, 1959.

Fruton, J. S., and S. Simmonds, *General Biochemistry*, 2nd ed., John Wiley & Sons, New York, 1958, Chs. 18–21.

Goodwin, T. W. (ed.), *The Metabolic Roles of Citrate*, Academic Press, New York, 1968. (A Biochemical Society Symposium in honor of Sir Hans Krebs.)

Greenberg, D. M. (ed.), *Metabolic Pathways*, vols. 1–2, Academic Press, New York, 1960–61.

Lehninger, A. L., *Biochemistry*, Worth Publishers, New York, 1970, Chs. 13–16.

Lehninger, A. L., *Bioenergetics*, W. A. Benjamin, New York, 1965.

Lowenstein, J. M. (ed.), *The Citric Acid Cycle*, vol. 13 of *Methods in Enzymology*, Academic Press, New York, 1969.

Lowenstein, J. M. (ed.), *Citric Acid Cycle: Control and Compartmentation*, Marcel Dekker, New York, 1969.

Tager, J. M., S. Papa, E. Quagliarello, and E. C. Slater, *Regulation of Metabolic Processes in Mitochondria*, American Elsevier Publishing Co., New York, 1966.

Articles

Axelrod, B., "Glycolysis," in *Metabolic Pathways*, 3rd ed., D. M. Greenberg (ed.), vol. 1, Academic Press, New York, 1967, p. 112.

De Duve, C., R. Wattiaux, and P. Baudhuin, "Distribution of enzymes between subcellular fractions in animal tissues," in *Advances in Enzymology*, F. F. Nord (ed.), vol. 24, John Wiley & Sons (Interscience Division), New York, 1962, p. 291.

Greville, G. D., "Pyruvate oxidation and the citrate cycle," in *Carbohydrate Metabolism and Its Disorders*, E. Dickens, P. J. Randle, and W. J. Whelan (eds.), vol. 1, Academic Press, New York, 1967, pp. 297–334.

Kornberg, H. L., "The coordination of metabolic routes," *Symp. Soc. Gen. Microbiol.*, **15**, 8–31 (1965).

Kornberg, H. L., and S. R. Elsden, "The metabolism of 2-carbon compounds by microorganisms," in *Advances in Enzymology*, F. F. Nord (ed.), vol. 23, John Wiley & Sons (Interscience Division), New York, 1961, p. 401.

Krebs, H. A., "The tricarboxylic acid cycle," in *The Harvey Lectures*, 1949–50, Academic Press, New York, 1950.

Lowenstein, J. M., "The tricarboxylic acid cycle," in *Metabolic Pathways*, 3rd ed., D. M. Greenberg (ed.), vol. 1, Academic Press, New York, 1967, p. 176.

Passoneau, J. V., and O. H. Lowry, "Phosphofructokinase and the control of the citric acid cycle," *Biochem. Biophys. Res. Commun.*, **13**, 372 (1963).

6 Energy conservation— nonphotosynthetic

It was said in Chapter 5 that a central consequence of the aerobic breakdown of hexose is ATP synthesis, which occurs at the oxidative steps of the Krebs cycle. Since ATP is involved in most of the energy-requiring reactions of the cell, clearly its synthesis is of great importance in the life of the cell, and the control of its formation would be expected to play a role in the control of numerous other events. This being the case, it is worthwhile to examine the matter of ATP formation in more detail and to pay special attention to the matter of its control.

ATP is formed by the esterification of inorganic phosphate with the terminal phosphate of ADP:

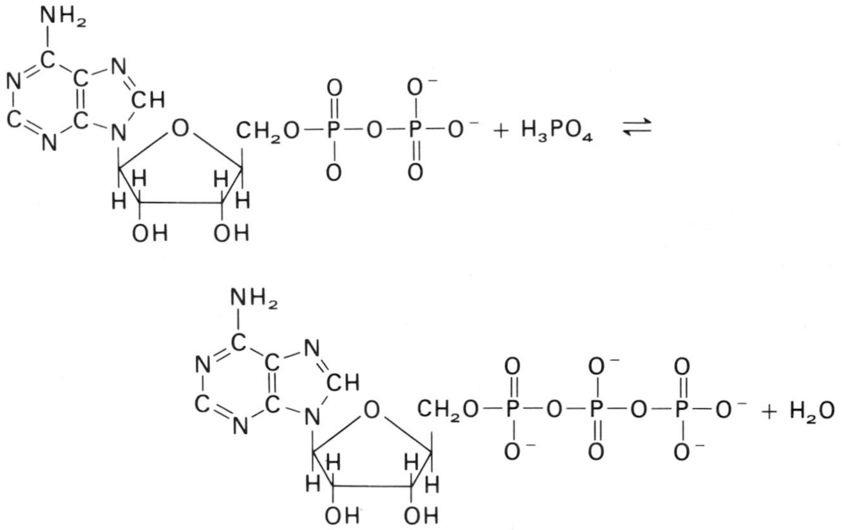

Cells have evolved three main routes to ATP synthesis, including substrate-linked phosphorylation (as in glycolysis), photosynthetic phosphorylation (see Chapter 7), and oxidative phosphorylation, which is the subject of this chapter. In addition, there are a number of ways ATP may be formed without a net increase in the available energy, an example being the myokinase reaction:

$$2\,ADP \;\rightleftharpoons\; ATP + AMP \qquad (AMP = \text{adenosine monophosphate})$$

Additional routes leading to ATP synthesis include a number of phosphate exchanges where phosphate is transferred from another nucleotide triphosphate (such as guanosine triphosphate) to ADP. Such reactions encountered earlier in the Krebs cycle do not result in an increase of "high-energy phosphate" but rather a change in its identity.

The reactions that lead to a net increase in high-energy phosphate compounds, of which ATP is the most important, have one central feature in common, and that feature will be stated here in the form of a central dogma: *Cells obtain energy only by moving electrons.* One can be a bit more explicit and say that cells really only capture energy by dropping electrons from a higher energy state to a lower one. A physiologist of a chemical persuasion would put it a bit differently, saying that energy transfer in cells is a matter of oxidation and reduction. It will be recalled that oxidation is the removal of an electron from something, while reduction is adding it to something else. Obviously, the two must be coupled together, because, when something is reduced, something else must be oxidized. The truth of the central dogma is seen in the fact that all respiratory ATP synthesis is coupled to oxidations and, specifically, to the transfer of electrons from a substrate, such as a Krebs cycle acid, where they may be said to have a high energy, to an electron acceptor such as oxygen. The difference between the energy of the electron in the substrate and in oxygen represents the amount of energy available to make ATP.

Photosynthesis will be seen in Chapter 7 to involve the passage of electrons from a high-energy state of the chlorophyll molecule to another acceptor, coupled to the synthesis of ATP. The applicability of the central dogma to the substrate-linked synthesis of ATP coupled to anaerobic glycolysis (see Chapter 5) is a bit more subtle. Suffice it to say that, in the breakdown of hexose, ATP formation is coupled to the passage of electrons from one part of the glycolytic pathway to another, carried in the form of NADH. Indeed, glycolysis, functioning in anaerobic cells, together with the reduction of pyruvate to lactate, is an example of the more general class of pathways called *fermentation*, which we define as a system for coupling ATP synthesis to electron flow, where the electron donor is an organic compound and the acceptor is a second organic compound derived from the first:

Hexose $\longrightarrow$ Triose—Ⓟ $\longrightarrow$ Pyruvate $\longrightarrow$ Lactate (or Ethanol)
NADH

It is easy to observe that cellular-energy capture invariably involves the movement of electrons, but the nature of the coupling between electron flow and ATP synthesis is a subject of intense research and remains, except in the case of substrate-linked phosphorylation, largely unexplained. The coupling process, which might be summarized

$$\text{Electron donor} \xleftarrow{\quad\underset{\text{ADP}+\circled{P}\qquad\text{ATP}}{}\quad} \text{Acceptor}$$

is unclear, both in general outline and in particulars. In fact, recent experimental work has been quite unable to distinguish between two widely differing theories of the coupling process, an impasse that is discussed below.

There is, however, a great deal of knowledge about the mechanisms for oxidation itself, and a large body of descriptive information exists about ATP synthesis, although the mechanism is, as we said, unclear. Much of what is known about the process is directly related to the matter of control of ATP production and is discussed shortly.

Oxidation-reduction enzymes

There are a huge number of known oxidative enzymes, and their study is a field of its own, with special techniques and language. For our purposes, we will consider all the enzymes as falling into one of two classes and consider the properties of the classes as a whole.

1. One class includes enzymes that catalyze oxidations by accepting electrons themselves and thus altering their own oxidation–reduction state.
2. Another class includes only enzymes that mediate the transfer of electrons between donor and acceptor without serving as electron donor or acceptor.

An example of the first sort of enzyme is succinate dehydrogenase, which catalyzes the oxidation of succinate by withdrawing two electrons, which reduce the flavin prosthetic group of the enzyme itself. The enzyme is then oxidized by another enzyme and is again ready to accept electrons from succinate. In a very real sense, such an enzyme is both a catalyst and a reactant.

A good example of the second class of enzymes is malate dehydrogenase, which catalyzes the oxidation of malate by NAD^+. The enzyme, in this case, is not oxidized or reduced and might be said to have the function of bringing malate and the pyridine nucleotide NAD^+ together in the right spatial arrangement to ensure rapid reaction. Both of these enzymes are called *dehydrogenases*, as both react with a substrate in an oxidation involving removal of two protons. The central event in the oxidation involves the removal of electrons, rather than hydrogen, and it is probably fair to say that these oxidations involve the transfer of hydrogen only in a secondary sense, as a consequence of electron withdrawal. This statement is based on the fact that the subsequent oxidation

167

of succinate dehydrogenase or NADH occurs by way of steps where only electron transfer appears to take place. The two processes are, however, very hard to separate, since hydrogen ion is involved in a fundamental way with ATP synthesis itself.

A general equation for oxidative enzyme action is given as

$$AH_2 + B \rightleftharpoons BH_2 + A$$

or

$$A \text{ (reduced)} + B \text{ (oxidized)} \rightleftharpoons A \text{ (oxidized)} + B \text{ (reduced)}$$

In our first class of enzymes, B is a part of the enzyme itself, whereas in the second it is a separate entity, often called a *coenzyme*, as in the case of NAD^+.

The enzymes that carry out cellular oxidation–reduction reactions are concentrated in specific regions of the cells that contain them. For example, in the case of prokaryotic cells, the plasma membrane contains most of the oxidative enzymes as well as those involved with ATP synthesis. In higher (eukaryotic) cells, the corresponding enzymes are located in the mitochondria and, in fact, comprise much of the structure of the inner mitochondrial membrane (see Figures 6-1 and 6-2).

Figure 6-1   Structure of a mitochondrion, illustrating mitochondrion configuration as found in pancreatic cells. Note that the *cristae* greatly enlarge the surface of the inner membrane. The granules shown could contain any one of a number of materials. In some instances they appear to contain high-molecular-weight polymers of sugars; under some circumstances they may be formed from insoluble salts of calcium. The matrix contains many enzymes in apparent solution, as well as DNA and RNA. The small specks represent (diagrammatically) ribosomes that are similar to those in bacteria and which are the sites of intramitochondrial protein synthesis.

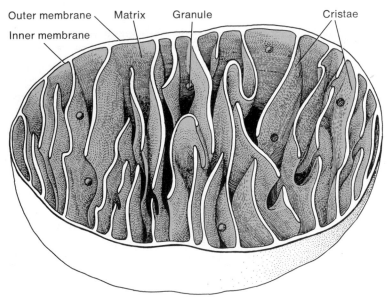

Outer membrane   Matrix   Granule   Cristae

Inner membrane

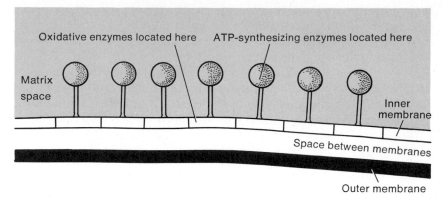

Figure 6-2 Structure of the inner mitochondrial membrane. The knobs where ATP is synthesized are identical to the F_1 coupling factor described later in this chapter. The stalks, when isolated from the remaining structure, appear to confer sensitivity to the inhibitor of ATP synthesis, oligomycin.

The respiratory chain

Let us turn our attention now to specific oxidations associated with the Krebs cycle. They are listed in Table 6-1, which also includes information about the quantity of ATP synthesized per mole of oxidizable substrate. In Table 6.1 it is seen that one of the four oxidations is linked to the reduction of the enzyme itself, while the others have one or the other of the pyridine nucleotides as acceptor. It is worth adding at this point that the reason for the four ATP's produced at the α-ketoglutarate dehydrogenase step is due to an extra substrate-linked phosphorylation similar to those described in Chapter 5. For our purposes, it is also worth noting that enzymes termed *transhydrogenases* catalyze the conversion of NADH to NADPH and vice versa, so that we may regard the two as largely equivalent, as far as the table is concerned.

The only remaining questions about the reactions listed in the table concern the reoxidation of either the enzyme, in the case of succinate

Table 6-1   The oxidations associated with the Krebs cycle

Reaction	Electron acceptor	Moles of ATP*
Isocitrate dehydrogenase	NAD+†	3
α-Ketoglutarate dehydrogenase	NAD+	4
Succinate dehydrogenase	flavin group on the enzyme	2
Malate dehydrogenase	NAD+	3

* Gives moles of ATP per mole of substrate oxidized.

† Classes of this enzyme exist that react, instead, with NADP+. These are found in the cytoplasm but not in the mitochondria, where the Krebs cycle enzymes are concentrated. It is likely that the NAD+ form of the enzyme is the most significant one in energy conservation.

dehydrogenase, or NADH in the others. The mechanism of such oxidations is known in considerable detail, and it must be said that the pathways for such electron transfer are surprisingly complex, even without considering the still more complex matter of how ATP is formed in the process. In animal or plant mitochondria the ultimate electron acceptor is molecular oxygen, while in bacteria it can be oxygen or an inorganic oxidant such as nitrate or sulfate. In the case of aerobic oxidations, the oxygen is reduced to water, and the overall process may be written

$$2\,AH_2 + O_2 \;\rightarrow\; 2\,A + 2\,H_2O$$

The 2:1 ratio of substrate to oxygen is a reflection of the fact that oxidation of the substrate involves removal of two electrons, whereas reduction of one molecule of oxygen to give water involves four. There are some oxidations in the cell that give rise to hydrogen peroxide (H_2O_2), but the enzyme catalase, which is an almost universal cellular constituent, converts it to water and oxygen, rendering the equation above still valid. Obscured in the arrow of the equation is a chain of oxidation-reductions of some complexity. This involves a series of different enzymes, each of which accepts electrons from its predecessor and passes them on to the following carrier. For example, one of the carriers, cytochrome c, now oxidized, picks up another electron from the preceding cytochrome c_1, and so on. These electron transfers may be summarized by writing

$$\text{Cytochrome } c_1 \;\rightarrow\; \text{Cytochrome } c \;\rightarrow\; \text{Cytochrome } a$$

where the arrows denote, not conversion of c_1 to c, but the transfer of an electron from one to the other. The complete system for aerobic electron transfer, which is known as the *respiratory chain*, is shown in Figure 6-3. It is seen to have two branches at the substrate end, one involved in NADH oxidation and the other in succinate oxidation. Some of the carriers are listed in Table 6-2, and it is seen that they include enzymes,

Figure 6-3 Path of electrons in the respiratory chain. The exact relation between cytochrome b, ubiquinone, and nonheme iron is presently unclear. There also appears to be nonheme iron associated with other regions of the respiratory chain. Finally, regions of the chain where ATP synthesis occurs are indicated by the dark arrows, although the precise locations and the exact nature of the interactions are subjects of considerable uncertainty.

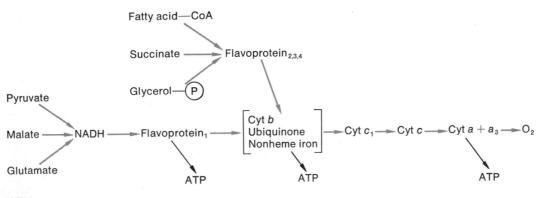

Table 6-2   Electron carriers of the respiratory chain

Carrier	Protein?	Prosthetic group	Number of electrons carried	$E°$
NAD$^+$	no	—	2	−0.31
Flavoprotein	yes	flavin	2	−0.05
Succinate	no	—	2	0.03
Cytochrome b	yes	heme	1	0.06
Ubiquinone	no	—	2	0.09
Cytochrome c_1	yes	heme	1	0.22
Cytochrome c	yes	heme	1	0.26
Cytochrome a	yes	heme	1	0.2
Oxygen	no	—	4	0.79

either flavin- or heme-containing, as well as small molecules. The reader will remember (from Chapter 3) that there is a relationship between the standard potential of a reaction and the free-energy change. It is quite literally true that the carriers of the respiratory chain are arranged in order of decreasing energy, which is expressed in Table 6-2 as an increasing potential. An electron from succinate, as it passes to oxygen via the chain, passes down an energy gradient, which is reflected in the difference in potential between succinate and oxygen. This energy difference is available for ATP synthesis.

The exact nature of the oxidation–reduction carriers has been a subject for study over the last half century, and our knowledge is quite detailed. A number of different flavoproteins have been studied that have a phosphate derivative of riboflavin as their prosthetic group,

This molecule consists of a substituted isoalloxazine ring system linked to D-ribitol. In this formula, R represents either phosphate, in the case of flavin mononucleotide (FMN), or AMP, in the case of flavin adenine dinucleotide (FAD). The oxidation or reduction of the flavoprotein takes place at the prosthetic group by a two-electron transition,

171

with the two electrons being transferred either together or sequentially.

Ubiquinone (also called coenzyme Q) is likewise oxidized by a two-electron transition and has the structure

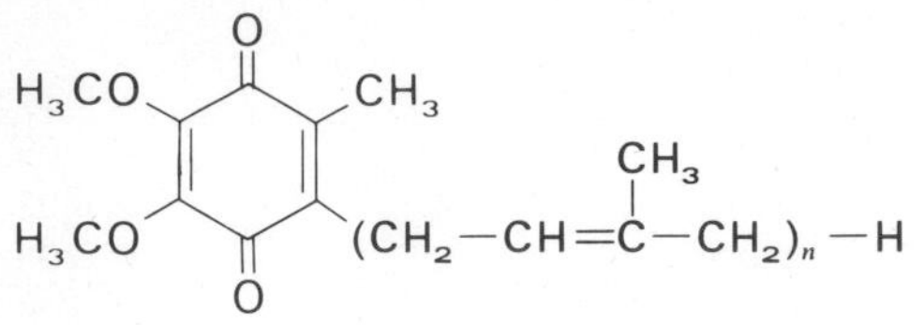

where the side chain may be of variable length. The exact location of ubiquinone in the respiratory chain is not clear, and it has even been suggested that it is not on the main line of the chain, because it is not oxidized and reduced at a rate high enough to be consistent with the rate of flow of electrons through the chain. A special role for ubiquinone in ATP synthesis has been suggested, but the evidence is quite inconclusive.

The cytochromes are all heme proteins whose oxidation–reduction occurs at the iron atom within the heme ring system:

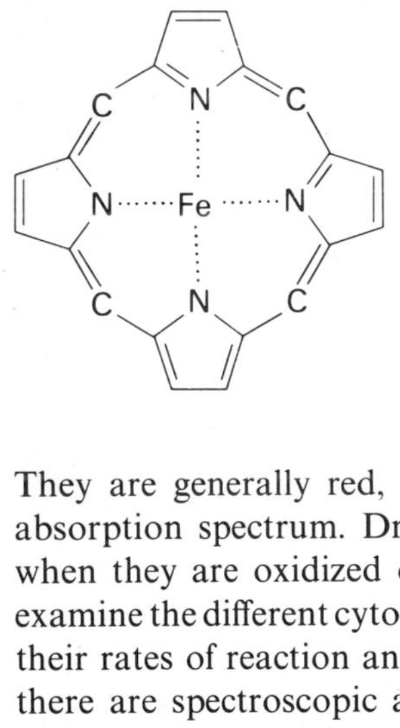

They are generally red, and each cytochrome exhibits a characteristic absorption spectrum. Dramatic changes in the spectrum are produced when they are oxidized or reduced, so that it is comparatively easy to examine the different cytochromes in intact mitochondria and to determine their rates of reaction and oxidation–reduction state. It is fortunate that there are spectroscopic approaches to the study of cytochromes, since, with the exception of cytochrome c, all are found to be integral parts of the mitochondrial membrane systems and have not been obtained in pure form without considerable alteration. Cytochrome c, on the other hand, is extremely easy to isolate, being apparently soluble in the intramitochondrial space. Cytochrome c is one of the relatively few proteins whose complete amino acid sequence is known, and its enzymic action has been studied in impressive detail.

It should be pointed out here that the cytochromes are quite unusual enzymes in that they react directly with other enzymes rather than substrates of lower molecular weight. Indeed, a case may be made for not including them under enzymes at all but rather considering them as a

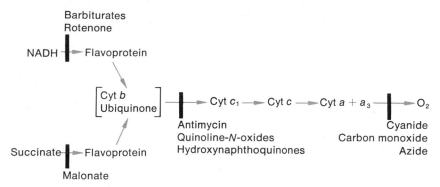

Figure 6-4 Sites of inhibition in the respiratory chain. In addition, a number of compounds interfere with electron transport in a nonspecific way. For instance, inhibitors that bind sulfhydryl groups inhibit the chain at several locations. In addition, excessive amounts of the inhibitors produce inhibition at more than one location; the strict specificity is true only when the inhibitors are used at relatively low levels.

separate category of electron-transferring proteins. Their action is so much a matter of electron flow down the chain that ordinary enzyme kinetics are not always easy to apply, and sometimes hydrodynamic analogies seem more valid.

Since the respiratory chain appears to form an integral part of the mitochondrial structure (or, in the case of bacteria, of the cell membrane), it has been a challenging object of study. The construction of the chain and especially the manner of coupling of electron flow to ATP synthesis have become clear largely through the use of inhibitors (see Figure 6-4). For example, the addition of an inhibitor such as antimycin in the presence of NADH has the effect of rendering cytochrome c_1 through a and a_3 more oxidized, while cytochrome b and flavoproteins are more

Figure 6-5 Artificial electron donors. Various compounds are able to feed electrons into the intact respiratory chain at points denoted by the gray arrows. In this figure, black arrows denote electron transfer that occurs normally in the passage of electrons from the substrate (succinate or NADH) to oxygen.

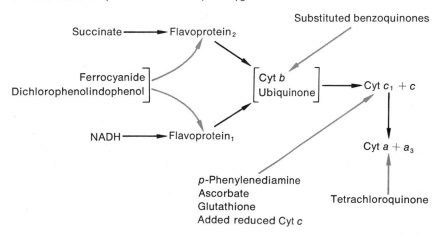

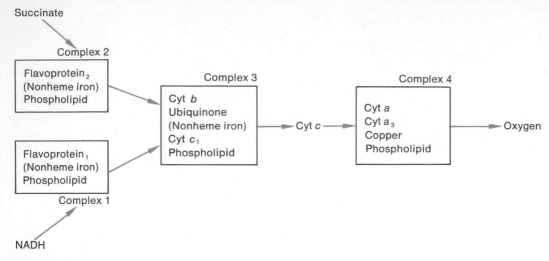

Figure 6-6 Resolution of the respiratory chain into components. When mitochondria are treated with specific detergents under defined conditions, the components of the inner mitochondrial membrane can be separated into four "complexes," each of which contains several proteins and is enzymically active. A functional respiratory chain may be reconstituted from the four complexes, together with cytochrome c and a substrate. These systems will carry out respiration at a rate similar to that of intact mitochondria but have never been shown to couple electron flow to ATP synthesis. Efforts to reconstitute ATP synthesis as well are underway in a number of laboratories.

reduced. By a number of such experiments, the order of the carriers in the chain has become known. The use of inhibitors, together with artificial electron acceptors (Figure 6-5), has led to the study of isolated portions of the chain and has greatly assisted in localizing the sites of ATP formation.

Oxidative ATP synthesis

As we said, the exact mechanism of coupling between oxidation and phosphorylation of ADP is not known. A large number of the opinions that exist about the subject have grown up through argument by analogy from substrate-linked phosphorylation or from studies using different inhibitors of the process. For example, there is a class of compounds, of which 2,4-dinitrophenol,

174

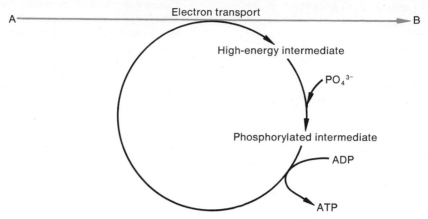

Figure 6-7   Outline of a mechanism for ATP synthesis. The final stage is the transfer of phosphate from the "phosphorylated high-energy intermediate" to ADP to form ATP. It must be said that this intermediate is hypothetical, never having been isolated or chemically identified. The final stage in the sequence of reactions, the transfer of phosphate to ADP, is sensitive to the antibiotic oligomycin, which, significantly, inhibits the ATPase described later in the chapter. Thus, a case may be made for the identity of the ATPase and the enzyme catalyzing the final phosphate transfer.

is the best known, that effectively uncouple oxidations from ATP formation. Addition of such compounds often stimulates electron flow while completely preventing any ATP synthesis. Studies of this kind have led to a picture of ATP synthesis, given in Figure 6-7, involving the synthesis of high-energy intermediates that lead to ATP. Although a large body of evidence supports this picture, the fact is that efforts to identify or isolate the intermediates have met with failure. The reason given is usually their extreme instability, but difficulty of isolation may also be a measure of the fact that they may not exist at all. Indeed, an alternative theory of ATP formation has been put forward in which high-energy intermediates in the sense of the figure *do not* exist; this theory is discussed below in the more appropriate context of Chapter 12.

Table 6-3   Uncoupling of oxidative phosphorylation by dinitrophenol*

Rat liver mitochondria were oxidizing succinate; oxidation was measured with a polarographic electrode, while ATP synthesis was estimated by following the incorporation of radioactive ^{32}P into ATP.

Conditions	Oxygen uptake (n atoms/min)	ATP/oxygen ratio
No uncoupler	460	1.6
+10 μM dinitrophenol	709	0

* The results are taken from an experiment in the author's laboratory.

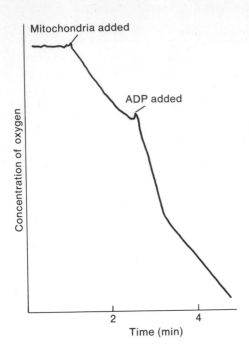

Figure 6-8 Respiratory control: the stimulation of mitochondrial respiration by ADP. This is a trace produced with a polarographic electrode sensitive to oxygen. A downward deflection of the trace indicates an increase in the rate at which oxygen is used up. The reaction mixture at the beginning of the experiment contains buffer, magnesium, phosphate, succinate, and water saturated with oxygen. When a suspension of mitochondria obtained by differential centrifugation of a tissue homogenate is added, oxygen uptake occurs as succinate is oxidized to fumarate. The addition of ADP stimulates oxygen uptake, which returns to the original rate when all the added ADP has been phosphorylated to ATP. The amount of oxygen used at the higher rate may serve as the basis for estimating the efficiency of ATP synthesis (the ATP/O ratio), since the amount of ATP synthesized is the same as the amount of ADP added. For additional information about the polarographic measurement of respiration, see Figure 6-10.

Respiratory control

Whatever the mechanism of the coupling process, there is one aspect of it with profound significance as far as the control of ATP production is concerned. If isolated mitochondria are incubated with a substrate (such as succinate) in the absence of ADP and phosphate, oxygen uptake is detected, even though no ATP can be produced. The subsequent addition of ADP and phosphate produces a marked increase in respiration (see Figure 6-8), which continues until the ADP (or phosphate) is used up. Thus, oxidation of succinate occurs at the maximum rate only under conditions where ATP may be synthesized, with lower rates of "wasteful" oxidation of substrate at other times. This phenomenon,

Figure 6-9 Respiratory control. Utilization of ATP produces ADP and inorganic phosphate according to the reaction

$$ATP + H_2O \rightleftharpoons ADP + PO_4^{3-}$$

The effect of the increase in ADP and PO_4^{3-} within the cell is a stimulation of respiration and consequent ATP synthesis.

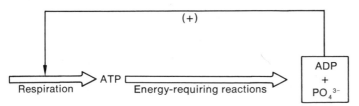

which is called *respiratory control*, is explained on the basis of the *high-energy intermediate theory* by saying that the intermediate, X, is itself inhibitory to respiration and that the addition of ADP and phosphate effects the removal of X and therefore stimulates electron flow. It is interesting that an uncoupling agent such as dinitrophenol, which is said to lead to the breakdown of intermediate, also stimulates respiration under similar conditions. It is also clear that respiratory control may form a sort of a positive feedback control for respiratory activity, since most processes leading to utilization of ATP produce ADP, which then stimulates respiration coupled to further ATP synthesis (see Figure 6-9). Respiratory control has also been observed with intact mammalian cells, which suggests that it is not simply a property of isolated mitochondria.

The reversibility of the respiratory chain

Recent evidence indicates that the reactions of the respiratory chain are reversible and that electrons may be made to flow in the direction of the substrate. Since the flow is up the energy gradient, the reaction requires energy, instead of leading to its production. For instance, electrons may be inserted into the chain at the cytochrome *c* locus and,

Figure 6-10 Measurement of cellular respiration with an oxygen electrode. In this diagram, a platinum electrode dips into a suspension of cells, mitochondria, or other enzyme systems, and records the oxygen concentration by a polarographic technique. To put it simply, a polarizing voltage is applied to the cell and current is measured. The current depends upon the degree of electrolysis of the atom being measured; in the case of oxygen, this is accomplished by passing about 0.5 volt across the cell. A virtue of the method is the range of polarizing voltages for different atoms, so that one may set the instrument to measure only the atom of interest, in this case oxygen.

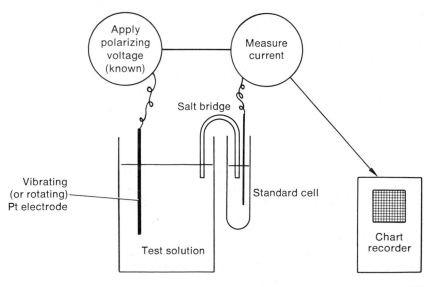

in the presence of cyanide, pass back along the chain to reduce cyto-chromes b, ubiquinone, or even NAD^+. This reaction requires ATP and so seems to involve a reversal not only of the chain but of the energy-coupling reactions shown in Figure 6-7.

A similar reversal reaction may be demonstrated using succinate as the electron donor, which, in the presence of antimycin (which inhibits the respiratory chain between cytochromes b and c) and ATP, leads to NAD^+ reduction. This reaction,

$$Succinate + NAD^+ + ATP \rightarrow Fumarate + NADH + H^+ + ADP + PO_4^{3-}$$

is thermodynamically unfavorable in the absence of ATP hydrolysis, since, as shown in Table 6-2, NAD^+ is of a higher energy than succinate. The reaction proceeds even in the absence of antimycin, under which condition ATP is not required, suggesting that high-energy intermediates are being formed in one part of the chain (between cytochrome b and oxygen) and are then used in the reduction of NAD^+. In fact, ADP inhibits the reduction, indicating that ATP synthesis is in competition with the reversal reaction for energy, presumably in the form of high-energy intermediates.

It has been suggested that the reversal of the respiratory chain might have implications for the control of cell function, as conditions leading to the availability of ATP would have the effect of sending electrons backward up the chain, leading to increased NADH and decreased net respiration. Since NADH (via NADPH) is used in many synthetic reactions, the effect would be to shift cell function in the direction of synthesis—an appropriate shift when ATP is in good supply.

The NADH–NADPH transhydrogenase

A final point related to the matter of control of respiratory activity concerns the transhydrogenase mentioned earlier. It is generally (but not universally) true that NADPH is used largely in reductive syntheses, whereas NADH is more closely associated with energy metabolism. Furthermore, the transhydrogenase is involved in the oxidation of isocitric acid, since NADPH is formed in that reaction, but only NADH may be oxidized by the respiratory chain. The transhydrogenase reaction

$$NADH + NADP^+ \rightleftharpoons NAD^+ + NADPH$$

represents a simple transfer of protons between two closely related com-pounds. There is little thermodynamic barrier to the reaction, as the potentials of the two are essentially the same. Since we have seen that there is a close relationship between potential, free-energy change, and equilibrium, it is clear that there is no net energy requirement for the reaction (ΔF is about 0) and that the equilibrium constant is about 1. This means that at equilibrium there should be about an equal partition between

NADH and NADPH, representing a balance that should be of importance to the cell. Recent measurements of the actual levels of reduction of NAD^+ and $NADP^+$ in intact mitochondria, however, reveal a very different picture. Under conditions where energy is available, NADPH is always found to be in a large excess over NADH, suggesting that energy is being used to push the reaction in that direction. Thus, under conditions where energy is available, hydrogen is shunted away from NADH, where it would mostly be used for production of further energy, toward NADPH where it is available for synthetic reactions. The transhydrogenase, then, may provide a means of coordinating two fundamental aspects of synthetic processes, the availability of energy and the generation of reducing power, while exerting a feedback inhibition on energy production itself.

The search for protein coupling factors

Since the exact mechanism of cellular-energy coupling remains obscure, efforts to understand it have ranged widely. Studies of intact mitochondria have produced a large array of data without, however, leading to a clear view of the overall mechanism. It appears that even isolated mitochondria are too complex to allow for the execution of simple experiments or for clear-cut analysis of results. Faced with an increasing mass of data, a number of workers have taken the different path of endeavoring to resolve energy coupling into individual reactions, catalyzed by individual enzymes, from which the net coupling process might be reassembled.

This approach has taken the form of isolation of a number of mitochondrial (and chloroplast) proteins that have in common the property of stimulating energy-linked processes, most notably synthesis of ATP. These proteins are called *coupling factors* and appear to be, in all cases, either enzymes or proteins that play a structural role. The best-characterized such protein is the mitochondrial ATPase, also called F_1, which has been extensively studied by Racker and co-workers (see Suggested Reading). This protein when isolated from mitochondria carries out the hydrolysis of adenosine triphosphate, that is,

$$ATP + H_2O \rightleftharpoons ADP + PO_4^{2-} + H^+$$

and it is believed that this reaction of the purified enzyme reflects the reversal of the final reaction of ATP synthesis as performed by the intact system. Such an opinion appears justified, as F_1 effectively stimulates ATP formation in certain kinds of submitochondrial preparations (see Chapter 1), and inactivation of its ATPase activity leads to parallel inactivation of its capacity to promote ATP synthesis.

Study of such proteins is prompted by the hope that it will eventually become possible to reassemble the entire process from the individual reactions. We have seen that such reconstitution has been of importance in verifying the reactions of such pathways as glycolysis. However, in

Table 6-4   Stimulation of ATP synthesis by coupling factors*

ATP synthesis was measured using mitochondrial fragments that had been treated to remove coupling factors, which could then be added back individually. The substrate for oxidation was NADH.

Conditions	ATP synthesis (% of complete system)
Mitochondrial fragments + F_1 + F_2 + F_3 + F_5 (complete system)	100
Minus F_1	5
Minus F_3	42

* Adapted from a paper by J. Fessenden-Raden, *J. Biol. Chem.*, **244**, 6662 (1969).

the case of cellular energetics, success appears to be rather distant, and, as we shall see in a subsequent chapter, there is at least one theory of energy coupling that would predict failure, on the grounds that it considers energy coupling to be a process not decomposable into a sequence of purely enzymic steps.

Anaerobic respiration

Since authors (and readers) of cell physiology books are members of the animal kingdom and, hence, obligate aerobes, great and understandable stress is placed upon the role of oxygen as an electron acceptor in respiration. Other electron acceptors exist, however, despite our inability to take advantage of them, and many bacteria are quite content to respire by alternative routes. For example, a number of microorganisms use nitrate as the terminal electron acceptor, reducing it to nitrite. Other organisms are able to reduce it further, to elementary nitrogen or, even nitric oxide. The bacterium *Escherichia coli* reduces nitrate by a mechanism strikingly similar to that employed in aerobic respiration, utilizing the respiratory chain with its cytochromes, flavins, and pyridine nucleotides. Passage of electrons along the respiratory chain to nitrate is coupled to synthesis of ATP by a mechanism similar again to that occurring when oxygen is the final electron acceptor, ATP formation being completely inhibited by the addition of uncoupling agents, such as dinitrophenol.

Certain other anaerobic bacteria also respire with sulfate as the electron acceptor, reducing it to sulfite. Again, the reaction is energy-linked and, again, the respiratory chain has been shown to be involved.

Fatty acid oxidation

We have been proceeding during the last two chapters as if all aerobic cellular-energy conservation occurred by means of the mitochondrial oxidation of carbohydrates and their derivatives. Indeed, we have adopted this apparent stance because of the central role that the glycolytic and citric acid pathways do, in fact, play. However, many eukaryotic cells utilize fatty acid oxidation as a major route for energy coupling and many store fats in large amounts as an energy reserve. The balance between the importance of fat and carbohydrate metabolism is quite variable and some cell types go rather far in one direction or the other. For example, in the muscle cells of migrating birds, fat oxidation is of paramount importance. On the other hand, brain cells depend largely upon the oxidation of carbohydrates, perhaps bringing into question the scientific validity of the useful term "fat head."

To avoid going into all the byways of fat metabolism, we shall confine our discussion to the oxidation of fats as they commonly occur in the form of *triglycerides*, or esters of glycerol and fatty acids. Glycerol

$$CH_2OH$$
$$|$$
$$CHOH$$
$$|$$
$$CH_2OH$$

is an example of a polyhydroxy alcohol and forms esters with fatty acids, which are simply organic acids with alkyl side chains of varying length and with varying degrees of unsaturation. The first stage of utilization of fats is the hydrolysis of triglycerides through the catalysis of enzymes called *lipases*, which are found in the cell cytoplasm:

$$CH_2OOC(CH_2)_n CH_3 \qquad\qquad CH_2OH$$
$$| \qquad\qquad\qquad\qquad\qquad\qquad |$$
$$CH-OOC(CH_2)_n CH_3 \xrightarrow{+H_2O} CHOH + 3\ {}^-OOC(CH_2)_n CH_3$$
$$| \qquad\qquad\qquad\qquad\qquad\qquad |$$
$$CH_2OOC(CH_2)_n CH_3 \qquad\qquad CH_2OH$$

This produces free glycerol, which can enter the glycolytic pathway after being phosphorylated by the *glycerol kinase reaction*:

$$\text{Glycerol} + \text{ATP} \rightleftharpoons \text{Glycerol phosphate} + \text{ADP}$$

The fatty acids that were cleaved from the glycerol may then be oxidized by the reactions described below. In fact, the removal of fatty acids from the cell by oxidation reactions must be a highly efficient procedure, as the concentration of free fatty acids in the cell is very low. This is important since fatty acids produce a number of toxic effects when their intracellular concentration becomes too high. For one thing, they are uncouplers of oxidative ATP synthesis, so that high concentrations lead to a diminished level of energy coupling.

Fatty acids are produced, then, by the action of lipases in the cytoplasm and are oxidized exclusively in the mitochondrion. Therefore, the first consideration is the passage of the acid into the mitochondrion, a process that begins with *activation* of the acid by means of the formation of a compound of the acid and coenzyme A:

$$RCOO^- + ATP + CoA-SH \rightleftharpoons R-\underset{\underset{O}{\|}}{C}-S-CoA + AMP + \text{\textcircled{P}}-\text{\textcircled{P}}$$

181

This reaction is catalyzed by an enzyme termed a *thiokinase* and forms the products acyl—CoA, AMP, and pyrophosphate. This last is then hydrolyzed by *pyrophosphatase* to give inorganic monophosphate. Since pyrophosphate cleavage is associated with a large negative free energy, it contributes to making the overall reaction of fatty acid activation a highly spontaneous one.

Transport of the fatty acid into the mitochondrion occurs through the formation of a carnitine derivative by transfer of the acyl portion of acyl—CoA to a hydroxy group of *carnitine*:

$$\begin{array}{c} CH_3 \\ | \\ H_3C - {}^+N - CH_2 - CH - CH_2 - COO^- \\ | \qquad\quad | \\ CH_3 \qquad OH \end{array}$$

The acylcarnitine ester is able to traverse the mitochondrial inner membrane readily, perhaps in part because of the combination of the polar carnitine and the nonpolar acyl portion (see Chapter 12). Once inside the mitochondrion, the acyl group is again transferred to CoASH so that the role of carnitine can be seen to be that of a shuttle (see Figure 6-11).

The first step in the actual oxidation of the fatty acid is a dehydrogenation wherein a double bond is formed *beta* to the site of attachment of CoA. This reaction is catalyzed by enzymes called *fatty acyl—CoA dehydrogenase*, which contain a FAD prosthetic group. When the acyl—CoA is oxidized, the FAD is electron acceptor. Reduced FAD, in turn, passes electrons to a second flavoprotein, then ubiquinone and, thence, through the remainder of the respiratory chain. The passage of electrons along the chain is coupled to the formation of ATP. The reaction catalyzed by fatty acyl—CoA dehydrogenase is the following:

$$R-CH_2-CH_2\overset{\displaystyle O}{\overset{\|}{C}}-S-CoA \quad \rightleftharpoons \quad R-\overset{\displaystyle H}{\underset{\displaystyle H}{C}}=\overset{}{C}-\overset{\displaystyle O}{\overset{\|}{C}}-S-CoA$$

+   +

FAD enzyme (oxidized)   FAD enzyme (reduced)

The second step in fatty acid oxidation is the addition of water across the double bond formed in the first step. This is catalyzed by *enolhydratase* and yields a hydroxyacyl—CoA as shown in the following reaction:

$$R-\overset{\displaystyle H}{\underset{\displaystyle H}{C}}=\overset{}{C}-\overset{\displaystyle O}{\overset{\|}{C}}-S-CoA + H_2O \quad \rightleftharpoons \quad R-\overset{\displaystyle HO}{\underset{\displaystyle H}{C}}-\overset{\displaystyle H}{\underset{\displaystyle H}{C}}-\overset{\displaystyle O}{\overset{\|}{C}}-S-CoA$$

The subsequent reaction is a second dehydrogenation wherein the hydroxy fatty acyl—CoA is oxidized to the corresponding ketone:

$$R-\overset{\displaystyle HO}{\underset{\displaystyle H}{C}}-\overset{\displaystyle H}{\underset{\displaystyle H}{C}}-\overset{\displaystyle O}{\overset{\|}{C}}-S-CoA + NAD^+ \quad \rightleftharpoons \quad R-\overset{\displaystyle O}{\overset{\|}{C}}-\overset{\displaystyle H}{\underset{\displaystyle H}{C}}-\overset{\displaystyle O}{\overset{\|}{C}}-S-CoA + H^+ + NADH$$

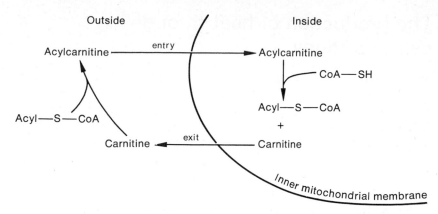

Figure 6-11 Acylcarnitine shuttle. Since acyl—S—CoA does not traverse the mitochondrial membrane, the acyl portion is first transferred to carnitine. It should be added that the mitochondrial barrier to acyl—S—CoA is the inner mitochondrial membrane, which, in fact, represents the main permeability barrier to most small molecules.

In this case, the electrons removed from the fatty acid derivative are also passed to the respiratory chain by way of NADH dehydrogenase with associated ATP synthesis.

Finally, the last step in a cycle of fatty acid oxidation is the cleavage of the acyl—CoA molecule in the beta position, and the transfer of the two carbon fragments to a second CoASH molecule:

$$R-\overset{\overset{O}{\|}}{C}-\overset{\overset{H}{|}}{\underset{\underset{H}{|}}{C}}-\overset{\overset{O}{\|}}{C}-S-CoA + CoASH \rightleftharpoons$$

$$R-\overset{\overset{O}{\|}}{C}-S-CoA + CH_3-\overset{\overset{O}{\|}}{C}-S-CoA$$

This reaction is catalyzed by an enzyme called a *thiolase* and leaves one molecule of acetyl—CoA and the CoA thioester of a fatty acid, which is now shortened by two carbon units. This molecule is now able to re-enter the cycle of oxidation, water addition, oxidation, and cleavage to yield a product shortened by two additional carbons, a process that can be repeated until only acetyl—CoA is left. The acetyl—CoA is able to contribute further to energy production by entering the Krebs cycle through condensation with oxaloacetate.

In order to gain an impression of the efficiency of energy coupling associated with fatty acid oxidation, note that three molecules of ATP are formed with each NAD^+-linked oxidation and two with each flavoprotein-linked oxidation. Thus, for each acetyl—CoA produced, there are five molecules of ATP. In addition, the oxidation of acetyl—CoA by way of the Krebs cycle produces an additional 12 ATP's per acetyl—CoA, raising the total to 17. Since a typical fatty acid might have a total of 16 or 18 carbons, this would have to be multiplied by 8 or 9 to give the total ATP production. In any case, the yield is clearly high.

The production of heat through cell oxidations

We have seen in this chapter that cellular oxidation reactions provide the negative free-energy changes necessary to drive the energy-requiring reactions of the cell. It will be recalled from Chapter 3 that the expression for free energy itself is composed of two terms, one relating to the entropy change and one to the change in heat produced by a reaction:

$$\Delta G = \Delta H - T \Delta S$$

In this equation, ΔG (as before) is the change in free energy, ΔH that in enthalpy, and ΔS that in entropy, while T is the absolute temperature at which the reaction occurs. We saw that the enthalpy is identical with the heat change at constant pressure, a condition usually obtained in the cell.

Normally, we are most concerned with the entire free-energy function, as it is what determines the spontaneity of reactions. However, in many organisms, including all birds and mammals and some invertebrates, the production of heat by biochemical reactions is of significance as well. The maintenance of body temperature (or hive temperature) depends on the existence of a balance between heat production by chemical reactions and heat loss by means of radiation or conduction. A number of mechanisms exist for regulating the loss of heat from animals, including sweating and panting as well as the adjustment of blood flow to bring more heat to the surface. In similar fashion, heat production is regulated by controls that sense the temperature and adjust the reactions accordingly.

It appears that a major site of heat production in mammals is the cells of *brown fat* that are distributed in several locations about the body. This tissue is most evident in newborn animals and in animals that undergo hibernation. The reason for this appears to be that adult animals normally increase their heat production by the mechanism of muscular shivering, a response that is not noted in newborn and is probably unsuitable during long hibernation. All the cells of an animal are probably capable of heat production, in that all carry out exergonic reactions that are able to evolve usually at least some heat. The unique feature of brown fat cells is that they appear specially modified for heat production and that their reactions are regulated by the cell temperature.

Brown fat tissue is heavily charged with droplets of fat, which we saw earlier to be a prime energy source. Brown fat differs however from the more widely distributed white fat in containing many more mitochondria and much higher concentrations of oxidative enzymes, including cytochromes, that lend it color. Thus, it is endowed with both substrate and catalysts for cellular oxidation. When young animals are chilled, their heat production (and oxidative metabolism) rises accordingly, and experiments with thermocouples implanted in the various tissues show that both functions are localized in the brown fat.

The chemical routes by which the heat is produced are presently under investigation, but some conclusions are possible. For one thing, it is clear that the major source of chemical energy is the fatty acids that are

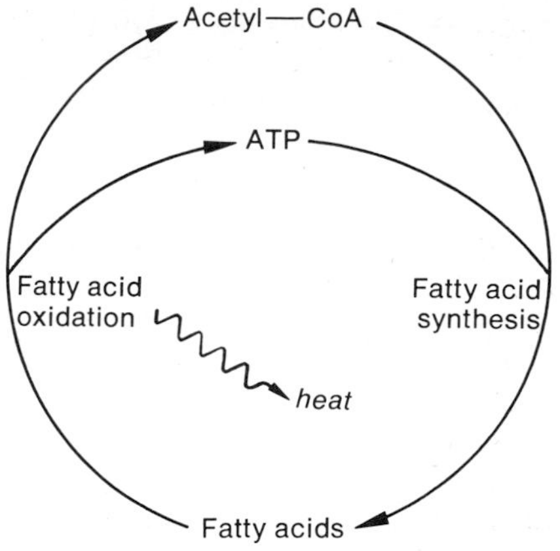

Figure 6-12 Production of heat through fat oxidation in brown fat tissue.

so abundant in these cells. When these are oxidized in the mitochondria, the discovery that ATP formation is low has suggested that respiration may be largely uncoupled. This observation makes sense if the energy from the oxidation is used directly for heat instead of biosyntheses requiring ATP. Since fatty acids are, themselves, uncouplers of oxidative phosphorylation, this sort of result may be trivial and may suggest only that the reactions are not being measured under ideal conditions. In any case, it has been suggested that the main route for heat evolution in these cells is an oxidative breakdown of fatty acids, by the pathways described earlier in this chapter, followed by resynthesis to the same fatty acids (and, ultimately, to triglycerides). It is probable that the ATP that is produced by the oxidation is reutilized at once in the resynthesis, so that the overall reaction is an endless cycle of oxidative degradation and reductive synthesis, the only net result of which is the production of heat. These relationships are illustrated in Figure 6-12.

The process whereby the cycle of fat breakdown and synthesis is initiated is not completely clear but the control appears to come from the central nervous system. Thus, temperature receptors in the skin of the animal sense the temperature and inform the appropriate center in the brain, which passes impulses to the brown fat cells by way of the sympathetic nervous system. The endings of the sympathetic system release the neural hormone *norepinephrine*, which appears to be the specific activator of fatty acid metabolism, possibly through the mediation of *cyclic adenosine monophosphate* (about which more in Chapter 11).

Suggested Reading

Books

Chance, B. (ed.), *Energy-linked Functions in Mitochondria*, Academic Press, New York, 1963.

Gran, F. C. (ed.), *Cellular Compartmentation and Control of Fatty Acid Metabolism*, Proceedings, Federation of European Biochemical Societies, 4th meeting, vol. 4, Academic Press, New York, 1967.

185

Keilin, D., *The History of Cell Respiration and Cytochrome*, Cambridge University Press, Cambridge, Mass., 1966.

Krebs, H. A., and H. L. Kornberg, *Energy Transformations in Living Matter*, Springer-Verlag, Berlin, 1957.

Lehninger, A. L., *Bioenergetics*, W. A. Benjamin, New York, 1965.

Papa, S., J. M. Tager, E. Quagliariello, and E. C. Slater (eds.), *The Energy Level and Metabolic Control in Mitochondria*, Adriatica Editrice, Bari, Italy, 1969.

Racker, E., *Mechanisms in Bioenergetics*, Academic Press, New York, 1965.

Singer, T. P. (ed.), *Biological Oxidations*, John Wiley & Sons, New York, 1966.

Slater, E. C. (ed.), *Flavins and Flavoproteins*, Elsevier Publishing Co., Amsterdam, 1966.

Articles

Garland, P. B., B. Chance, L. Ernster, C. Lee, and D. Wong, "Flavoproteins of mitochondrial fatty acid oxidation," *Proc. Nat. Acad. Sci. U.S.*, **58**, 1698 (1967).

Green, D. E., "The mitochondrial electron transfer system," in *Comprehensive Biochemistry*, M. Florkin and E. Stotz (eds.), vol. 14, American Elsevier Publishing Co., New York, 1966.

Greville, G. D., and P. K. Tubbs, "The catabolism of long-chain fatty acids in mammalian tissues," in *Essays in Biochemistry*, P. N. Campbell and G. D. Greville (eds.), vol. 4, Academic Press, New York, 1968. Excellent and readable review.

Harris, R. A., J. T. Penniston, J. Asai, and D. E. Green, "The conformational basis of energy transformations in membrane systems: II. Correlation between conformational change and functional states," *Proc. Nat. Acad. Sci. U.S.*, **59**, 830 (1968).

Hayaishi, O., and M. Nozak, "Nature and mechanisms of oxygenases," *Science*, **164**, 389 (1969).

King, T. E., "Reconstitution of the respiratory chain," in *Advances in Enzymology*, F. F. Nord (ed.), vol. 28, John Wiley & Sons (Interscience Division), New York, 1966, p. 155.

Kröger, A., and M. Klingenberg, "On the role of ubiquinone," in *Current Topics in Bioenergetics*, D. R. Sanadi (ed.), Academic Press, New York, 1967, p. 152.

Lardy, H. A., and S. M. Ferguson, "Oxidative phosphorylation in mitochondria," *Ann. Rev. Biochem.*, **38**, 991 (1969).

Okunuki, K., M. D. Kamen, and I. Sekuzu, *Structure and Function of Cytochromes*, University of Tokyo Press, Tokyo, 1967.

Racker, E., "The membrane of the mitochondrion," *Sci. Am.*, **218**, 32 (Feb. 1968).

Slater, E. C., "Oxidative phosphorylation," in *Comprehensive Biochemistry*, M. Florkin and E. Stotz (eds.), vol. 14, American Elsevier Publishing Co., New York, 1966.

7 Energy transfer involving light

Light is the ultimate source of most of the energy entering the living world. Members of the plant kingdom (as well as some bacteria) possess highly organized systems for the absorption of light and the utilization of the energy thus absorbed for the processes of life. The process whereby radiant energy is converted to a chemical form of energy, leading to the synthesis of cell material, is called *photosynthesis*. Historically, the study of photosynthesis has passed through a number of phases, each centering on one aspect of the process believed to be fundamental at the time. Early studies of photosynthesis saw it as the light-induced production of carbohydrates and the process was usually summarized by the equation

$$CO_2 + H_2O \xrightarrow{\text{light, chlorophyll}} CH_2O + O_2$$

a formulation based on the fact that CO_2 was known to be taken up, oxygen was observed to be evolved, and carbohydrate (here abbreviated CH_2O) was the chief storage product in green plants. This view, according to which photosynthesis was a special light-requiring type of sugar synthesis, was entirely consistent with the information then available.

However, in the 1930's additional information led to the questioning of the universal applicability of the above equation. For one thing, photosynthesis was discovered in several groups of bacteria, often involving quite different reactants. For example, some photosynthetic bacteria utilize hydrogen sulfide instead of water, as in the case of green plants, and yield elementary sulfur as an end product. In such a case, the equation is written

$$CO_2 + 2 H_2S \xrightarrow{\text{light, chlorophyll}} CH_2O + 2 S + H_2O$$

It will be observed that both sorts of photosynthesis include two basic processes, one being the formation of carbohydrate from CO_2, the other being an oxidation, in one case that of water to form oxygen, in the other that of H_2S to elementary sulfur. Furthermore, from everything that we

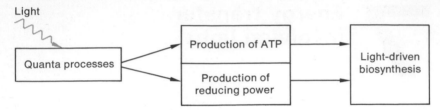

Figure 7-1 Photosynthesis: a flow diagram.

know about synthetic reactions, the production of carbohydrate is likely to involve two separate requirements, that of energy and that of reducing power (electrons). The apparent complexity of photosynthesis may be resolved by considering the total process to be the sum of several interconnected, equally fundamental events, illustrated in the flow diagram of Figure 7-1. In the diagram, "quanta processes" are those aspects of light absorption and energy transfer that are not chemical. A useful test is the ability of the truly quantum processes to take place at very low temperature, such as that of liquid nitrogen, conditions under which chemical reactions would be infinitely slow. It will be seen shortly that ATP production and the formation of reducing power are aspects of the same thing and that the two are equally necessary for the synthesis of carbohydrates.

Chlorophyll and the quantum processes

The compound responsible for most of the energy-trapping light absorption of both plants and photosynthetic bacteria is *chlorophyll*. Like the cytochromes, it is a porphyrin compound, but instead of the iron of cytochromes it contains a magnesium atom bound in the center of the ring system. A number of different types of chlorophyll occur in plant and photosynthetic bacterial cells, all having the basic Mg–porphyrin ring system shown in Figure 7-2 but with differing substituents attached to it. In addition, in many plants, *carotenoids* and, in red and blue-green algae, *phycobilins* are also able to absorb light energy in photosynthesis and are

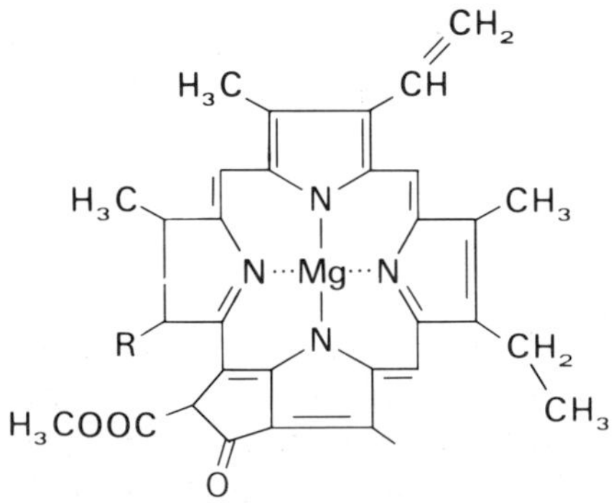

Figure 7-2 Basic structure of chlorophyll. R denotes the long *phytol* side chain, which is built of terpenoid units and which confers lipid solubility on the molecule. The ring structure that is shown is that of chlorophyll *a*; other chlorophylls exhibit modification as to the side groups or (in the case of bacterial chlorophyll) the degree of reduction of the ring system itself.

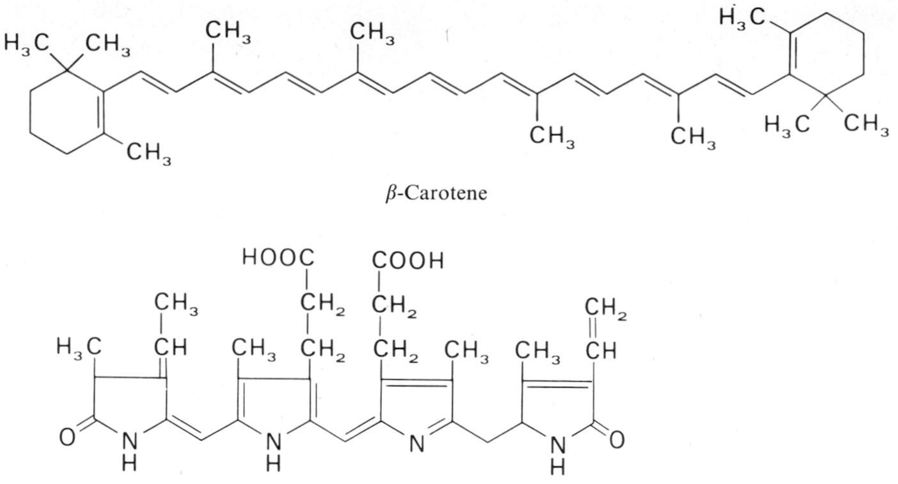

β-Carotene

Phycoerythrobilin

Figure 7-3 Examples of accessory photosynthetic pigments.

termed *accessory pigments* (see Figure 7-3). These molecules act in a secondary manner with respect to chlorophyll in that the energy must first be transferred to a chlorophyll molecule before further "fixation" of the energy can take place. The importance of carotenoid light absorption is probably related to the fact that green chlorophyll and yellow carotenoids absorb in different regions of the spectrum and so enable the cell to use a wider wavelength range in energy trapping.

It is obvious that, for light to produce an effect on cells, it must be absorbed. The region of the spectrum of interest in photosynthesis is that of visible light (or near-infrared in the case of bacteria), which represents interaction between quanta and the electronic energy levels of molecules. When such absorption takes place, a quantum of light produces an excitation in the molecule, wherein an electron is raised to a higher energy level. The energy available is given by the familiar expression

$$E = h\frac{c}{\lambda}$$

where E is the energy of a light quantum, c the velocity of light (a constant), h Planck's constant, and λ the wavelength of the light quantum. Thus, the amount of energy available in photosynthesis is dependent on the wavelength, and it is likely that most of the infrared region, because of its lower energy, is not important in photosynthesis. Similarly, the greater energies of the shorter wavelength ultraviolet region are not photosynthetically important, inasmuch as they produce such dramatic changes in a large number of organic molecules as to be lethal to the cell. It is for this reason that UV light is used to kill microorganisms, and cells, generally, do well to be as opaque to UV quanta as possible.

The absorption of light by chlorophyll leads to an electron being raised to a higher energy level, representing the primary excitation of

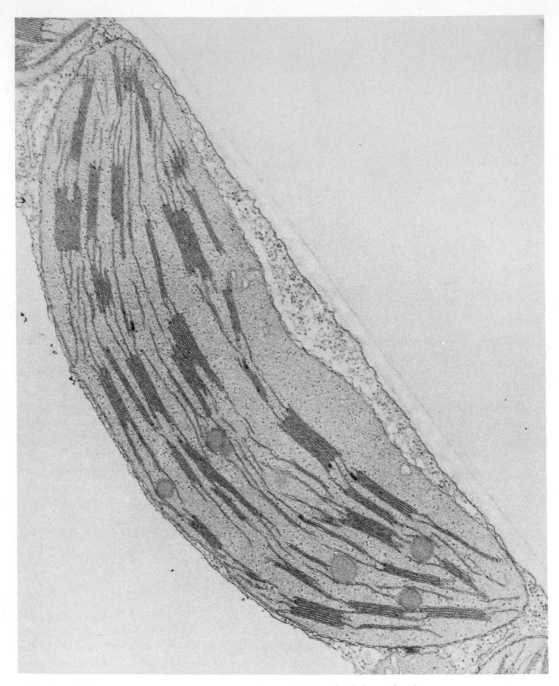

Figure 7-4 Chloroplast from tobacco plant. In plants, photosynthesis is confined to membranous organelles, the chloroplasts. There, the photochemical centers as well as proteins and other molecules involved in electron transfer are located in the thylakoid membranes. This electron micrograph shows grana, where the internal membranes are folded back on themselves, as well as the outer chloroplast membrane. This particular chloroplast is seen to be closely applied to the cell wall. The micrograph was prepared by fixation with glutaraldehyde, followed by osmium tetroxide. The magnification is × 37,500. (Courtesy of Dr. Herbert W. Israel.)

190

photosynthesis. The energy may then be conserved in some chemical reaction—say, using the electron in a reduction—or the energy may become dissipated as heat or some form of reradiation (such as fluorescence) as it returns to its ground state. The considerable variety of ways whereby the electron may return to its lower state in the chlorophyll molecule makes it a useful system for studying absorption and reemission processes in general, a fact that probably accounts for the recent influx of molecular physicists into the area.

Energy radiated from photosynthetic pigments by fluorescence is lost to the organism, and biologists are understandably more interested in

Figure 7-5 High-magnification view of tobacco chloroplast. Observe the stacking of membranes to form grana. Magnification is × 92,000, with fixation and staining as described in the legend to Figure 7-4. (Courtesy of Dr. Herbert W. Israel.)

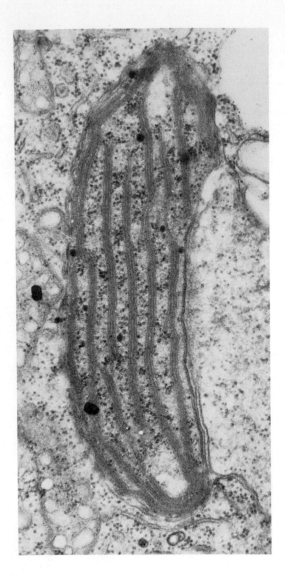

Figure 7-6 Algal chloroplast. Note that in this chloroplast from *Ochromonas danica* (*Chrysophycophyta*), the thylakoid membranes are not divided up into grana. Indeed, one can say that they compose one, continuous granum. Note also that ribosomes (small specks) are distributed both within and outside the chloroplast. The chloroplast ribosomes are implicated in intrachloroplast protein synthesis. Magnification is × 40,700; fixation was with glutaraldehyde, followed by osmium tetroxide. (Courtesy Dr. Sarah P. Gibbs.)

the energy that is conserved in the formation of high-energy bonds or biological work such as transport. However, the study of fluorescence has been of importance in coming to understand the mechanisms of energy transfer, representing, in a sense, a partial reaction of photosynthesis. For example, carotenoids and certain types of chlorophyll fluoresce strongly when isolated from the chloroplast, whereas in their natural location in the intact chloroplast they do not. It appears that their integration with the energy transfer apparatus within the chloroplast provides other routes for energy dissipation, so that the nonproductive route of fluorescence is suppressed or "quenched." When light is absorbed by a photopigment, the pigment remains in the activated (or excited) state for an average of about 1 nanosecond (10^{-9} second), at the end of which time it returns to the ground state with fluorescent emission of light. If, on the other hand, the energy is utilized by the photosynthetic organelle, it will be transferred from the excited pigment much more rapidly, within an average of 1 picosecond (10^{-12} second), and this much greater speed of energy transfer enables the productive pathway to prevail over the more sluggish fluorescence.

Electron transfer in photosynthesis

The reader is reminded of our central dogma about energy capture in cells: It always occurs by moving electrons. We have just said that the photosynthetic absorption of light (or indeed any photochemical process) involves the excitation of an electron to a higher energy level. It is consistent with everything we know to say that the electron may be regarded as a reductant, and that it passes along a series of oxidation–reduction compounds in a manner quite similar to the oxidation of a substrate by the respiratory chain.

Indeed, there are a number of striking similarities between the respiratory chain of mitochondria on the one hand, and the photosynthetic electron transfer system on the other. Pyridine nucleotides, flavoproteins, quinones, and cytochromes are involved in both cases, and the chains in both cases are interrupted by a number of inhibitors, including *antimycin* and *quinoline-N-oxides*. In both instances, the movement of electrons along the chain is coupled to synthesis of ATP, and, in both, electron flow is under regulation according to the availability of ADP and inorganic phosphate.

Figure 7-7 illustrates the photosynthetic electron transfer pathway in chloroplasts. It is clear that light interacts at two places in the chain, the so-called *photosystems I and II*, which have been identified by their different spectra. The two photosystems may be thought of as electron pumps, where excitation leads to electron transfer from the photosystem to the next carrier in the chain. This transfer leads to a deficiency in electrons, which is neutralized by passage of an electron from the previous carrier. Thus, the effect of light might be said to enable the photopigment to reduce the next carrier while oxidizing the previous carrier in the electron transfer chain.

Components of the photosynthetic chain include two cytochromes, *cytochrome b_6*, which is spectroscopically similar to cytochrome *b* of mitochondria, and *cytochrome f*, which is, spectroscopically, a *c*-type cytochrome. Prior to cytochrome b_6 is *plastoquinone*, which has the structure

$$H_3C \underset{O}{\overset{O}{\underset{\|}{\overset{\|}{\bigcirc}}}} H \quad (CH_2-CH=\overset{\overset{CH_3}{|}}{C}-CH_2)_n H$$

and the two cytochromes are followed by *plastocyanin*, a copper protein. It is significant that this portion of the photosynthetic chain is similar to

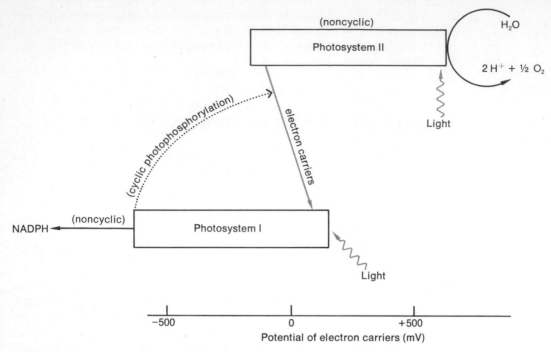

Figure 7-7 Outline of electron transfer in chloroplast photosynthesis. The effect of light is to elevate an electron in the photosystems to a redox state sufficiently high that it can cascade back to ground state by way of the several electron carriers (see also Table 7-1). This electron movement is coupled to ATP synthesis in a manner analogous to that in oxidative phosphorylation.

the respiratory chain of mitochondria (and many bacteria) in including the following sequence: *quinone, b-type cytochrome, c-type cytochrome,* and *copper protein.* It will be recalled that the final cytochrome of the oxidative system, cytochrome *a*, contains copper; it is likely that, in both cytochrome *a* and plastocyanin, the copper takes part in the oxidation and reduction of the enzyme. An additional similarity between oxidative and photosynthetic electron pathways is that both are interrupted between *b* and *c* cytochromes by a set of inhibitors, including antimycin and

Table 7-1   Some components of the photosynthetic electron transfer pathway

Arranged in order of increasing oxidation–reduction potential

Component	Molecular weight	Prosthetic group (if applicable)	Potential (mV)
Ferredoxin	12,000	nonheme iron	−430
NADP$^+$	743	—	−320
Cytochrome b_6	—	iron–heme	−30
Plastoquinone	748	—	113
Cytochrome f	ca. 20,000	iron–heme	360
Plastocyanin	21,000	copper	400

quinoline-*N*-oxides. Finally, it can be seen in Figure 7-7 that one branch of the pathway involves the reduction of NADP (to form NADPH) by *ferredoxin*. Ferredoxin is an iron-containing protein of unusually low oxidation–reduction potential and, unlike the cytochromes, which also contain iron, the iron is not associated with heme. It might be mentioned that there are nonheme iron proteins in mitochondria as well, although in general their function is not well understood.

Cyclic and noncyclic ATP synthesis

There are two basic mechanisms for ATP synthesis in green plant photosynthesis, and they differ in the pathway of associated electron flow as well as in inhibitor sensitivity. *Cyclic photophosphorylation* appears to involve photosystem I only and to function as a closed loop, without giving rise to *net* reduction of NADP. Cyclic photophosphorylation is sensitive to neither antimycin nor to *dichlorophenyldimethylurea* (DCMU), both of which inhibit noncyclic photophosphorylation. Cyclic photophosphorylation utilizes only part of the available electron transfer apparatus and might be said to be pure photophosphorylation in that its only product is ATP and there is no generation of reduced compounds. Finally, it appears that bacterial photosynthesis includes only cyclic photophosphorylation; light energy leads to the production of ATP only. Since these organisms frequently are found to grow in anaerobic situations, where components of the external medium are highly reduced, they obtain "reducing power" directly from the medium and, in their photosynthesis, concentrate on ATP production.

Cyclic photophosphorylation is well named, since there is neither a first electron donor nor a final acceptor. The path of electrons describes a closed loop and, as we have seen, there can be no net production of

Figure 7-8 Cyclic photophosphorylation. Note that this pathway comprises the loop (including dotted line) in Figure 7-7. (a) and (b) denote alternative pathways where ferredoxin is either included or omitted. The exact locus of ATP synthesis is not completely certain.

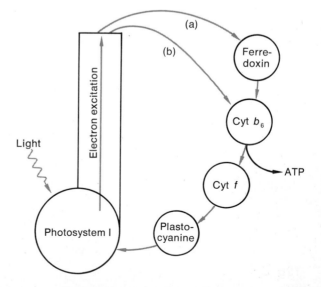

reduced compounds. On the other hand, plants that carry out photosynthesis do so in the presence of oxygen and thus in a relatively oxidized environment. Since, in such a situation, "reducing power" is unavailable from the environment directly, and since many of the biosyntheses that plants carry out are reductive, the photosynthetic mechanism has evolved to include a source of reduced compounds. The method whereby this can occur is called noncyclic photophosphorylation and is shown in Figure 7-9. The ultimate source of electrons is water, and withdrawal of electrons from water leads to production of oxygen gas. The occurrence of noncyclic photophosphorylation in plants accounts for oxygen production in such organisms (as opposed to photosynthetic bacteria) and is of obvious significance in maintaining the level of oxygen in the environment and hence the balance between photosynthetic and oxidative organisms. It should be added that the identity of water as the primary source of electrons in noncyclic photophosphorylation is consistent with tracer studies where addition of water labeled at oxygen gave rise to labeled gaseous oxygen.

The final electron acceptor in noncyclic photophosphorylation is $NADP^+$, the electrons traversing the photosynthetic electron transfer chain, as shown in Figure 7-9. In this system, as in the case of cyclic photophosphorylation, chlorophyll acts as a pump, raising electrons to an energy level at which they can reduce the next component in the chain. Another way of saying this is that chlorophyll in the presence of light is a relatively strong reducing agent (and a less strong oxidizing agent). The reduced product of noncyclic photophosphorylation, NADPH, is used by cells as a reducing agent in a number of biosynthetic reactions. It may also give rise to NADH by way of the transhydrogenase reaction

Figure 7-9   Noncyclic photophosphorylation. Note that oxygen (from water) is produced at the very beginning of the process and that the reduction of $NADP^+$ to form NADPH is, in a sense, the final result. The dotted arrows denote cyclic photophosphorylation "short circuits."

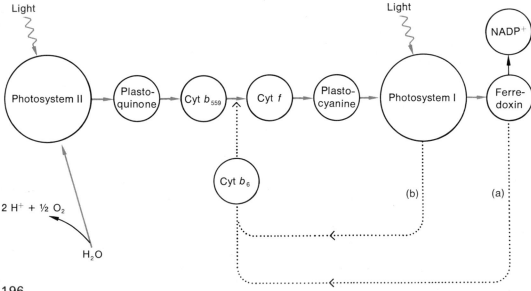

described in Chapter 6. Since plant cells contain mitochondria, as well as chloroplasts, it is possible for any NADPH not needed in biosynthesis to be converted to NADH and then oxidized with coupled ATP synthesis. This is probably a minor pathway, since mechanisms exist that can shift the balance in photosynthesis between cyclic and noncyclic processes as the need for energy (in the form of ATP) or reducing compound (in the form of NADPH) predominates. The molecular basis for this regulation is not yet understood.

An important distinction between cyclic and noncyclic photophosphorylation is the sensitivity of the latter to DCMU. This compound inhibits flow of electrons from water to NADP without influencing cyclic photophosphorylation. In the presence of DCMU, electrons can still reach NADP if they are inserted into the chain by an artificial electron donor, such as a reduced dichlorophenolindophenol dye, in which case only photosystem I is involved.

The relation of ion transport to photosynthesis

Not only can light energy be conserved through photosynthetic ATP synthesis, but it can provide the driving force for energy-linked ion transport across membranes associated with plant and bacterial photosynthesis. For example, chloroplasts have been shown to take up a variety of cations, such as calcium and potassium, and, under suitable conditions, the uptake can compete with ATP synthesis for energy—that is, can produce uncoupling of photophosphorylation.

It appears likely that movement of ions must be considered as more than an alternative route for photosynthetic energy dissipation and that it forms an essential feature of light-induced energy trapping. Thus, transfer of protons (hydronium ion) appears to be obligatory for energy conservation. When light shines on chloroplasts in an unbuffered medium, the pH of the medium rises, owing to the transfer of protons into the chloroplasts. The inward transfer of protons continues for a considerable period of time, suggesting that an anion must also be taken up. Otherwise, the uptake of protons would quickly lead to formation of a large membrane potential (positive on the inside), which would form an electrostatic barrier opposing additional proton uptake (see Chapter 12). Indeed, it have been said to include so-called "high-energy intermediates," but protons, thus maintaining electrical neutrality.

The mechanism of photosynthetic ATP synthesis, like that of oxidative ATP synthesis, is not known. The pathways of energy in both processes have been said to include so-called "high-energy intermediates," but these have been neither isolated nor observed. There is an alternative point of view that regards the high-energy state as an electrical potential across a chloroplast (or mitochondrial) membrane, and this approach will be taken in the more appropriate context of membrane processes (see Chapters 11 and 12). However, in the present discussion it is quite relevant to observe that the photo-induced migration of protons (and other ions) across the chloroplast membrane may be viewed in the light of the idea that a membrane potential may be central to energy coupling. Thus, ion movement across the chloroplast membrane may represent not the use of energy as an alternative to ATP synthesis but rather as an intimate feature of the ATP-synthesizing mechanism itself. This approach has become especially appealing in view of recent experiments where chloroplasts in the dark have been induced to synthesize ATP from ADP and phosphate by rapidly shifting the pH of their medium, so that a rapid flux of protons occurred across the chloroplast membrane (see Figure 7-10).

Figure 7-10 "Acid bath" experiment, in which chloroplasts make ATP in the dark when a proton gradient is imposed across their membrane. The first step is the equilibration of protons at pH 4.0, where they penetrate the chloroplast until the internal pH is also at that pH. Then (in the dark) the chloroplasts are placed in a basic medium (pH 8.5) that also contains phosphate and ADP. Synthesis of ATP is coupled to the efflux of protons, a result suggesting that proton movement may underlie light-induced ATP synthesis as well. Indeed, it is important to add that this proton-coupled ATP synthesis is prevented by known inhibitors of photophosphorylation. A complete description of this experiment can be found in the original paper by A. T. Jagendorf and E. Uribe, *Proc. Nat. Acad. Sci. U.S.,* **55**, 170 (1966).

Step 1: Equilibration of chloroplasts at pH 4

Reaction medium at pH 4.0

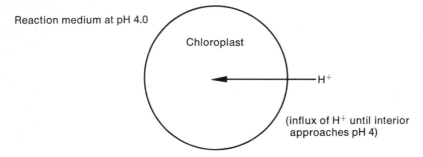

(influx of H^+ until interior approaches pH 4)

Step 2: Addition of base (pH 8.5), ADP, and PO_4^{3-} to medium

Reaction medium at pH 8.5

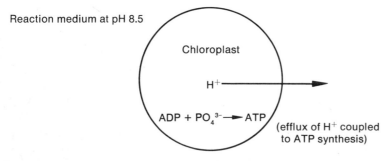

(efflux of H^+ coupled to ATP synthesis)

An additional form of experimental evidence for the importance of membrane potentials in chloroplast energy trapping has come from studies of the spectra of chloroplast pigments, such as carotenoids. Under conditions of energy coupling, the spectroscopic peaks of the carotenoids shift slightly toward the red region of the spectrum, and the most reasonable explanation for such a "red shift" appears to be the influence of membrane potential upon the electronic structure of the carotenoid located within the membrane. It should be added that such effects of external electrical fields on spectra are well known and obey predictable laws, so that it is possible to estimate the electrical field within the chloroplast by the degree of spectroscopic displacement. Indeed, estimation of the membrane potentials made in this fashion are in agreement with those made by other techniques.

Light reactions and dark reactions

Both the primary light adsorption events and the electron transport reactions leading to $NADP^+$ reduction and ATP synthesis require light for their occurrence. The actual synthesis of hexose is a "dark reaction" —light is not required if there are sources of energy and reducing equivalents (for example, NADPH). As far as mechanism is concerned, carbohydrate synthesis is a secondary event in photosynthesis, but in terms of physiology (and ecology) it is of the utmost importance. It is such an obvious and important result of photosynthesis that early students of photosynthesis saw nothing but the "fixation" of carbon dioxide to form sugar and assumed this reaction to be the whole story.

The pathway for carbohydrate synthesis by green plants is largely a reversal of glycolysis. The early steps of the series were demonstrated in elegant experiments in which algae were incubated with radioactive CO_2 and then plunged into boiling ethanol after a short interval in the light. The plant extracts were then examined by paper chromatography to see which intracellular compounds were the ones first labeled. It turned out that 3-phosphoglycerate was the earliest compound to become radioactive, but that a number of sugars were labeled soon thereafter. On the basis of such experiments, a cyclic process for CO_2 incorporation was described, involving as its initial reaction the fusion of CO_2 and the 5-carbon sugar ribulose diphosphate:

$$
\begin{array}{l}
CH_2O-\text{\textcircled{P}} \\
| \\
C=O \\
| \\
H-C-OH \\
| \\
H-C-OH \\
| \\
CH_2O-\text{\textcircled{P}}
\end{array}
\quad
\xrightarrow{H_2O + CO_2}
\quad
\begin{array}{l}
CH_2O-\text{\textcircled{P}} \\
| \\
2\ CHOH \\
| \\
COOH
\end{array}
$$

The product of this reaction, two molecules of 3-carbon phosphoglycerate, will be familiar as a member of the glycolytic pathway. The reaction does

199

not require light nor any of the products of photosynthesis (ATP or NADPH). The subsequent use of phosphoglycerate to synthesize sugars proceeds by way of a pathway that requires both ATP and NADPH—illustrating the obligatory nature of the connection between photosynthesis and CO_2 incorporation (see Figure 7-11). Without becoming involved in the complexity of the pathway, it is worth noting that most of its reactions are found in animal tissue as well, and the reason the pathway is associated with photosynthesis is because of the requirements for ATP and NADPH. Indeed, it used to be thought that CO_2 fixation was a unique property of green plants, but recently a number of different reactions incorporating CO_2 into organic compounds in animal cells have been discovered. These are, however, rather minor routes in animal cells, whereas CO_2

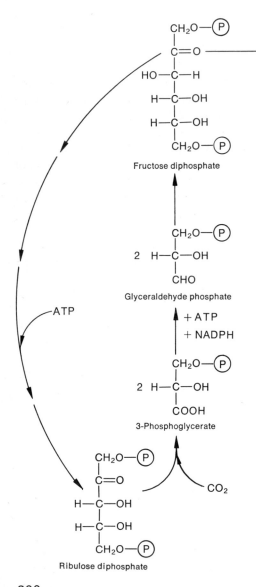

Figure 7-11 Outline of photosynthetic carbon dioxide incorporation. The cycle is greatly simplified by the omission of the extremely complex series of rearrangements leading to regeneration of ribulose diphosphate. It is likely that there is a six-carbon intermediate, formed by the addition of CO_2 to ribulose diphosphate, which is rapidly cleaved to form two molecules of phosphoglycerate. Note that the reactions between phosphoglycerate and glucose are essentially those of glycolysis run in reverse.

200

accounts for all carbon in the biosyntheses of many plants under many conditions.

The control of photosynthesis through regulation of pigment synthesis

Since the photosynthetic pigments are at the heart of the whole affair, it is reasonable that the formation of the pigments themselves might be a locus of control. Photosynthetic organisms live under a variety of conditions of nutrient and illumination, and it is necessary that the rate of photosynthetic energy trapping reflect the situation at hand. For example, it is well known that seedlings grown under conditions where photosynthesis is impossible (in the dark) do not form the photosynthetic apparatus, which includes chlorophyll and accessory pigments. If such *etiolated* plants are placed in the light, the concentration of pigments soon reaches a normal value.

Much of the most compelling information about control of photosynthesis comes from studies of microorganisms, which are useful subjects because of the possibility of controlled culture and homogeneous populations. Unicellular algae such as *Chlamydomonas* or *Euglena* may be grown in pure culture, and the concentration of pigment in such cells is proportional to light intensity over a considerable range. Similar results are noted with the photosynthetic bacterium *Rhodopseudomonas*, and it is interesting that there appears to be an optimum light intensity for pigment synthesis. Thus, cells grown aerobically in the dark make no pigment, cells grown in dim light do, and in bright light synthesis again is suppressed. This should be regarded as an adaption whereby cells contain maximum pigment under those conditions where the absorption of radiant energy is most critical—conditions of minimal illumination.

Cells of *Rhodopseudomonas* are able to grow aerobically in the dark in the presence of organic substrates using respiratory electron transport for energy production. It is interesting that the presence of oxygen suppresses pigment production completely, so that dividing cells soon exhibit much lower concentrations of chlorophyll and carotenoids. Thus, in this instance, the availability of oxygen—which allows oxidative phosphorylation—serves to eliminate the apparatus for photosynthetic ATP synthesis, a situation analogous to the Pasteur effect, where the fermentative route to ATP synthesis is turned off. One might suppose that oxidative ATP formation represents the most efficient form of energy trapping by cells and that nature prefers to use it where possible. Obviously, oxygen does not eliminate photosynthesis in the case of higher plants, because, unlike the photosynthetic bacteria, green plants are commonly found in aerobic situations (such as my flower box). Indeed, the aerobic character of our planet is largely caused by the fact that green plants evolve oxygen as a photosynthetic by-product—and it is obvious that oxygen-induced repression of photosynthesis in higher plants would lead to ecological disaster.

Phototaxis

Many photosynthetic organisms, including plants and bacteria, exhibit the ability to orient themselves with respect to a source of light. Such *phototaxis* may be regarded as a regulation device associated with photosynthesis, whereby organisms maximize their photosynthetic capability by positioning themselves in an optimal fashion. For example, it has been known for a half-century that photosynthetic motile bacteria, such as *Rhodospyrillum*, congregate in a light spot on a microscope slide. In fact, these bacteria possess a relatively primitive response mechanism wherein they reverse direction of swimming if they experience a diminution of light intensity. Thus, when the bacterium swims across a light–dark boundary in a direction that produces a drop in intensity, it changes direction and returns to the lighter region. This form of taxis, or sensitivity to temporal changes in light intensity, is called *phobophototaxis* and does not include the ability to sense direction of incident light.

On the other hand, many motile, photosynthetic organisms, including algae and gametes of higher plants, are sensitive to the direction of light flux (*topophototaxis*) and have highly developed systems for sensing it. For example, the unicellular alga *Euglena* responds to light direction by swimming toward it, and analysis of the effectiveness of light of different wavelengths suggests that the light-sensitive pigment involved is a carotenoid. In the case of *Euglena*, this is of special interest because of the occurrence of a carotenoid-containing eyespot, which is situated near the flagellum and appears to be closely involved with phototactic response. On the other hand, an eyespot-less mutant of *Chlamydomonas* exhibits phototaxis, although of lesser sensitivity than normal. Finally, it is interesting that *Euglena* is sensitive not only to the direction from which light comes, but also to the degree of polarization of light.

Since phototaxis represents a form of photosynthetic control, it is not surprising that there are close metabolic links with photosynthesis. For example, *Euglena* grows in the dark by means of respiration, and dark-grown cells are perfectly motile but do not exhibit phototaxis. Likewise, uncouplers of photosynthetic phosphorylation inhibit phototaxis to a much greater extent than they do motility itself.

The biological emission of light

Cells generally are able to give off electromagnetic radiation, and its release is coupled to the chemical reactions within the cell. In most instances, the radiation is in the infrared region and might be described as metabolic heat production. In some cases, widely dispersed in the plant and animal kingdoms, the radiation produced is of sufficiently high energy to fall within the visible region of the spectrum and is termed *bioluminescence*. Bioluminescence does not in any sense involve a reversal of photosynthesis, since it is found in nonphotosynthetic organisms (such as fireflies), but it shows some interesting similarities.

What is known of the mechanism of light production suggests that it should be thought of as a special case of *chemiluminescence*, which is the release of energy in the form of light in the course of a chemical reaction. As most chemical reactions in the cell are catalyzed by enzymes, it is no surprise that the luminescent reaction requires the presence of an enzyme, which is called, somewhat poetically, *luciferase*. Luciferase is actually a class of enzymes, the members of which catalyze the light-producing oxidation of any one of a number of substrates, depending on the phylogenetic source of the system. In the firefly, the substrate is a complex organic molecule,

called *luciferin*. The oxidation of luciferin to yield light and the involvement of ATP in the whole affair is summarized by the following reactions:

(1)   Luciferin + ATP   →   Luciferin—AMP + ℗—℗

(2)   Luciferin—AMP + $\frac{1}{2}O_2$ →   Oxidized luciferin—AMP + H_2O

(3)   Oxidized luciferin—AMP   →   Oxidized luciferin + AMP + light

For the process to be truly cyclic, a means must be at hand whereby oxidized luciferin can become reduced again. In bacterial luminescence, the reducing agent appears to be a flavoprotein which, as seen in Chapter 6, receives electrons from the primary substrates of respiration, such as succinate and NADH:

It is interesting to note that, although the details of mechanism are not yet completely worked out in the various luminescent organisms, the general features are suggestive of a process rather like photosynthesis run backward. Photosynthesis is light absorption coupled to the formation of ATP and reducing power, while the light-producing systems appear to *require* ATP (or energy-coupled respiration) and reducing power (in the form of a reduced flavin in bacteria) for light to be emitted. The details of the two processes, however, differ significantly in a number of obvious

203

ways, including the nature of the quantum events, structures involved, and their phylogenetic distribution.

Photorespiration

Higher plant cells contain mitochondria and are therefore able to respire. Indeed, in the dark, when photosynthesis does not occur, cells respire with the products of photosynthesis as substrates. This respiration is associated with oxidative phosphorylation and is sensitive to the usual inhibitors of respiration, such as cyanide and carbon monoxide.

There is a second type of respiration that occurs specifically in the light and that differs from ordinary respiration in several ways. For one thing, this *photorespiration* is not inhibited by compounds that inhibit mito-chondrial respiration. It is not coupled to the synthesis of ATP and so is, in this sense, wasteful. It appears that photorespiration should be regarded as the loss of some reduced products of photosynthesis through their oxidation by molecular oxygen. It has been claimed that this respiration plays a regulatory role in plant cells by adjusting the net production of reducing power by photosynthesis, although little evidence is presently available.

Although the enzymic details of photorespiration are not yet clear, it does appear that a major substrate for the process is glycolic acid

$$HOH_2C-COOH$$

It is not known by what means glycolic acid is formed from the reduced products of ordinary photosynthesis, but it appears that it can be oxidized by the following reaction:

$$HOH_2C-COOH + O_2 \rightleftharpoons \overset{\overset{\textstyle H}{|}}{O=C}-COOH + H_2O_2$$

This is catalyzed by the enzyme *glycolate oxidase*, a flavoprotein that is located in the peroxisomes (see Chapter 1). The H_2O_2 is decomposed to water and molecular oxygen by the catalysis of the enzyme *catalase*:

$$H_2O_2 \rightleftharpoons H_2O + \tfrac{1}{2}O_2$$

a reaction that also occurs in the peroxisomes.

Nitrogen fixation

Nitrogen fixation is the process by which molecular nitrogen (N_2) in the atmosphere is incorporated into organic compounds. Owing to the importance of amino, nitrate, and other nitrogen-containing groups in the living world, the process of nitrogen fixation is of great importance. Although lower plants are often able to carry out nitrogen fixation, it is not, strictly speaking, an aspect of photosynthesis. The reason that we are considering it here is that the enzymic machinery involved includes components similar to the photosynthetic electron transfer chain.

Cells that fix nitrogen include all the photosynthetic bacteria, certain soil bacteria, such as *Azotobacter*, certain anaerobic bacteria, such as *Clostridia*, blue-green algae, and symbiotic systems, including higher plants in association with certain bacteria. In the latter case, the association is a true symbiosis in that cultured bacteria of the *Rhizobium* group isolated from suitable plants are, alone, unable to carry out fixation, while the plants (which include legumes and alders) are also unable to carry it out when the bacteria are absent. Clearly, both plant and bacterium play an essential role in the process. In the case of this symbiosis, the bacteria invade the roots of suitable plants and cause them to form nodules that have a characteristic structure. Interestingly, these contain the heme protein, hemoglobin, otherwise found only in animals, and whose function in this case may be to regulate the concentration of oxygen in the nodule tissue.

Nitrogen fixation is a reductive process: molecular nitrogen is reduced to ammonia. It appears that the reducing agent is commonly molecular hydrogen (H_2), so that one should probably refer to the net process as *nitrogen–hydrogen fixation*. Clearly, too, ATP is required for the fixation to proceed, although we do not know at which step it is needed.

$$N_2 + 3\,H_2 + 12\,ATP \rightleftharpoons 2\,NH_3 + 12\,ADP + 12\,PO_4^{3-}$$

Indeed, the individual steps in the process are not yet completely worked out, in part because of difficulties in obtaining cell-free systems that retain activity. It is clear that a sort of electron transfer chain of enzymes is operative and that, at the least, it contains *ferredoxin*, a *hydrogenase*, a second nonheme iron protein (ferredoxin being the first), and, finally, a protein that contains both iron and molybdenum. It is clear that the hydrogenase is the first carrier in the series in that it reacts with H_2. It is also likely that the final reduction of N_2 is carried out by the iron–molybdenum protein, but the order of the other two components is not certain. It can, however, be predicted that these matters will be worked out in the very near future, since increasingly active cell-free systems are being obtained as well as improved assays for activity. In this last connection, it should be mentioned that the assay of nitrogen incorporation has always been something of a problem, since the available tracer isotope of nitrogen is ^{15}N, which is not radioactive. Thus its incorporation must be followed by means of a mass spectrometer, a technique both expensive and cumbersome. Recently, it has been found that the enzyme system can be assayed by allowing it to "fix" acetylene instead of nitrogen, acetylene having the advantage of being detectable by simple means. Indeed, the enzyme system can also reduce nitrous oxide, azide, and cyanide. In all cases, the system is inhibited by carbon monoxide.

Suggested Reading

Books

Calvin, M., and J. A. Bassham, *The Photosynthesis of Carbon Compounds*, W. A. Benjamin, New York, 1962.

Clayton, R., *Molecular Physics in Photosynthesis*, Xerox College Publishing, Lexington, Mass., 1965.

Heath, O. V. S., *The Physiological Aspects of Photosynthesis*, Stanford University Press, Stanford, Calif., 1969.

Hill, R., and C. P. Whittingham, *Photosynthesis*, Methuen & Co., London, 1957.

Kamen, M. D., *Primary Processes in Photosynthesis*, Academic Press, New York, 1963.

Kirk, J. T. O., and R. A. E. Tilney-Bassett, *The Plastids*, W. H. Freeman and Co., San Francisco, 1967.

McElroy, W. D., and B. Glass (eds.), *A Symposium on Light and Life*, Johns Hopkins Press, Baltimore, 1961.

Shibata, K., *Comparative Biochemistry and Biophysics of Photosynthesis*, University of Tokyo Press, Tokyo, 1968.

Vernon, I. P., and G. R. Seely (eds.), *The Chlorophylls*, Academic Press, New York, 1966.

Articles

Arnon, D. I., "Photosynthetic activity of isolated chloroplasts," *Physiol. Rev.*, **47**, 317–58 (1967).

Avron, M., "Mechanism of photoinduced electron transport in chloroplasts," in *Current Topics in Bioenergetics*, D. R. Sanadi (ed.), vol. 2, Academic Press, New York, 1967, p. 1.

Boardman, N. K., "The photochemical system of photosynthesis," in *Advances in Enzymology*, F. F. Nord (ed.), vol. 30, John Wiley & Sons (Interscience Division), New York, 1968, p. 1.

Gross, E. L., and L. Packer, "Ion transport and conformational changes in spinach chloroplast grana, *Arch. Biochem. Biophys.*, **121**, 779 (1967); **122**, 237 (1967).

Hendricks, S. P., "How light interacts with living matter," *Sci. Am.*, **219**, 174 (Sept., 1968).

Levin, R. P., "The mechanism of photosynthesis," *Sci. Am.*, **221**, 58 (Dec., 1969).

Olson, J. M., "The evolution of photosynthesis," *Science*, **168**, 438 (1970).

Pfennig, N., "Photosynthetic bacteria," *Ann. Rev. Microbiol.*, **21**, 285 (1967).

Walker, D. A., and A. R. Crofts, "Photosynthesis," *Ann. Rev. Biochem.*, **39**, 389 (1970).

8

Some pathways for the biosynthesis of small molecules

In the past three chapters, we have seen that a number of mechanisms exist whereby organic molecules are degraded by cells into simpler products. In general, these processes lead to a release of energy, which is conserved by cells largely in the form of adenosine triphosphate (ATP). We also observed that energy conservation was associated with the movement of electrons—that is, oxidation or reduction. In the case of respiratory ATP synthesis, the energy-yielding breakdown of organic molecules, such as sugars, involved oxidations or removal of electrons. We turn now to a set of reactions that are, in a sense, the opposite of those mentioned above in that they are generally reductive (on the one hand) and in that they require ATP instead of producing it (on the other). We will now become concerned with biosynthetic processes that are not inherently spontaneous in that they must be coupled to ATP hydrolysis in order to proceed.

It is worth thinking about biological syntheses in a rather general sense before proceeding to specific examples. As we have said, ATP hydrolysis is generally the source of energy, so that syntheses are coupled to the processes of oxidative, fermentative, or photosynthetic phosphorylation. When biosynthetic reactions require chemical reduction of an intermediate compound, the reducing agent is often (but not always) NADPH, which we have seen to be produced directly by photosynthetic electron transport and indirectly, via a transhydrogenase, by oxidative electron flow. Thus, in general, the coenzyme NADPH is chiefly involved in synthetic processes, while NADH plays a greater role in energy production; but this rule is not always completely valid and should only be regarded as a trend. The importance to the cell of separating coenzymes that function in energy-producing (degradative) and energy-requiring (synthetic) pathways will become obvious in the following discussion.

It should be clear that syntheses and degradations are, in a real sense, opposite. For example, certain cells can synthesize the sugar glucose from carbon dioxide and water and, at the same time, or under different conditions, break down glucose to the same two components. Likewise, cells often can, on the one hand, synthesize specific amino acids from simpler precursor molecules and, on the other hand, break them down into the same, or similar, simple molecules. Since cells tend to carry out their activities in the most economical fashion, one might reasonably expect the enzymes and pathways for synthesis and breakdown to be the same. Indeed, there are many enzymes that do function in both directions, but when one considers whole pathways, a rather different rule emerges: *Pathways for synthesis and breakdown are seldom (or never) identical.* A little reflection will suggest the importance of this generalization. If the same completely identical pathways were performing both syntheses and degradations, the cell would be without the means of controlling the two processes. There would be a simple equilibrium between small precursors and the more complex molecules, and the position of this equilibrium would be determined by the free-energy differences between the two classes of compounds. Unless a continuous supply of ATP were consumed, the net reaction would be heavily in the direction of small molecules, since we have seen that the breakdown direction is spontaneous, whereas the synthesis is not. To put it differently, the occurrence of nonidentical pathways permits different control mechanisms to operate in the two directions, and this, in turn, permits the balance between precursor and product to be regulated according to the needs of the cell and without excessive expenditure of energy in the form of ATP.

Another feature of the regulation of biosynthetic reactions in relation to energy-producing ones is the frequent separation of the two in different regions of the cell. Thus, the product of a synthesis is often located in a compartment of the cell where enzymes that might effect its degradation are excluded. For example, fatty acid synthesis is, in eukaryotic cells, localized in the endoplasmic reticulum, while their breakdown by means of oxidations occurs within mitochondria.

Regulatory mechanisms in biosynthesis

We turn now to the types of regulation that cells employ in biosynthetic pathways. In the first place, it is generally the case that all reactions in a pathway are not under specific regulatory control beyond that imposed by the kinetic features of the enzyme, such as its maximum velocity and the Michaelis constant. Rather, there are generally a few control points in a pathway, and usually the first step in understanding the regulatory processes is the identification of these reactions. Sometimes a useful clue comes from a comparison of the corresponding biosynthetic and degradative pathways. Often these will employ many of the same enzymes, with a few differing in such properties as kinetic constants, coenzymes, or substrate specificity. The latter enzymes usually turn out to be the place to look for possible regulation. When identical enzymes are employed in both kinds of pathway, they are most unlikely to be sites of control, for reasons stated above.

Regulation may be imposed on either the activity of an enzyme or on its concentration in the cell. The former includes either stimulation or

Table 8-1 *Regulatory mechanisms in*
          metabolic pathways

I. Acting on enzyme synthesis	stimulation	= induction
	decrease	= repression
II. Acting on enzyme activity	stimulation	= activation
	decrease	= inhibition

inhibition of the rate of an enzyme-catalyzed reaction, while the latter includes stimulation or suppression of enzyme synthesis. These various possibilities are summarized in Table 8-1, and, in the ensuing discussion, it will be important to be familiar with the distinctions between the several possibilities. We shall not be concerned here with the mechanisms of these forms of regulation; control of enzyme activity is discussed in some detail in Chapter 4, and control of enzyme synthesis is discussed in the context of protein synthesis in Chapter 16.

There are a few additional remarks that can be made about general patterns of regulation that occur in synthetic pathways. For one thing, the regulating molecule is often the final product of the pathway, and, in cases where it leads to allosteric inhibition of an enzyme in the pathway, that target enzyme is often the first enzyme in the sequence. This pattern of regulation is called *end-product inhibition* and is exceedingly common in syntheses. In instances when a pathway is branched—that is, when a single precursor gives rise to more than one end product—the site of allosteric inhibition is commonly the first enzyme after the last branch point. In other words, the product of a pathway is usually involved in the regulation of only that portion of the pathway that leads uniquely to it.

Examples of biosynthesis: sugars

Since we have examined the breakdown of sugars in detail in the chapter on glycolysis (Chapter 5), it seems productive now to consider their synthesis with special reference to the ways in which the two pathways are different. We observed the central importance of the glycolytic pathway by which six-carbon sugars were transformed, finally, to pyruvate. In a similar fashion, the major route for the synthesis of sugars proceeds from pyruvate to hexoses but in a somewhat different manner, as it employs some different reactions.

The overall pathway for the synthesis of sugars, a process also called *gluconeogenesis*, is outlined in Figure 8-1. It is clear that several of the same intermediates occur both here and in glycolysis, but it is also apparent that important differences exist as well. In general, these differences may be understood in the need for separate regulation of synthesis and degradation. They also are related to the quite different thermodynamic situation that obtains in the two directions. Thus, the glycolytic pathway includes several reactions that are associated with so great a negative free-energy

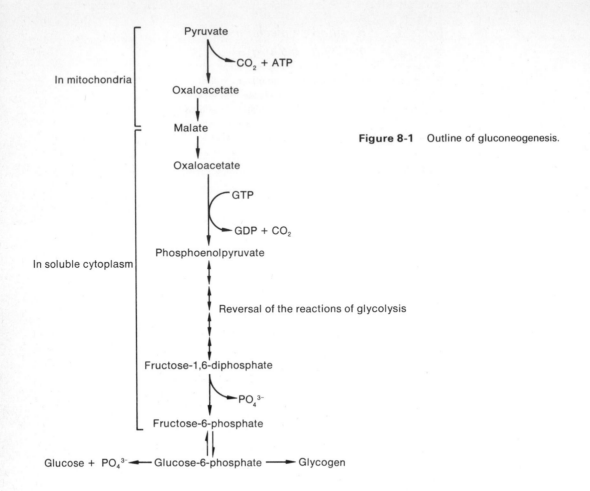

Figure 8-1 Outline of gluconeogenesis.

change that they are, for all practical purposes, irreversible. For example, the glycolytic reaction catalyzed by the enzyme *pyruvate kinase*, where phosphoenolpyruvate is dephosphorylated to form pyruvate,

Phosphoenolypyruvate + ADP $\rightleftharpoons$ Pyruvate + ATP

exhibits a free-energy change of over -7 kcal/mole, so that a simple reversal is quite out of the question. In fact, it appears that oxaloacetate is an intermediate in the formation of phosphoenolpyruvate from pyruvate, so that a quite different mechanism must be in existence.

Indeed, the route from phosphoenolpyruvate to pyruvate requires a single enzyme, the kinase, while the reverse net reaction occurs in four steps, which are summarized in Table 8-2. It is also clear that some of the reactions in this more complex pathway are localized within the mitochondrion, while the glycolytic reactions are found to occur entirely in the nonmembranous cytoplasm.

There are several features of this set of four reactions that are instructive in comparing glycolysis and gluconeogenesis. For one thing, the dephosphorylation of phosphoenolpyruvate (to yield pyruvate) leads to the

*Table 8-2   Synthesis of phosphoenolpyruvate from
pyruvate in hexose synthesis (gluconeogenesis)*

Reaction	Cellular location
1. Pyruvate + CO_2 + ATP $\rightleftharpoons$ Oxaloacetate + ADP + P_i	Mitochondrion
2. Oxaloacetate + NADH + H^+ $\rightleftharpoons$ Malate + NAD^+	Mitochondrion
3. Malate + NAD^+ $\rightleftharpoons$ Oxaloacetate + NADH + H^+	Cytoplasm*
4. Oxaloacetate + GTP $\rightleftharpoons$ Phosphoenolpyruvate + CO_2 + GDP	Cytoplasm*

* That is, the nonmembranous cytoplasm.

synthesis of adenosine triphosphate (ATP), while the final phosphorylation that yields phosphoenolpyruvate (from oxaloacetate) employs guanosine triphosphate (GTP). The first reaction of the four appears to be
influential in the control of the overall process. It is catalyzed by the
enzyme *pyruvate carboxylase*, which requires as a cofactor (or activator)
acetyl—S—CoA. Since the intramitochondrial concentration of acetyl—
S—CoA would be expected to be high when there is a favorable supply
of precursors for gluconeogenesis, this means that, when the precursors
are available, the pathway would be turned on.

It is also important to comment on the apparent identity of reactions
2 and 3 in Table 8-2. The conversion of oxaloacetate to malate is a spontaneous one (with $\Delta G^{\circ\prime} = -6.7$ kcal/mole), but, because of the compartmentation, it is effectively even more strongly pulled in the direction
of malate, owing to the removal of malate from the mitochondrion. The
migration of malate from the mitochondrion to the cytoplasm that must
occur between reactions 2 and 3 also provides an additional locus for
possible control, since the availability of energy in mitochondria can
significantly alter the permeability to anions.

When phosphoenolpyruvate is converted to pyruvate, there is a synthesis of a single molecule of ATP from ADP and phosphate. Even
including the formation of ATP, this reaction is highly spontaneous. On
the other hand, the reverse (gluconeogenetic) reaction can be seen to
require two molecules of "high-energy phosphate compounds" in the
form of one of ATP and one of GTP. In addition, energy is provided for
this net reaction in more subtle ways. For example, the reaction is favored
by the relatively high $NADH/NAD^+$ ratio that occurs in the mitochondrion and the correspondingly low ratio that occurs in the soluble cytoplasm. In addition, energy may be expended in the movement of malate
from the inside of the mitochondrion, since it has been demonstrated that
such anion transport can be stimulated by the energy-linked transport of
cations in the same direction.

Phosphoenolpyruvate is converted to fructose-1,6-diphosphate by a
series of reactions that are identical to those involved in glycolysis.
Fructose-1,6-diphosphate is then hydrolyzed to fructose-6-phosphate by
a reaction that is not the reverse of the glycolytic one. It will be recalled

that glycolysis includes the phosphorylation of fructose-6-phosphate to yield the diphosphate:

Fructose-6-phosphate + ATP $\rightleftharpoons$ Fructose-1,6-diphosphate + ADP

a reaction that is associated with a negative free-energy change. Thus, a simple reversal of the reaction would not be thermodynamically favorable, as it would have to lead to the synthesis of ATP, which we have seen requires considerable energy. For this reason, the reaction in the synthetic direction is a hydrolysis, catalyzed by the enzyme *fructose-1,6-diphosphate phosphatase*:

Fructose-1,6-diphosphate + H_2O $\rightleftharpoons$ Fructose-6-phosphate + P_i

This enzyme is important in the regulation of the overall synthetic pathway, since it is subject to allosteric inhibition by AMP. This may be thought of as a sort of indirect case of end-product inhibition, since it is sensitive to the ratio between ATP and AMP. Because the total adenine nucleotide content of cells is relatively constant, changes in the amount of any one of the three, AMP, ADP, or ATP, will result in changes in the others. It happens that, in cells, when the ATP concentration is high, the concentrations of both ADP and AMP are usually depressed, so that one effect of a high ATP level, where energy will be available for gluconeogenesis, is that the pathway by which this occurs will be activated. When the biosynthesis has proceeded to the extent that the ATP/AMP ratio begins to decline, the increasing AMP levels will inhibit further activity of the pathway until the balance is restored.

Fructose-6-phosphate is next converted to the isomer glucose-6-phosphate by the reversal of the glycolytic isomerase reaction. Glucose-6-phosphate, in turn, can serve as a source of the hexose glucose as well as of various polymers, such as starch or glycogen. It is interesting that, when glucose-6-phosphate is converted into glucose, the path is not the reversal of the glycolytic reaction

Glucose + ATP $\rightleftharpoons$ Glucose-6-phosphate + ADP

which would be thermodynamically unfavorable. Instead, another hydrolysis is employed, so that inorganic phosphate is released:

Glucose-6-phosphate + H_2O $\rightleftharpoons$ Glucose + P_i

In a similar way, the synthesis of glycogen from glucose-6-phosphate does not follow the pathway used when glycogen is utilized, a breakdown that is essentially a phosphorolysis by inorganic phosphate and is catalyzed by *glycogen phosphorylase*, about which more later. Instead, glucose-6-phosphate is converted into glucose-1-phosphate, which reacts with uridine triphosphate (UTP) as follows:

Glucose-1-phosphate + UTP $\rightleftharpoons$ UDP—glucose + PP_i

The pyrophosphate (PP_i) produced in this reaction is hydrolyzed to two molecules of inorganic monophosphate,

$$PP_i + H_2O \rightleftharpoons 2 P_i$$

a reaction that is accompanied by a large negative free-energy change and is therefore spontaneous. The free-energy change is sufficient, indeed, to "pull" the preceding one toward the right as well.

The actual synthesis of the polymer glycogen occurs with the transfer of the glucose from a UDP—glucose molecule to the end of the growing glucose polymer chain:

$$UDP-glucose + (glucose)_n \rightleftharpoons UDP + (glucose)_{n+1}$$

This reaction is catalyzed by *glycogen synthetase*. In vertebrates, UDP—glucose gives maximum activity in this reaction, while ADP—glucose is much less effective. On the other hand, in lower organisms, this preference is reversed, with ADP—glucose being most active. It may be recalled that glycogen is a branched polymer of glucose, and so a separate category of enzymes called *transglycosidases* is required to form the branches (by synthesizing $1 \rightarrow 6$ bonds instead of the $1 \rightarrow 4$ bonds formed by the preceding reaction).

Finally, we should mention that the pathway described for the synthesis of sugars from pyruvate allows entry from a number of directions. For example, some, but not all, amino acids can give rise to a net formation of glycogen and enter the synthetic pathway by being converted first to Krebs cycle intermediates and thence to oxaloacetate and pyruvate by the route outlined earlier. On the other hand, animals cannot synthesize glycogen from fatty acids, while many plants and bacteria are able to do so. This is because of the occurrence of the glyoxylate cycle (see Chapter 6), by which these organisms are able to effect the net synthesis of Krebs cycle intermediates from acetyl—S—CoA. Thus, it is clear that acetyl—S—CoA is a necessary intermediate in the formation of sugars from fatty acids and that Krebs cycle intermediates, such as oxaloacetate, are obligatory intermediates as well.

The regulation of gluconeogenesis by hormones

Many cellular processes are, in animals, subject to control by the action of molecules called *hormones*. These compounds, which vary in size from small organic molecules to proteins, serve as intercellular communicating messengers, being produced in one part of an organism and received in another. Although this sort of communication has little meaning in the case of a unicellular organism, some of the underlying mechanisms behind the action of hormones are found in altered form there as well. For example, we shall shortly examine the role of a molecule called cyclic AMP in hormonal mechanisms, while the same molecule will be encountered in Chapter 16 playing a role in the regulation of protein synthesis in bacteria.

Hormones are synthesized in the *endocrine glands*, from which they are released into the bloodstream to find their way to the target tissue. It has long been known that removal of certain glands leads to changes in carbohydrate metabolism and that such effects can often be reversed by administering the hormone secreted by that gland. For example, beta cells of the pancreas produce a protein hormone called *insulin*. When the pancreas loses this function through disease (diabetes), surgical removal, or administration of alloxan

$$
\begin{array}{c}
\text{H} \\
| \\
\text{O} \diagup \text{N} \diagdown \text{O} \\
\text{O} \diagdown \text{N} \diagup \\
| \\
\text{O} \quad \text{H}
\end{array}
$$

which prevents insulin secretion, the most dramatic metabolic effect that is noted is a sharp rise in the concentration of blood sugar (hyperglycemia). Moreover, glucose uptake by liver and muscle is inhibited, while biosynthesis of fatty acids and other cell components is also lowered. It appears that the site of action of insulin is at the level of glucose transport across the cellular membranes of liver, muscle, and other tissues and that the rise in blood glucose is a response to this inability—an effort, one might say, to overcome the block. Administration of insulin to animals that are deficient and exhibit the symptoms described above leads to increased uptake of glucose by the tissues and a lowering of the blood sugar concentration.

There are other hormones that play a direct role in regulation of carbohydrate metabolism by influencing not the transport of carbohydrates but rather key enzyme activity in the pathways for glycogen synthesis. For example, *glucagon*, which is synthesized by the pancreas, and *epinephrine*, which is synthesized in the adrenal medulla, both influence the biosynthesis of glycogen from glucose-1-phosphate, as well as the reverse reaction. We say that the synthesis and breakdown of glycogen occur by different pathways and that the control of these two processes occurs at different enzymes—but with a coordinated mechanism, so that inhibition of synthesis of glycogen will be associated with a stimulation of its breakdown and vice versa.

In understanding this control, it is necessary to identify the enzymic steps where the control takes place. In the case of glycogen breakdown, the crucial enzyme is *glycogen phosphorylase*, which catalyzes the phosphorolysis of glycogen by inorganic phosphate. This enzyme has been obtained in high purity and has a molecular weight of 380,000 daltons. The enzyme exists in two forms, *phosphorylase a*, which is active, and *phosphorylase b*, which is much less so. It develops that the active form of the enzyme is a tetramer, while the inactive one is a dimer. Transition from phosphorylase b to a is, itself, a phosphorylation reaction,

$$\text{Phosphorylase } b + 4\,\text{ATP} \rightleftharpoons \text{Phosphorylase } a + 4\,\text{ADP}$$

and is catalyzed by an enzyme, *phosphorylase kinase*. This reaction is essentially irreversible, and the inactivation of phosphorylase a is catalyzed by a second enzyme, *phosphorylase phosphatase*:

$$\text{Phosphorylase } a + 4\,H_2O \rightleftharpoons \text{Phosphorylase } b + 4\,P_i$$

Obviously, the action of these two enzymes provides a means whereby phosphorylase itself can be regulated. Equally obviously, the search for a regulatory mechanism shifts to these enzymes whose activity must therefore be under some form of regulation.

Phosphorylase kinase is similar to *phosphorylase* in that it exists in two forms, one active and one inactive. In this case, the inactive form of the enzyme is activated by the important regulatory molecule *cyclic adenosine monophosphate* (cyclic AMP):

$$\text{Kinase}_{\text{inactive}} \xrightarrow{\text{cyclic AMP}} \text{Kinase}_{\text{active}}$$

This reaction pushes the level of regulation one step farther back, and the next question is, obviously: What regulates the formation of cyclic AMP? Cyclic AMP is formed in the plasma membranes of a variety of cells by the action of yet another enzyme, *adenyl cyclase*:

$$\text{ATP} \rightleftharpoons \text{Cyclic AMP} + PP_i$$

It is this reaction that is under direct influence of the hormones that regulate glycogen synthesis and breakdown. Thus, the formation of cyclic AMP is dramatically stimulated by both epinephrine,

and glucagon, which is a peptide with 29 amino acid residues. Thus, the action of these two hormones is to stimulate formation of cyclic AMP, which stimulates production of active kinase, which, in turn, stimulates the production of active (phosphorylated) glycogen phosphorylase. Thus, breakdown of glycogen and synthesis of glucose-1-phosphate are favored by the action of these hormones.

The same hormones also regulate the activity of the enzyme glycogen synthetase, and again cyclic AMP plays a role. Glycogen synthetase is like phosphorylase in existing in two forms, one phosphorylated. In this case, however, phosphorylation does not confer (simple) activation.

215

Rather, the phosphorylated enzyme's kinetic properties are changed, so that it is active only in the presence of rather high concentrations of glucose-6-phosphate. The stimulation is prevented by UDP. The phosphorylated enzyme can be dephosphorylated by a second enzyme, *glycogen synthetase phosphatase*, and the product enzyme is then fully active even in the absence of glucose-6-phosphate. The phosphorylation of this form of the enzyme is catalyzed by yet another enzyme, a *glycogen synthetase kinase*, which is activated by cyclic AMP. Thus, the same chain of enzyme activation leads from the relevant hormones via cyclic AMP finally to glycogen synthetase. The net effect of the process is, however, generally opposite from the stimulation of phosphorylase, since under most conditions the action of the sequence leads to lesser activity, owing to the control by glucose-6-phosphate. Not only, then, do glucagon and epinephrine stimulate the breakdown of glycogen by activating glycogen phosphorylase, but they usually inhibit the reverse reaction, catalyzed by glycogen synthetase. The interrelations between these effects are illustrated in Figure 8-2 and, although complicated, represent a control mechanism of considerable delicacy.

The *adenyl cyclase* activity localized in the cell plasma membrane represents the site of interaction between the intercellular regulatory device represented by the hormones and the intracellular communication channel represented by cyclic AMP. This system has the effect of being a receptor of the hormone signal as well as a sort of chemical amplification device, since the direct action of the hormone is on an enzyme. This means that the hormone can lead to the production of cyclic AMP on a much more than one-to-one stoichiometry. Thus, the very low concentrations of hormones that usually are effective when in the bloodstream can have influences that are magnified by the catalytic action of the enzyme. It should be added, finally, that the activity of adenyl cyclase is influenced by a number of other hormones and that its product, cyclic AMP, regulates the activity of a number of other metabolic processes. The effect of both the hormones and cyclic AMP differs greatly in different tissues, so that a great deal of specificity of action is possible.

Figure 8-2 Regulation of glycogen metabolism under the influence of hormones. The fundamental reaction (interconversion of glycogen and glucose-6-phosphate) is in the middle and is denoted by capital letters. The various enzymes involved are indicated by italics. Additional information about the processes is given in the text.

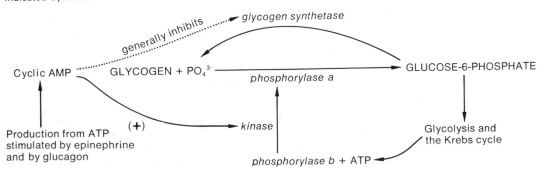

Examples of biosyntheses: lipids

When the synthesis of fatty acids is compared with their breakdown by oxidation, one finds again that the reverse reaction does not occur by the same steps and that the locations of the processes within the cell are quite different, oxidation occurring in the mitochondrion and synthesis in either the endoplasmic reticulum or the soluble fraction. We observed (in Chapter 6) that fatty acid oxidation yields acetyl—S—CoA, and it is the case also that fatty acid synthesis proceeds from the same compound. In the case of synthesis, however, there is a second CoA derivative, malonyl—S—CoA,

$$
\begin{array}{l}
COO^- \\
| \\
CH_2 \\
| \\
C-S-CoA \\
\| \\
O
\end{array}
$$

that actually serves as the immediate precursor. The reader will note that this is a derivative of malonic acid, which was seen in Chapter 4 to be a competitive inhibitor of the mitochondrial enzyme succinate dehydrogenase. Malonyl—S—CoA is, however, not formed from malonate, but rather from acetyl—S—CoA and carbon dioxide by the following net reaction:

$$CO_2 + ATP + acetyl{-}S{-}CoA \rightleftharpoons Malonyl{-}S{-}CoA + ADP + P_i$$

which is catalyzed by an enzyme called *acetyl—S—CoA carboxylase*. This enzyme, like others catalyzing a single carbon transfer, contains *biotin* as a prosthetic group, a nitrogen atom in the biotin serving as site of attachment of the CO_2. This enzyme reaction appears to be the major regulatory step in the whole synthetic process. The reaction is stimulated by citrate, an effect that enables the fatty acid-synthesizing pathway to "sense" the concentration of Krebs cycle intermediates. Since these intermediates, via acetyl—S—CoA, are precursors of fatty acids, this mechanism is important in adjusting the synthetic apparatus to its supply of raw material.

It will be recalled from Chapter 4 that fatty acid synthesis in yeast is carried out by a multienzyme complex containing six separate enzymes and an additional protein called the *acyl carrier protein* (ACP). This protein binds acyl groups to a sulfhydryl group located on a pantothenic acid prosthetic group, and subsequent reactions involving the acyl residues occur with them in the bound form. Thus, the first steps in the synthetic pathway are binding of acetyl—S—CoA and malonyl—S—CoA to the acyl carrier protein (ACP):

(1) Acetyl—S—CoA + ACP—SH $\rightleftharpoons$ Acetyl—S—ACP + CoA—SH

(2) Malonyl—S—CoA + ACP—SH $\rightleftharpoons$ Malonyl—S—ACP + CoA—SH

These then react together to form acetoacetyl—S—ACP:

(3) Acetyl—S—ACP + malonyl—S—ACP $\rightleftharpoons$ Acetoacetyl—S—ACP + ACP—SH + CO_2

Thus, fusion has led to the formation of the four-carbon acetoacetyl—S—ACP,

$$CH_2\overset{\overset{\displaystyle O}{\|}}{C}CH_2C-S-ACP$$

with the elimination of CO_2. Thus, the role of CO_2 is catalytic: It is required for the synthesis of malonyl—S—CoA and it is regenerated in reaction (3). As a general rule, decarboxylations are associated with large negative free energies, so that this step tends to make the pathway for the synthesis of fatty acids largely unidirectional.

We stated earlier that biosyntheses were often reductive in nature and that the reducing agent was often NADPH, a generalization borne out in the following reduction:

(4) Acetoacetyl—S—ACP + NADPH + H^+ $\rightleftharpoons$ β-Hydroxybutyryl—S—ACP + $NADP^+$

This is followed by a dehydration, leaving a carbon–carbon double bond:

$$(5)\quad CH_3\overset{\overset{\displaystyle OH}{|}}{\underset{\underset{\displaystyle H}{|}}{C}}-CH_2-\overset{\overset{\displaystyle O}{\|}}{C}-S-ACP \rightleftharpoons CH_3\overset{\overset{\displaystyle H}{|}}{C}=CH-\overset{\overset{\displaystyle O}{\|}}{C}-S-ACP + H_2O$$

$\quad\quad\quad \beta$-Hydroxybutyryl—S—ACP $\quad\quad\quad\quad$ Crotonyl—S—ACP

Finally, the double bond undergoes another reduction to yield the fully saturated butyryl—S—ACP,

$$CH_3CH_2CH_2\overset{\overset{\displaystyle O}{\|}}{C}-S-ACP$$

In the case of the yeast fatty acid-synthesizing complex, the reducing agent in this second reduction is reduced flavin coenzyme (flavin mononucleotide, $FNNH_2$—reaction 6a), while in the case of animal tissue (and the bacterium E. coli), the reducing agent is again NADPH (reaction 6b):

(6a) Crotonyl—S—ACP + enzyme—$FMNH_2$ $\rightleftharpoons$ Butyryl—S—ACP + enzyme—FMN

(6b) Crotonyl—S—ACP + NADPH + H^+ $\rightleftharpoons$ Butyryl—S—ACP + $NADP^+$

This completes a cycle of synthesis that has seen the formation of a four-carbon fatty acid bound to the acyl carrier protein (ACP). This reacts then with a second molecule of malonyl—S—ACP, after which a second cycle of reduction, dehydration, and reduction occurs, leading to a fatty acyl chain enlarged by two more carbon atoms. This continues until the chain reaches the full length of the fatty acid being synthesized, at which time the free acid is released from the carrier protein in a hydrolysis catalyzed by a *deacylase*.

It is apparent that there are important similarities between the synthesis of fatty acids described above and their oxidation outlined in Chapter 6. The alternation of *oxidation-reduction* and *hydration-dehydration* reactions is the same in the two cases, although reductions and dehydrations are confined to synthesis, and oxidations and hydrations are found in the oxidative pathway. The differences in mechanism are also apparent. The reactions occur in different compartments of the cell and involve quite different enzymes. Malonyl—S—CoA plays a central role in the synthesis of fatty acids and none at all in their oxidation. These differences appear to be most important in the regulation of the two processes. For example, synthesis appears to be regulated by the availability of malonyl—S—CoA, so that it can be adjusted without interfering with oxidation. We saw that this control of malonyl—S—CoA synthesis was effected by an allosteric control of *acetyl—S—CoA carboxylase* by citrate or iso-citrate. These activators appear to convert an inactive protomer form of the enzyme to the enzymatically functional polymeric form, thereby stimulating activity.

Examples of biosynthesis: amino acids

The synthesis of amino acids is an important topic in cell biology both because of their essential role as the building blocks of proteins and because the study of their biosynthesis has revealed regulatory mechanisms of high effectiveness. Many of the first studies of end-product inhibition of metabolic pathways were performed with bacterial amino acid synthesis, of which we shall now look at some examples.

Cells exhibit greatly different ability to synthesize amino acids. For example, some bacteria, including *E. coli*, can synthesize all the 20 or so amino acids that they require from simple inorganic sources. On the other hand, animal cells, as well as those of many species of micro-organisms must receive about half of the 20 common amino acids in their diet, as they do not possess the enzymatic machinery to synthesize them. This fact leads to a grouping of amino acids into two nutritional categories for any given organism, the *essential* and *nonessential* amino acids. Essential amino acids are those that the organisms are unable to synthesize themselves, and the list is different for various organisms. Table 8-3 illustrates the two classes in the case of the laboratory rat. It should be added that there are some amino acids that are nonessential for most organisms that have been studied. In general, these are those amino acids that are synthesized by means of relatively short pathways from common metabolic intermediates, such as those of the Krebs cycle. There are cases where the distinction between essential and nonessential is not completely exact. For example, rats require arginine in their diet but can, in fact, synthesize it at a low (and insufficient) rate. Moreover, different cell types

219

Table 8-3 *Nutritional classification of amino acids in the case of the laboratory rat*

Essential (not synthesized)	Nonessential (synthesized)
Arginine	Alanine
Histidine	Aspartate
Isoleucine	Cystine
Leucine	Glutamate
Lysine	Glycine
Methionine	Hydroxyproline
Phenylalanine	Proline
Threonine	Serine
Tryptophan	Tyrosine
Valine	

from the same organism may have different synthetic abilities, and it is commonly observed that when cells are isolated under culture conditions, their list of essential amino acids becomes longer, owing to their change in environment.

An example of an almost universal nonessential amino acid that is metabolically close to a Krebs cycle precursor is glutamate. This compound is readily formed from the Krebs cycle acid, α-ketoglutarate, by a single reaction catalyzed by the enzyme *glutamate dehydrogenase*:

$$NH_3 + \alpha\text{-Ketoglutarate} + NADPH + H^+ \rightleftharpoons Glutamate + NADP^+$$

This reaction is reversible, so that glutamate can serve as a carbon and energy source by entering the Krebs cycle at the level of α-ketoglutarate. This reaction turns out to be a very important one in the synthesis of amino acids other than glutamate, since it is a major route for the introduction of amino groups into amino acids in general. This occurs by means of a *transamination* reaction where the amino group from one amino acid, in this case glutamate, changes places with a carbonyl group on another molecule. This can be illustrated by the synthesis of alanine from pyruvate plus glutamate:

$$\begin{array}{l}
\text{COO}^- \\
| \\
\text{CHNH}_3 \quad \text{CH}_3 \\
| \qquad\quad | \\
\text{CH}_2 \;\; + \; \text{C}=\text{O} \\
| \qquad\quad | \\
\text{CH}_2 \qquad \text{COO}^- \\
| \\
\text{COO}^-
\end{array}
\rightleftharpoons
\begin{array}{l}
\text{COO}^- \\
| \\
\text{C}=\text{O} \quad\; \text{CH}_3 \\
| \qquad\quad | \\
\text{CH}_2 \;\; + \; \text{CHNH}_3{}^+ \\
| \qquad\qquad | \\
\text{CH}_2 \qquad\; \text{COO}^- \\
| \\
\text{COO}^-
\end{array}$$

By a similar reaction, glutamate serves as amino donor in the synthesis of aspartate from oxaloacetate.

220

In contrast to the simple routes for the formation of glutamate and alanine, essential amino acids are synthesized by long pathways, which have as their sole feature in common their almost universal regulation by allosteric end-product inhibition. For example, the synthetic pathway for valine, which is one of the least complicated ones, is shown in Figure 8-3. It is characteristic of such pathways that allosteric inhibition by the product of the pathway occurs at the first reaction of the sequence. Note also that the final reaction is a transamination where the amino donor might be glutamate, as seen earlier.

A number of other amino acid-synthesizing pathways are similar to that for valine in being unbranched and in exhibiting end-product inhibition of the first reaction. Histidine, leucine, and threonine are examples of such amino acids, and we shall not show the pathways here, as they can be found in any modern textbook of biochemistry. Somewhat more complicated pathways and regulatory mechanisms are noted in the cases of other amino acids. As an example, the synthesis of tryptophan, phenylalanine, and tyrosine all include a portion of a common pathway, so that simple feedback inhibition is inadequate. The relations between these amino acids are outlined in Figure 8-4, and it is apparent that a complicated system exists for the control of the pathways. Several general points about control networks are illustrated by this example. For one thing, tryptophan can be seen to serve as allosteric inhibitor of the first reaction that leads uniquely to its own formation. Tyrosine and phenylalanine produce a similar effect on the first reaction leading to their formation and serve to inhibit the initial reaction in the whole sequence as well.

Figure 8-3 Synthesis of valine. The dotted line denotes allosteric inhibition.

221

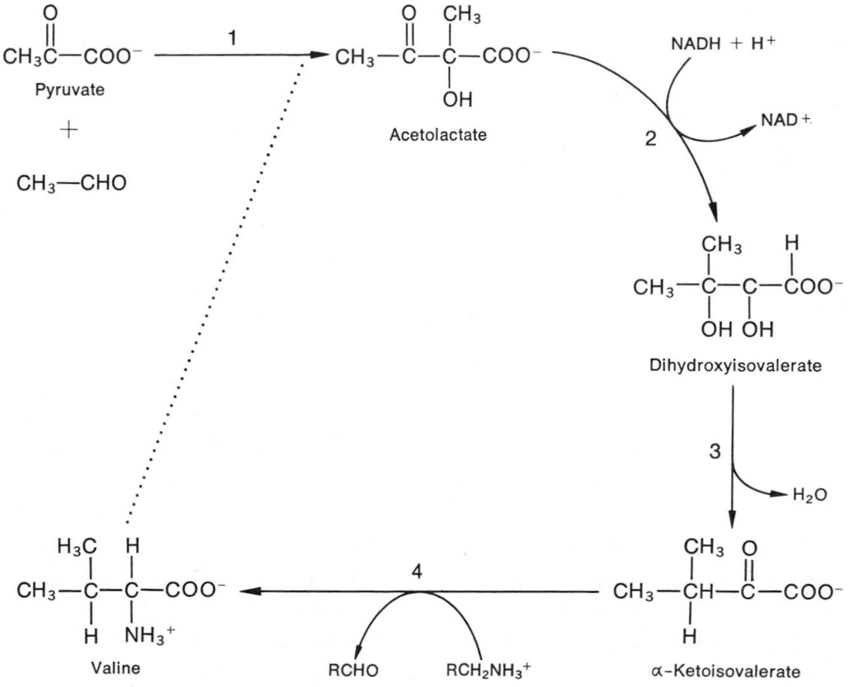

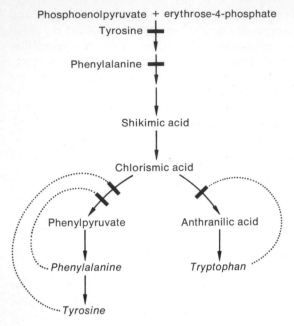

Phosphoenolpyruvate + erythrose-4-phosphate

Figure 8-4   Outline of synthetic routes for the aromatic amino acids, phenylalanine, tyrosine, and tryptophan. Inhibition is denoted by dotted lines.

In this situation, it is clear that tyrosine and phenylalanine exert a sort of primacy over tryptophan in that they are able to regulate its rate of synthesis, while it is unable to affect theirs.

In addition, it is clear that both phenylalanine and tyrosine are able to produce allosteric inhibition of the same enzymes. This implies that these particular enzymes must have more than one regulatory site (as opposed to a single catalytic site) with at least one each able to bind tyrosine and phenylalanine. This situation is termed *multivalent* inhibition and is of general occurrence where a single allosteric enzyme catalyzes a reaction leading to more than one final product. In some cases, multivalent inhibitors are additive in their inhibition and can inhibit independently. On the other hand, cases are known where both inhibitory end products are required to produce any inhibitory effect. This latter situation is termed *concerted multivalent inhibition* and would have adaptive value in situations where the several end products were all required in reasonable amounts for continued cell function, since it ensures that no one component will be suppressed by an excess of another.

The mechanism of allosteric inhibition was discussed in Chapter 4, but it seems worthwhile here to return to it in a more specific context. It will be recalled that the essence of allostery is the occurrence of two classes of enzyme site, the active *catalytic site* and a *regulatory site* (or sites). Binding of the substrate molecule to the catalytic site normally leads to the enzymic reaction, while binding of a regulatory molecule to the regulatory site leads to a modification of the level of activity. This modification, in turn, is the result of a change in the tertiary (conformational) or quaternary (aggregation) structure of the enzyme. An excellent example of allosteric regulation is found in the control of the synthesis of glutamine,

222

$$
\begin{array}{l}
\text{COOH} \\
| \\
\text{CHNH}_2 \\
| \\
\text{CH}_2 \\
| \\
\text{CH}_2 \\
| \\
\text{C}-\text{NH}_2 \\
\| \\
\text{O}
\end{array}
$$

which is important as a precursor of a number of other compounds, including the coenzyme NAD^+ and purines, as well as being a component of proteins.

Glutamine is synthesized from a glutamic acid by the following reaction, which is catalyzed by *glutamine synthetase*:

$$\text{Glutamate} + \text{NH}_3 + \text{ATP} \rightleftharpoons \text{Glutamine} + \text{ADP} + \text{P}_i$$

This enzyme has been most rigorously characterized in the case of *Escherichia coli*, from whose cells it has been isolated in a quite pure state. The enzyme is composed of 12 subunits (or protomers), each with a molecular weight of 53,000 daltons. Electron micrographs indicate that the 12 subunits are arranged in two hexagonal layers, and this aggregate can be easily dissociated into subunits by treatment with guanidine, urea, or a detergent. Chelating agents also promote dissociation. Divalent cations such as Mg^{2+} appear to be involved in the binding of the subunits together.

As we said, glutamine is an important precursor of a number of compounds, including tryptophan, adenylic acid, cytidylic acid, glucosamine, alanine, glycine, histidine, and carbamyl phosphate (the last being the precursor of numerous other compounds). In fact, in cells of *E. coli*, as well as in those of a number of organisms studied, all these compounds exert allosteric inhibition of glutamine synthetase. The inhibition has the interesting property of not being very great with even large amounts of any one inhibitor and requiring a mixture of inhibitors to approach complete inhibition. Indeed, one can show that inhibition with mixtures of these compounds approaches levels that are more than additive, so that some sort of concerted mechanism must be in operation. Analysis of the properties of the enzyme under different conditions reveals that there are at least eight separate sites to accommodate the several inhibitory molecules, so that this system must represent an extreme case of multivalent inhibition.

To see the full array of controls that can occur in a single metabolic reaction, we now point out that the *synthesis* of the enzyme glutamine synthetase is also subject to feedback control. The role of glutamine as a precursor of other molecules largely reflects its ability to serve as a metabolic donor of amino groups. Since it turns out that the need for amino-group donation is altered by the availability of nutritionally obtained ammonium ion (NH_4^+), it is not unexpected that, when bacterial cells are grown in the presence of ammonia, the synthesis of glutamine

Table 8-4   Some examples* of regulation of amino acid synthesis in E. coli by means of allosteric inhibition and repression of enzyme synthesis

End product	Enzyme affected
A. Allosteric inhibition by end products	
Histidine	Pyrophosphorylase
Isoleucine	Threonine deaminase
Threonine	Homoserine dehydrogenase
Tryptophan	Anthranilate synthetase
B. Repression of enzyme synthesis by end products	
Arginine	Several
Histidine	Several
Isoleucine	Threonine deaminase
Leucine	Several
Valine	Acetoacetate synthetase

* The list is by no means exhaustive.

synthetase is *repressed*. The mechanism of such enzyme repression will be considered later in the context of protein synthesis. It is likely that the roles of allosteric inhibition and enzyme repression are rather different. Inhibition of activity appears to serve to regulate the flow of material through a pathway and assist in the interrelation of different metabolic transformations, while the function of repression is much more crude, serving to protect the cell from the uneconomical synthesis of enzymes that it does not need because components are provided nutritionally instead. See Table 8-4 for additional examples of inhibition and repression.

Finally, the example of glutamine synthetase is fortunate, since it exhibits yet another sort of regulation that is not included in either repression or allostery. When cells of *E. coli* are grown in a medium into which NH_4^+ ion is introduced, not only does the synthesis of glutamine synthetase cease (repression) but the molecules that already exist in the cell are inactivated by the action of a specific enzyme. It turns out that the reaction catalyzed by this inactivating enzyme is the addition of an adenylate unit by a covalent bond to the synthetase molecule, the adenylate originating in a molecule of ATP. This reaction leads to a total of 12 adenyl groups being bound to the enzyme or 1 per subunit. The changes that occur in the properties of the enzyme upon this modification turn out to be reversible, since there appears to be a second enzyme able to de-adenylate the enzyme under conditions when ammonium ion is not available.

Thus we see that the control of synthetic pathways (as illustrated in the case of amino acid synthesis) occurs on several levels and that the combination of available mechanisms enables the pathways to adjust

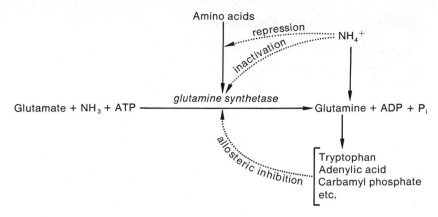

Figure 8-5   Regulatory mechanisms influencing glutamine synthetase.

themselves as the cell's requirements for the end products of the pathways change.

The several sorts of regulation that we have noted in the case of the glutamine synthetase system are summarized in Figure 8-5.

Suggested Reading

Books

Cohen, G. N., *Biosynthesis of Small Molecules*, Harper & Row, New York, 1967.

Cohen, G. N., *The Regulation of Cell Metabolism*, Holt, Rinehart and Winston, New York, 1968.

Dawson, R. M. C., and D. W. Rhodes (eds.), *Metabolism and Physiological Significance of Lipids*, John Wiley & Sons, New York, 1964.

Lehninger, A. L., *Biochemistry*, Worth Publishers, New York, 1970, Chs. 22–24.

Masoro, E. J., *Physiological Chemistry of Lipids in Mammals*, W. B. Saunders Co., Philadelphia, 1968.

Meister, A., *Biochemistry of the Amino Acids*, vol. II, 2nd ed., Academic Press, New York, 1965.

Articles

Scrutton, M. C., and M. F. Utter, "The regulation of glycolysis and gluconeogenesis in animal tissues," *Ann. Rev. Biochem.*, **37**, 249 (1968).

Stadtman, E. R., "Allosteric regulation of enzyme activity," in *Advances in Enzymology*, F. F. Nord (ed.), vol. 28, John Wiley & Sons (Interscience Division), New York, 1966, p. 41.

Stadtman, E. R., "Enzyme multiplicity and function in the regulation of divergent metabolic pathways," *Bacteriol. Rev.*, **27**, 170 (1963).

9 The special physiology of microorganisms

We have had frequent occasions in the preceding chapters to refer to physiological topics that are best (or often) studied in microorganisms. Such topics include induced enzymes, aspects of growth, aspects of protein synthesis, and a number of others. Indeed, it appears that microorganisms, and especially bacteria, have been extremely useful and influential experimental organisms; the purpose of this chapter is the review of some of the reasons for this utility and, in the process, to summarize some of what is known about this area of physiology. Usually, bacterial physiology is considered in the context of a course in microbiology rather than physiology, but there is no significance to this tactic other than custom, and in this book we need not be bound by such convention.

We saw in Chapter 1 that cells fell quite naturally into the two categories of *prokaryotic* and *eukaryotic*, based, among other things, on the presence of a nuclear membrane and certain intracellular organelles. Similarities among cell types within the two groups are much greater than between the groups. Thus, *prokaryotic cells*, which include only bacteria and blue-green algae, share a similar cellular organization as well as many common features of function, although the physiological similarities are somewhat blurred by the wide variation in bacteria.

Cell structure

The reader is reminded of the salient features of prokaryotic structure by examining Figure 9-1, which is a highly generalized diagram of such a cell. The figure stresses the simplicity of such cells and does not include structures, such as mesosomes, flagella, or photosynthetic membranes, found only in restricted groups within the prokaryotic category. As we shall see, this very simplicity lends bacteria and blue-green algae much of their attractiveness as experimental subjects, especially in investigations that require the disruption of cells and resolution into morphological components.

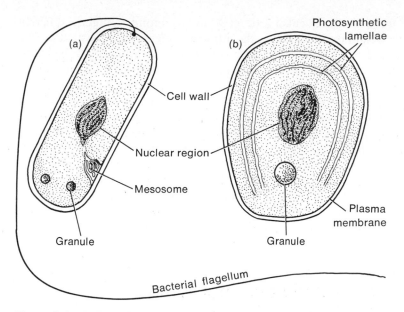

Figure 9-1   Prokaryotic cells. (a) A bacterium. (b) A blue-green alga. The granules might be composed of food-storage material such as glycogen or, in the case of some bacteria, of inorganic material such as polyphosphate. Flagella are not found in all bacteria, but where they occur, they are of the relatively simple sort, as compared to the 9 + 2 configuration in eukaryotic cells. No blue-green algae are flagellated. Finally, lamellae are characteristic of a number of types of blue-green algae but are also observed in some photosynthetic bacteria.

Although prokaryotic cells have many features in common, it is worth going a bit beyond Figure 9-1 by enumerating some of the special features of these cells and, especially, by mentioning some of the differences between the two subclasses of prokaryote, namely the bacterial and blue-green algal cells. For one thing, the gross form of cells of the two subclasses can be quite different. Bacteria tend to be smaller and range from spherical (coccus) to elongated rods (bacillus) while blue-green algae are often larger, with a great deal more variation in shape. Blue-green algae often grow in filaments comprising a number of cells, while such a growth habit is exceedingly uncommon in bacteria.

Blue-green algae are similar to bacteria in possessing an outer capsule (or sheath) layer with a rigid cell wall between the layer and the plasma membrane. In both cases, the cell wall is unlike that of higher plants in lacking cellulose and containing muramic acid and mucopolysaccharides. In both instances, the cell walls do not constitute osmotic barriers, and, in both, the walls can be attacked by the enzyme lysozyme.

Blue-green algae and bacteria exhibit a plasma membrane of the usual bilayer type, the function of which will be discussed in the following section. The interior of the cell is much less complex than that of a eukaryotic cell and, with the exception of certain specific invaginations of the plasma membrane (also discussed in the following section), no organelles are to be found. In both cases, examination with the electron microscope reveals the cytoplasm to contain numerous ribosomes, which

227

differ from those of eukaryotic cells in exhibiting a slightly different particle weight (sedimenting in the ultracentrifuge with a constant of 70 S as opposed to 80 S for eukaryotic ribosomes). In addition, prokaryotic ribosomes are free in the cytoplasm and not attached to any membrane system such as the endoplasmic reticulum of eukaryotic cells (which is, in any case, lacking in the prokaryotic cell interior). Finally, prokaryotic cells of both types frequently contain a variety of cellular inclusions, including droplets of fluid and solid granules, such as those of polyphosphate and highly polymerized carbohydrates.

It will also be recalled that the prokaryotic nuclear region is quite different from that of eukaryotic cells. It not only lacks a nuclear membrane, but its DNA exists unbound to basic proteins, such as histones, and, in both bacteria and blue-green algae, contains methylated nucleotide bases, such as methylcytosine, not ordinarily found in higher organisms.

The prokaryotic cell membrane

When a eukaryotic cell is disrupted and subjected to differential centrifugation, the fraction containing cellular membranes normally contains membranes that are from a variety of locations and that differ greatly in their function and composition. For example, unless special precautions are taken, a membrane fraction will contain inner and outer mitochondrial membranes (which are quite different), endoplasmic reticulum, nuclear membrane, and the cell membrane itself. This presents enormous problems in obtaining homogeneous preparations. For example, quite erroneous conclusions have been drawn from studies employing mitochondria that were contaminated with endoplasmic reticulum. In contrast, the situation in prokaryotes is much more easily examined, since there is basically only a single membrane, the cell membrane, which carries on the several functions of mitochondria, chloroplasts, and cell membrane of eukaryotic cells.

It appears likely that the prokaryotic cell membrane should be thought of as a structural and functional *mosaic*, with specific regions associated with respiration, photosynthesis, transport, and synthesis of cell wall components. How closely the units of the mosaic may be associated remains an open question. For example, there is evidence that the process of cell respiration in blue-green algae employs some of the same cytochromes as photosynthetic electron transfer, a situation that cannot occur in higher plants where the oxidation and photosynthetic respiratory chains are located in mitochondria and chloroplasts, respectively.

This simple picture of prokaryotic cell membranes requires, however, some amplification, as many instances occur where the membrane is seen to be *invaginated* so as to penetrate the cell interior with characteristic structures. For example, the bacterial *mesosome*, mentioned in Chapter 1, is such an infolding of the cell membrane. It is believed to serve as a point of attachment for the DNA of the nuclear region during cell division, and it has been suggested that the close proximity of membrane and DNA is of significance in DNA replication, perhaps because ATP is required for replication and is synthesized at the membrane by the process of oxidative (or photosynthetic) phosphorylation. In the case of blue-green algae, the plasma membrane is seen likewise to form structures within the cell, which are apparently associated with photosynthesis. Such *thylakoid* membranes vary considerably in their penetration of the cell and in their complexity, ranging from small buds and vesicles to complex folds of membrane. Similar variation is observed in the case of photosynthetic bacteria, which range from simple photosynthetic vesicles to systems of parallel membranes.

Finally, the individual components of prokaryotic membranes often exhibit wide variation when compared to those in eukaryotic cells. For example, the mitochondria of the great variety of higher organisms that have been examined contain cytochromes c and c_1, with absorption maxima at 550 and 552 nanometers, which appear functionally identical no matter what the source. This consistency of higher organisms is in striking contrast to the situation in prokaryotic cells. There, a great variety of cytochromes of the c type are found in different organisms, with more than one often found in a single cell. In some cases, the cytochrome contains a flavin prosthetic group as well as heme, a situation not obtaining in higher organisms. Molecular weights of c-type cytochromes vary greatly in bacteria, and whereas cytochrome c from higher cells is loosely bound to membranes so that it can be removed by salt extraction, in prokaryotic cells a complete range of tightness of binding is observed. Similar variation occurs in the other classes of bacterial cytochromes. While a single cytochrome b is found in mitochondria from a wide selection of higher cells, there are numerous bacterial cytochrome b's with quite differing properties. Similarly, prokaryotic cells contain classes of respiratory enzyme not found at all in higher forms (such as cytochrome o or cytochrome d),

*Table 9-1   Some bacterial cytochromes**

Example	Absorption maximum (nm)	Organisms (among others)
Cytochrome a	605	Many
Cytochrome a_1	595	*Acetobacter*
Cytochrome b_1	558	*Corynebacterium, Escherichia coli*
Cytochrome b	554	Many
Cytochrome b-574	574	*Micrococcus*
Cytochrome c-551	551	*Rhizobium*
Cytochrome c-552	552	*Rhodospirillum*
Cytochrome c-554	554	*Acetobacter*
Cytochrome d	629	*Pseudomonas*
Cytochrome o	565	*Micrococcus, Rhodopseudomonas*

* This table is included to illustrate the variety found among bacterial cytochromes. Cytochrome *d* appears to be a terminal oxidase analogous to cytochrome *a*. Cytochrome *o* is detected by its ability to bind carbon monoxide and is of unknown function.

and photosynthetic forms contain photosynthetic pigments of considerably more variety than those observed in eukaryotic plant cells.

Nutrition

Bacteria and, to a lesser extent, blue-green algae are noted for the variety of environmental conditions under which they are able to exist and grow. Bacteria grow in the range of environments extending from marine and fresh waters, where relatively low concentrations of organic compounds are available, to such rich media as human tissues. One feature of bacteria that enables them to fill such a diversity of niches is the great variety in their nutritional requirements. Thus, some bacteria (and blue-green algae) grow in the virtual absence of organic carbon and require only carbon dioxide and a variety of inorganic salts. In contrast, other bacteria require numerous organic compounds, often including sugars, amino acids, nucleotide bases, and vitamins. In the latter case, the nutritional requirements often approximate those of animals.

The environmental advantages that bacteria enjoy are furthered by their ability to adapt to new nutritional situations through the mechanism of enzymic induction mentioned earlier. Further flexibility is found on the genetic level; the relatively rapid generation time of microorganisms renders possible rapid mutation to new nutritional requirements. Thus, during a time interval, a population of bacteria is likely to contain at least some cells with differing synthetic capabilities and, therefore, differing requirements. If a change in the environment occurs that favors one of these, natural selection will enable it to grow until it becomes the dominant cell type in the population. This genetic flexibility in bacterial nutrition is utilized in the important technique of *enrichment culture*, where the experimenter seeks to obtain a pure culture of an organism capable of metabolizing a specific nutrient by providing it as a growth factor, while limiting alternative sources.

Table 9-2 illustrates the components of a medium able to sustain growth in the case of certain strains of *E. coli*. It is provided as an example of the

Table 9-2   *Medium for culture of Escherichia coli*

Component	Final concentration (molar)
KH_2PO_4	1×10^{-1}
NH_4Cl	2×10^{-2}
Glucose	1.2×10^{-2}
Na_2SO_4	1×10^{-3}
$MgCl_2$	8×10^{-4}
$CaCl_2$	1.4×10^{-5}
$FeCl_2$	8.5×10^{-6}

rather limited number of materials required, at least in the simplest case. The reader should note that the components listed must provide the necessary atoms to allow the synthesis of all cellular components, and he should reflect on the probable sources of the several elements required for, say, proteins and nucleic acids.

The nutritional classification of cells

In considering the relationships between individual cells and the living world as a whole, it is useful to classify cells as to their nutritional requirements. This can be quite complex, as cells must be described with regard to energy and carbon sources as well as numerous other components. In general, it is well to omit consideration of many of these since a single, point mutation can totally change a cell's requirement for a particular vitamin or amino acid. On the other hand, the nature of the carbon and energy sources depends, in general, on whole pathways and, often, on the gross structure of cells as well. For example, the conversion of a cell accustomed to obtaining energy from the aerobic oxidation of sugar could not be converted to photosynthesis, save by a host of mutational steps involving complex biochemical pathways and the morphological structures and pigments associated with photosynthetic energy conversion.

Since the carbon and energy sources for cells are frequently different, it is well to consider classifying nutritional types on the basis of both. Based upon the source of cell energy, it is thus possible to distinguish among *phototrophs*, *chemolithotrophs*, and *chemoorganotrophs*. Phototrophs obtain energy by means of the photosynthetic absorption of visible radiation. Green plants are obviously in this category, as are virtually all blue-green algae and certain photosynthetic bacteria. It will be recalled that green plants use water as the original source of reducing power (electrons) and that the overall reaction

$$H_2O + CO_2 \xrightarrow{\text{light}} CH_2O + O_2$$

leads not only to the synthesis of cell material, "CH_2O," but also to the production of oxygen. Bacteria use other compounds as a source of electrons and do not give rise to molecular oxygen. Indeed, many are *obligate anaerobes*, which is to say that they are poisoned by molecular oxygen. For instance, green sulfur bacteria use hydrogen sulfide as a source of electrons,

$$2\,H_2S + CO_2 \xrightarrow{\text{light}} CH_2O + H_2O + 2\,S$$

while nonsulfur purple bacteria employ an organic reducing source such as isopropanol:

$$2\,CH_3CHOHCH_3 + CO_2 \xrightarrow{\text{light}} CH_2O + 2\,CH_3C{-}CH_3 + H_2O$$

It should be mentioned that blue-green algae are most frequently similar to higher plants in employing water as an electron donor and producing molecular oxygen. In a few instances, however, it appears that molecular hydrogen can serve as the source of reducing power,

$$2\ H_2 + CO_2 \xrightarrow{\text{light}} CH_2O + H_2O$$

a technique also employed by some photosynthetic bacteria.

Chemolithotrophs employ an inorganic energy source that is oxidized by molecular oxygen. Numerous soil and water bacteria are in this category, as are a few blue-green algae that can utilize an inorganic energy source when kept in the dark. For example, it is well known that the reaction

$$H_2 + \tfrac{1}{2} O_2 \rightleftharpoons H_2O$$

between hydrogen and oxygen is highly spontaneous—that is, can be accompanied by the release of considerable energy. The hydrogen bacteria have evolved ways of coupling this reaction to the synthesis of ATP and so utilize it for primary energy production. Bacteria of the genus *Nitrosomonas* obtain energy by the oxidation of ammonia to nitrite,

$$NH_3 + \tfrac{3}{2} O_2 \rightarrow NO_2^- + H^+ + H_2O$$

while cells of the genus *Nitrobacter* oxidize nitrite to nitrate,

$$NO_2^- + \tfrac{1}{2} O_2 \rightarrow NO_3^-$$

These two reactions, taken together, constitute the process of *nitrification*, by which is meant the conversion of nitrogen to increasingly oxidized forms. Nitrification is, in a sense, a sequel to those nitrogen-fixing reactions (carried out by certain bacteria and blue-green algae) in which molecular nitrogen is reduced to yield ammonia. The relationships between these reactions are summarized in Figure 9-2.

All organisms that are neither phototrophs or chemolithotrophs fall into the category of *chemoorganotrophs*, which obtain energy from reactions involving organic compounds. Many bacteria and all animals fall within this class. Among the bacteria, chemoorganotroph energy conservation occurs either by means of fermentation (to be discussed shortly) or respiration, which can employ either oxygen or some other inorganic material as the final electron acceptor. Thus, respiration can be aerobic, so that if AH_2 is the organic electron donor, respiration takes the form

$$AH_2 + \tfrac{1}{2} O_2 \rightarrow A + H_2O$$

On the other hand, the electron acceptor may be such an inorganic substance as sulfate (which is reduced to form sulfide) or nitrate (to yield nitrogen gas, ammonia, or nitrous oxide). In such cases, the mechanisms of both oxidation and energy coupling are similar to the situation that

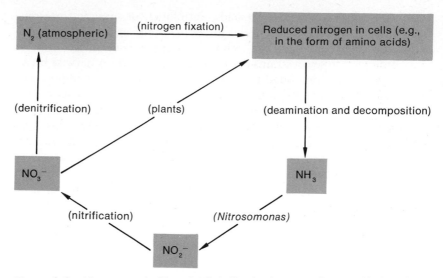

Figure 9-2 Nitrogen cycle. The role of nitrification is seen to be an oxidation of nitrogen to nitrate, in which form it can re-enter the cellular world, being converted by plants (and certain bacteria) into reduced, organic nitrogen.

obtains when oxygen is the electron acceptor, a matter discussed in detail in Chapter 5.

A separate nutritional grouping of organisms is often made on the basis of carbon source. Thus, *autotrophs* are able to "fix" CO_2; that is, CO_2 is able to serve as their sole carbon source. Such organisms obtain energy either from photosynthesis or from oxidation of inorganic energy sources and so comprise the photolithotrophs and the chemolithotrophs. In contrast, *heterotrophic* organisms must rely on organic sources of carbon. This category includes most bacteria and all animals.

Sources of nutritional carbon

Heterotrophic bacteria are able to grow on an enormous range of carbon sources and, as we have seen, can even adapt to changes in the composition of the medium in which they grow. Organisms that carry out fermentations use a variety of hexoses, pentoses, as well as such smaller molecules as lactate. Cells that carry out oxidations of organic compounds can also utilize many sugars, including hexoses, pentoses, disaccharides, polysaccharides, and amino sugars. Other cells oxidize such polyols as sorbitol and ribitol,

$$
\begin{array}{c}
CH_2OH \\
| \\
H-C-OH \\
| \\
H-C-OH \\
| \\
H-C-OH \\
| \\
CH_2OH
\end{array}
$$

while still others utilize amino acids, antibiotics, or even testosterone. Indeed, there appears to be no end to the list of the organic compounds oxidizable by bacteria, and the technique of enrichment culture enables microbiologists to select strains of cells able to oxidize (or otherwise break down) nearly any compound tried. This ability is of obvious importance in the chemical industry, which often relies on bacteria to carry out chemical transformations or to remove wastes from solution. The "bio-degradability" of many materials, such as pesticides, detergents, and industrial waste, clearly depends on the degree to which the nutritional requirements of bacteria can be extended. So, indeed, does one aspect of the endeavor to increase the world's food reserve: Bacteria able to oxidize, and thus utilize, hydrocarbons are likely to be of increasing importance through their ability to convert petrochemicals (such as crude oil) into nutritionally valuable protein.

Bacterial fermentations

We considered fermentations earlier in the context of the glycolytic pathway, which provides a major route for fermentative energy production in yeast and many bacteria. The purpose of this section is to give an idea of the range of prokaryotic organisms that obtain energy by a fermentative mechanism and to illustrate some of the different pathways involved. Since no blue-green alga that has been studied has proved to obtain a major part of its energy from a fermentation, this section will deal solely with bacteria.

Fermentations have been described by Pasteur as *life without air*, a definition rather too inclusive in light of modern knowledge of anaerobic respiration and photosynthesis. We observed in Chapter 5 that the essence of a fermentation was that, as in the case of respiration, energy is conserved by transport of an electron from one molecule to another but with the distinction that the electron acceptor is derived by chemical transformation of the donor:

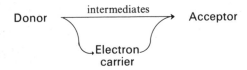

Thus, one can think of a fermentative pathway as a linear path where, at an early phase, there is oxidation of a metabolite (removal of an electron) and, at a later point, a reductive (electron-adding) step occurs. It will be recalled that, in the glycolytic pathway, the electron carrier mediating between the two oxidation-reduction reactions is the important coenzyme nicotinamide adenine dinucleotide (NAD^+). The mechanism whereby the fermentative pathway yields high-energy terminal phosphate bonds of ATP was discussed in Chapter 4. Here, it will only be noted that, in contrast to most respiratory and photosynthetic ATP production, fermentative ATP synthesis is not coupled directly to electron flow along a chain of redox proteins. Rather, it occurs at the level of structural re-arrangements in the pathway itself, which lead to formation of phosphorylated compounds (such as phosphoenolpyruvate) with high free energies of hydrolysis.

An important general property of fermentative energy conservation is its relative inefficiency when compared to respiration. Thus, the yield of ATP per mole of starting material (such as sugar) is low and the pathways must consume a great deal of such material to supply the energy needed by a growing cell. This massive rate of breakdown means that relatively little of the material serves as a carbon source, so that there is usually virtually complete recovery of carbon in the product or products of the fermentation. In addition, since aerobic oxidations do not play a role in fermentations, it is also the case that the net oxidation-reduction state of the products cannot differ from those of the source.

Fermentative pathways

A number of types of bacteria obtain energy by fermentations based on pathways similar to the glycolytic route studied most extensively in yeast and muscle tissue. The reader is reminded of the main features of this pathway by examination of Figure 9-3; a more detailed account will be found in Chapter 5. Bacteria employ a number of variations upon this general theme and several of them are described in this section.

For example, many bacterial species, including those of the genera *Streptococcus* and *Lactobacillus*, ferment glucose and produce lactate as the main product, although often with small amounts of ethanol, formic acid, and acetic acid. This pattern of source and product is known as a *homolactic fermentation* and appears to occur completely by means of the reactions of the glycolytic pathway. Thus, when glucose labeled with ^{14}C in the 1 position is fermented by such cells, lactate may be isolated with the label in the methyl carbon:

$$
\begin{array}{c}
^{14}CH_3 \\
| \\
H-C-OH \\
| \\
COO^-
\end{array}
$$

The reader should satisfy himself that this result is consistent with the assumption that homolactic fermentations employ the glycolytic pathway in conversion of glucose to lactate.

Other sets of fermentation products are observed even when the enzymic basis is the glycolytic pathway. For instance, a number of bacteria, including species of *Clostridium*, ferment glucose and produce hydrogen, carbon dioxide, butyric acid, and acetic acid. Although the detailed mechanism of this is not clear, it appears that glycolysis provides the basic

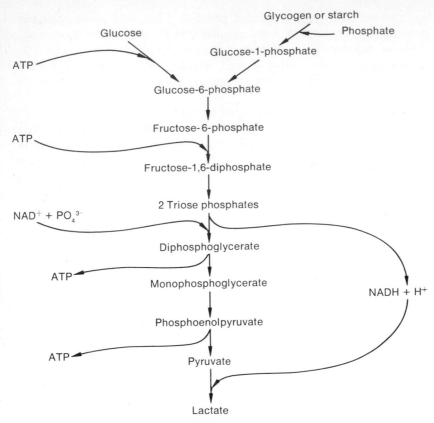

Figure 9-3 Summary of glycolysis. Where ATP is consumed, it is understood that ADP is produced, and vice versa.

route. Similar organisms apparently employ a similar mechanism to produce, instead, butyl or isopropyl alcohol, and at least one bacterium, *C. acetobutylicum*, converts glucose to acetone. It is believed that, in these cases, glucose is converted to pyruvate by glycolysis, after which pyruvate is cleaved to CO_2 and the two-carbon acetylcoenzyme A, which, in turn, can yield various four-carbon products by means of condensation to form acetoacetate,

$$CH_3-\overset{\overset{\displaystyle O}{\|}}{C}-CH_2-COO^-$$

The role of the glycolytic pathway in these fermentations has been also confirmed by studies with ^{14}C-labeled glucose.

It should not be thought that glucose is the only sugar fermentable by bacteria. Indeed, we have noted earlier that bacteria tend to be quite catholic in their taste in fermentable sugars. Thus, hexoses appear to be readily interconvertible by many bacteria in the form of their monophosphates, which enter the glycolytic route as glucose or fructose

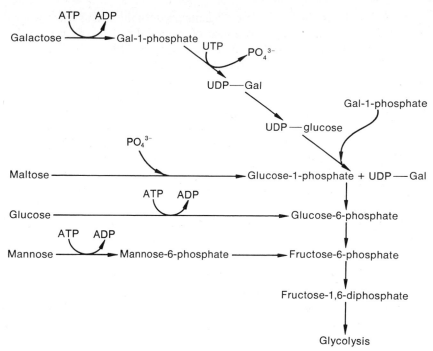

Figure 9-4   Routes by which several hexoses can contribute to glycolytic fermentation.

phosphate. Some reactions that enable hexoses other than glucose to be fermented by glycolytic pathway are shown in Figure 9-4.

Sensitivity to inhibitory compounds

Bacterial cells frequently differ from eukaryotic ones in their sensitivity to a number of chemicals, a sort of distinction that is of great importance in pointing up differences in function between the two categories of cell. The distinction is of perhaps greater practical significance in providing the basis for the chemical treatment of bacterial diseases, a strategy that has led to enormous success in eliminating many diseases and associated human suffering. Of course, this very success has also led to a profound alteration of the human population structure, the consequences of which are just now being fully appreciated (and feared). In any case, the chemical approach to bacterial infections is widely employed and is based on the selectivity of many toxic substances, wherein certain substances kill or prevent growth of bacterial cells while remaining relatively harmless to the cells of the host. We shall see that the selectivity of many such compounds is great, but not perfect, and that their rational use requires an intelligent estimate of the severity of the disease and the gravity of possible side effects resulting from only partial selectivity of the drug employed.

Early efforts to find specific antibacterial agents were largely unsuccessful, owing to limited knowledge of bacterial (and host) metabolism and to

237

the high toxicity to the host of most agents tried. In the 1920's there was a flurry of interest in the possible use of bacterial viruses as medically useful agents and it is true that such bacteriophage have the conspicious virtue of not attacking nonbacterial cells. Unfortunately, it developed that bacteria, which often live in tenuous equilibrium with populations of bacteriophage, are able to mutate to nonsensitive forms so that the only effect of viruses is to select the nonsensitive mutants, which rapidly take over the population and which are obviously insensitive to subsequent introduction of that particular virus. This illustrates an added boundary condition on the selection of an antibacterial drug: The rate of mutation to an insensitive form must be relatively low. If a drug is unable to wipe out the entire population, owing to a residue of insensitive cells, it is very likely to be utterly worthless.

Probably the first important class of antibacterial drug to be developed was the *sulfonamides*, which were available in time to significantly reduce battle fatalities during World War II. The first of these employed was *prontosil*, which had a double ring structure:

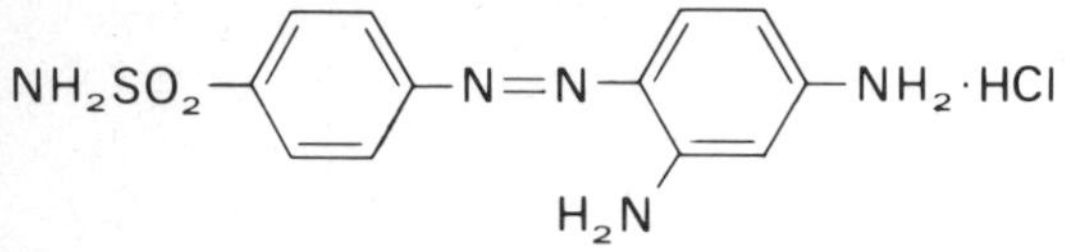

It was soon learned that prontosil was cleaved in the body and that its activity as an antibacterial compound reflected that of one of the cleavage products, *sulfinilamide*,

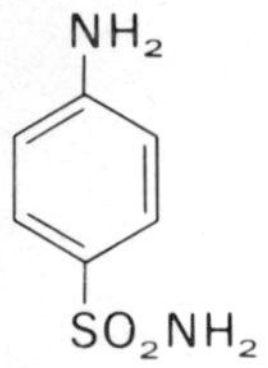

which was subsequently used directly. A large number of other derivatives of sulfinilamide have been synthesized and some exhibit sufficient variation in target organism to be of use. In general, these drugs remain useful in dealing with infections resulting from traumatic injury as well as with certain pneumonias, meningitises, and infections of the urogenital system.

The basis for antibacterial action of such drugs is instructive, as is the reason for its low toxicity for humans. These compounds are examples of a larger class of *antimetabolites*, called *folic acid antagonists*. The vitamin folic acid,

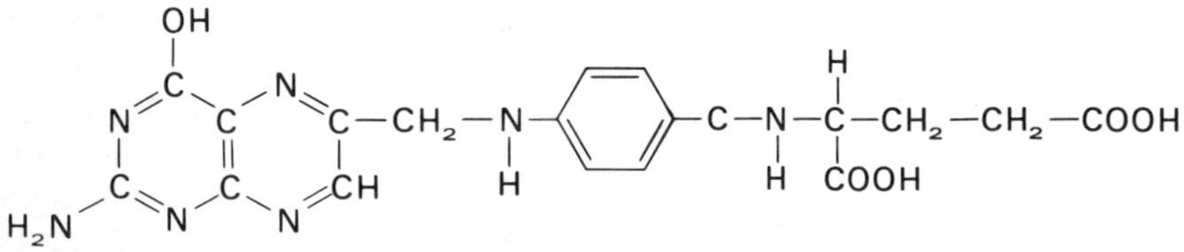

is a cofactor in various syntheses that involve transfer of a single carbon atom, examples of which include formation of various nucleotide bases.

238

Many bacteria are able to synthesize folic acid, and one stage in the synthesis requires *para-aminobenzoic acid*,

which ends up in the obvious location in the folic acid structure. It develops that the sulfanilimides are competitive inhibitors (see Chapter 4) of enzymes concerned with incorporation of *p*-aminobenzoic acid into the folic acid, and their toxicity for bacteria is a direct result of that inhibition and consequent inability to make the vitamin. This example of competitive inhibition should not be surprising in view of the close similarities of the two structures. The reason that sulfanilamides do not kill human (or other mammalian) cells is that they are unable to synthesize folic acid, to begin with, and require the vitamin in their diet. Thus, there is no incorporation of *p*-aminobenzoic acid with which the sulfanilamide can compete.

It should be added here that other folic acid antagonists are of importance also as antitumor drugs. Thus, *aminopterin* (4-aminofolic acid),

is an inhibitor of the reduction of folic acid to the tetrahydro form that is active as a coenzyme. This compound is a useful drug in the treatment of certain cancers, notably leukemias, the cells of which are clearly more sensitive to it than are the normal cells of the same individual.

Antibiotics

An antibiotic is defined as a substance that is produced by one organism and is detrimental to the function of another. Obviously, antibiotics of practical utility are mostly those that inhibit bacteria, and all are produced by various microorganisms, including bacteria and fungi. Again, a usable antibiotic is one that is effective in harming the invading microorganism but not the host. This requirement is often met by *penicillin*, since its action is at the level of bacterial cell wall synthesis, and mammalian cells are not troubled with having to make bacterial cell walls for themselves. Penicillin is actually a class of compounds of which one example is penicillin G, whose structure is shown:

One difficulty with the biologically produced penicillins is the ability of certain bacteria to mutate to resistant forms, and much effort has gone into the chemical modification of the penicillin structure to circumvent this regrettable adaptability. A number of bacteria are able to synthesize an enzyme, *penicillinase*, which, as the name implies, breaks down the antibiotic. Indeed, some bacteria are able not only to render the drug harmless in this manner but also to utilize it as a source of nutritional carbon. This ability, which may be called "feeding upon the hand that bites you," is not conducive to effective chemotherapy, and it is fortunate (for us) that insertion of dimethoxyphenyl groups into the penicillin structure renders the drug resistant to penicillinase.

Since microorganisms differ in their sensitivity to antibiotics, many being quite unaffected by, say, penicillin, it is fortunate that a wide variety of antibiotics with different mechanisms of action are available. For example, *streptomycin* (Figure 9-5) acts by inhibiting protein synthesis and causing erroneous reading of the mRNA code at the ribosome. This appears to occur through alteration of the 30-S ribosomal structure. In addition, the class of *tetracycline* compounds, including tetracycline itself,

and chlorotetracycline (also called Aureomycin),

acts by inhibiting protein synthesis in the prokaryotic ribosome. Still another inhibitor of protein synthesis is the important drug *chloramphenicol*,

which is known to inhibit protein synthesis in prokaryotic ribosomes, interacting with the 50-S subunit, perhaps competing with mRNA for binding sites on the ribosome. Antibiotics that interfere with protein

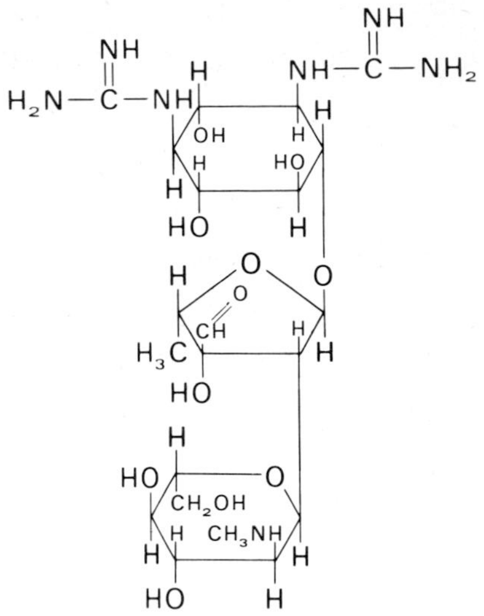

Figure 9-5   Streptomycin.

synthesis clearly must not inhibit protein synthesis in the host organism, and it is interesting that the ribosomes of bacteria differ from
those of higher organisms in sufficiently important ways to make such
specificity possible. Table 9-3 demonstrates the difference between the
two classes of ribosome with regard to inhibitor sensitivity, size, and
location. It is especially interesting that the ribosomes being studied in
mitochondria and chloroplasts are much more similar to those of bacteria
than to those attached to the endoplasmic reticulum of the same cells in
which they occur. This similarity has led many people to surmise that
bacteria and chloroplasts may have evolved from some sort of microorganisms living in a symbiotic fashion in eukaryotic cells. This hypothesis
maintains that such intracellular bacteria (or whatever) gradually took
over the energy-producing role in the host cell and, in return, gave up much
of its synthetic ability. This point of view is also appealing with regard
to many similarities between bacteria and mitochondria (and chloroplasts), such as the occurrence of circular DNA in all three locations. On
the other hand, there is very little direct evidence that would enable one

Table 9-3   The two classes of ribosomes

	Size (Svedberg units)	Location	Compounds that inhibit protein synthesis
Class I	80 S	Endoplasmic reticulum of eukaryotic cells	Cyclohexamide*
Class II	70 S	Prokaryotic cells, and mitochondria and chloroplasts from eukaryotic cells	Chloramphenicol

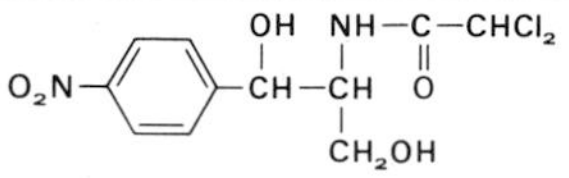

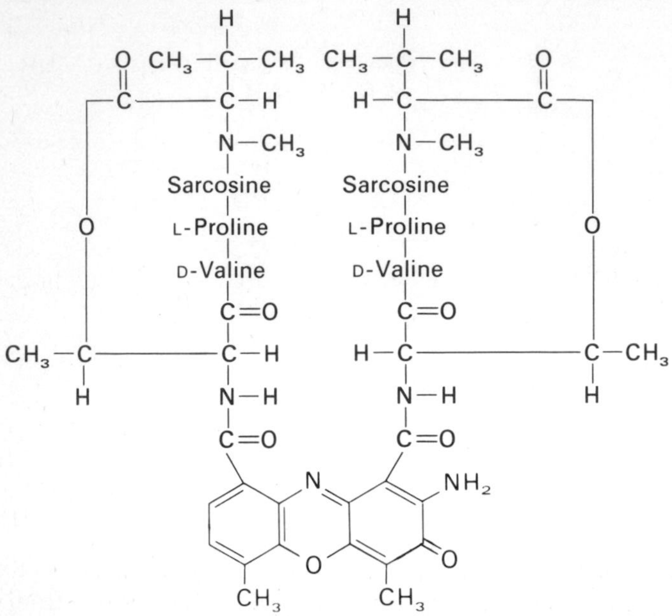

Figure 9-6 Actinomycin D.

to accept the view in any definitive fashion and, in fact, the view is some-what difficult to relate to the double-membrane structure of mitochondria (that is, what is the origin of the outer mitochondrial membrane?). Also, the nuclear genetic control of some features of mitochondrial and chloro-plast structure renders the overall picture rather more ambiguous. While these organelles show autonomy, they may not be quite autonomous enough to be consistent with a bacterial origin.

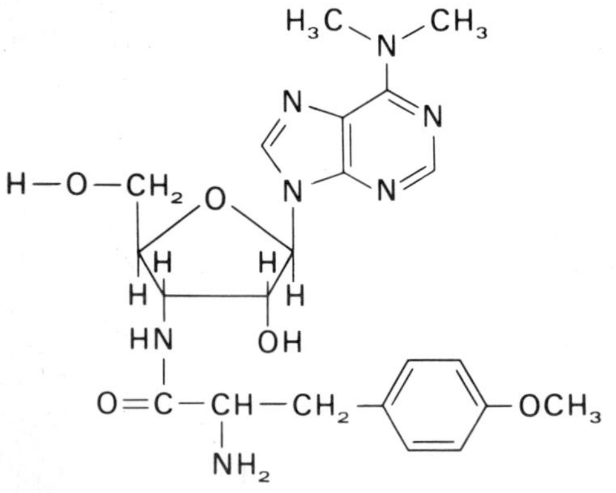

Finally, other antibiotics damage bacterial cell function at the level of nucleic acid synthesis. *Actinomycin D*, which acts against bacteria and also some tumors, has the structure shown in Figure 9-6. This compound acts by inhibiting mRNA synthesis at the DNA template (see Chapter 16) and has been shown to bind directly to DNA. One imagines that the DNA of microorganisms is more susceptible to such binding by being relatively more exposed than that of the eukaryotic nucleus, where DNA is already bound to such basic proteins as the histones. *Puromycin*, which is not a particularly useful antibiotic to the clinician, is nonetheless of interest as another nucleic acid-level inhibitor. It has the structure and owes its inhibitory properties to its similarity to the adenosine end of the aminoacyl tRNA molecule:

Suggested Reading

Books

Bonner, J., and J. E. Varner, *Plant Biochemistry*, Academic Press, New York, 1965.

Cheldelin, U. H., *Metabolic Pathways in Microorganisms*, John Wiley & Sons, New York, 1961.

Davis, B., R. Dulbecco, H. Eisen, H. Ginsberg, and B. Wood, *Principles of Microbiology and Immunology*, Harper & Row, New York, 1968.

Gunsalus, I. C., and R. Y. Stanier (eds.), *The Bacteria*, vols. 1–5, Academic Press, New York, 1964.

Kluyver, A. J., and C. B. van Niel, *The Microbe's Contribution to Biology*, Harvard University Press, Cambridge, Mass., 1956.

Microbial Reaction to Environment (Eleventh Symposium of the Society of General Microbiology), Cambridge University Press, Cambridge, 1961.

Oginsky, E., and W. Umbreit, *An Introduction to Bacterial Physiology*, W. H. Freeman and Co., San Francisco, 1959.

Rose, A. H., *Chemical Microbiology*, Butterworth & Co., London, 1965.

Sokatch, J. R., *Bacterial Physiology and Metabolism*, Academic Press, New York, 1969.

Stephenson, M., *Bacterial Metabolism*, MIT. Press, Cambridge, Mass., 1966. (A paperback reprint of the earlier 1949 hardcover edition.)

Articles

Echlin, P., "The blue-green algae," *Sci. Am.*, **214**, 74 (June, 1966).

Gibson, F., and J. Pittard, Jr., "Pathways of biosynthesis of aromatic amino acids and vitamins," *Bacteriol. Rev.*, **32**, 465 (1968).

Hegeman, G., and S. L. Rosenberg, "The evolution of bacterial enzyme systems," *Ann. Rev. Microbiol.*, **24**, 429 (1970).

Iino, T., "Genetics and chemistry of bacterial flagella," *Bacteriol. Rev.*, **33**, 454 (1969).

Lang, N., "The fine structure of blue-green algae," *Ann. Rev. Microbiol.*, **22**, 15 (1968).

Payne, W. J., "Energy yields and growth of heterotrophy," *Ann. Rev. Microbiol.*, **24**, 197 (1970).

Salton, M. R. J., "Structure and function of bacterial cell membranes," *Ann. Rev. Microbiol.*, **21**, 417 (1967).

Sharon, N., "The bacterial cell wall," *Sci. Am.*, **220**, 92 (May, 1969).

Part III

Membranes and the integration of cell function

10 Studying the intact cell

The preceding chapters of this book have dealt largely with the parts of cells rather than the cells themselves and often with activities that, at least in principle, could occur in a test tube. Such an approach is largely dictated by the enormous complexity of cells, where, to make a beginning, it is necessary to isolate as completely as possible the part of cell function that one wishes to study. Frequently, this means a physical isolation of a part of the cell, so that it can be studied alone. For instance, many of the reactions involving energy coupling that we have considered are known from studies not of intact cells but of mitochondria or chloroplasts, or even from studies employing enzyme systems derived from them. Such an approach leads often to information about the mechanism of a reaction without, however, the means of deciding whether the reaction has any role in the entire cell.

However, it should not be thought that the divide-and-conquer approach to cell physiology is without virtues; it is obviously of interest to know what reactions and physical activities are associated with a given part of the cell. But, once these activities are known, the questions immediately become more complex and relate frequently to the relation between the organelle and the rest of the living cell. Thus, it no longer is suitable to remove the organelle from the cell; ways must be developed to study it *in situ*, indeed to study the entire cell in such a manner as to make clear the functioning of each part. This is clearly a very difficult matter, and it should not come as too much of a surprise that we appear to be only on the threshold of viewing the whole cell with any sophistication. In considering the intact cell, one has not only to cope with its great complexity but one must design experiments or make measurements that are not likely to produce so much damage as to render the measurements meaningless. In other words, the cell physiologist is confronted not only by the great subtlety of cells but also by their fragility.

This chapter serves as introduction to the remainder of the book, which will be concerned with increasingly integrated aspects of cell function. The purpose of the present chapter is to note some of the difficulties in approaching the intact cell while permitting it to remain intact, and to consider some special techniques that are often useful.

Cellular indeterminacy

In physics, it is impossible to obtain simultaneous measurements of, say, position and momentum of a particle beyond a certain limit of accuracy. This "indeterminacy" is not a measure of the difficulty of making physical measurements but an underlying property of the world as disclosed by quantum mechanics. An intuitive grasp of this impossibility is allowed by considering that the only way to locate a subatomic particle precisely is by hitting it with another particle, which, however, produces an unspecified change in the target's momentum. Thus one must, in principle, disturb the particle's momentum to measure its location and, indeed, vice versa. To put it more simply, the act of measurement perturbs the system being measured and produces, in this case, a fundamental limit to the information that can ever be obtained about that specific system.

Consider in similar fashion the case of the cell. Imagine that one wishes to know, with some accuracy, the internal pH of a particular living cell. There are two techniques that might be used and they are, in principle (but not in practice), extremely simple. One can employ a pH indicator dye such as bromothymol blue or perhaps a pH electrode. If one chooses the dye, it is necessary to use one that is as little toxic as possible (bromothymol blue is a good choice). The dye must either be injected into the cell with a micropipet (in which case, one does well to choose a large cell for the experiment), or it may be allowed to penetrate the cell through the cell membrane, provided, of course, the cell is permeable to the dye. Then it is only necessary to somehow measure the color that the dye takes on when inside the cell, and using that, to compute the pH.

Two problems need to be kept in mind throughout this sort of measurement. In the first place, the intrusion of dye into the cell must produce as little deleterious effect as possible. Clearly, if the cell is killed, either by the process of injection or by the harmful action of a dye, measurements obtained are very nearly worthless. But, even if the cell survives the treatment, how near is it to normality? Indeed, the problem frequently is a circular one, since the criterion of normality is often related to the measurement being performed.

In addition, it is obviously necessary that the measurements be susceptible to interpretation—that is, that they mean what they appear to. Even if we satisfy ourselves that the cell is in reasonably good condition after its shot of bromothymol blue dye (or whatever), how are we to measure the color change accurately, and is it really analogous to the same change observed in a beaker when we carry out a standard curve? For example, some charged dyes become closely associated with the cell membrane, which, as we shall see, can itself bear a sizable charge. Perhaps all the dye is really located within a few nanometers of the cell membranes, in which case the measurements hardly represent conditions in the rest of the cell interior. Perhaps all the dye has been taken up by some closed organelle, and one actually measures the pH of *its* interior. Numerous possibilities exist and must be kept in mind.

Nondestructive measurements of cells

It is useful to consider a few of the techniques that may be applied to analysis of cells without producing important changes in them. Obviously every sort of cell study requires some alteration of the cell, if only the passage of light through it. In applying a given technique, it is necessary to satisfy oneself that the perturbation is minimal and will be unlikely to influence the results. It should be clear that what represents a minimal perturbation to one cell type might be intolerable for another. For instance, spectrophotometry, which involves the transit of light through a cell, is often quite innocuous in the case of an animal cell, while quite difficult in that of a photosynthetic plant, where the light, at some wavelengths, may stimulate energy production and produce profound effects in the intracellular state.

In many cases, a technique is satisfactory only if one takes care not to overdo it. For example, spectrophotometry, as we shall see, may work quite well at low light intensities, while producing damage when too much light is used. Frequently, one treads a thin line between adequate light for required accuracy and too much, which can fry and otherwise disrupt cellular structure. Likewise, the essence of the tracer technique, which has been applied with dramatic success to cell study, is that only trace amounts of radioactive substance may be used. Larger dosages can give rise to extensive damage, owing to a high intracellular flux of ionizing radiation produced on decay of the radioactive substance. In these (and all other) cases, a light touch is required.

Spectrophotometric methods

Spectrophotometric examination of intact cells has often been used to analyze cell content chemically and to follow reactions going on *in vivo*. There are two general approaches that may be used. In the first place, one can use a *microspectrophotometer* (Figure 10-1) to examine specific regions of a single cell and determine their spectrum, which, in turn, can tell a great deal about their chemical composition. This technique has been used, for example, in the localization of deoxyribonucleic acid (DNA), which has a characteristic spectrum, in the cell nucleus. Likewise, mitochondria exhibit the characteristic composite spectrum of the cytochromes of the respiratory chain.

A second form of spectrophotometry is that applied not to single cells via a microscope but to suspensions of cells. Usually cells are examined

Figure 10-1 Microspectrophotometer. The combination of a high-resolution optical microscope and visible-ultraviolet spectrophotometer permits the spectroscopic analysis of cell components. (Courtesy of the Olympus Corporation of America.)

at wavelengths in the visible or near-ultraviolet region where specific chemical constituents absorb. For example, one can set a spectrophotometer on 340 millimeters, where reduced pyridine nucleotides (NADH and NADPH) absorb, and thus keep track of the degree of reduction of these important compounds. In the past, such observations of intact cells (or even intact organelles) have been technically difficult, owing to the high degree of light scattering produced by a suspension. To obtain a high-enough cell density to enable one to measure a particular component, it was necessary to have a very turbid suspension whose light scattering blanked out the changes being studied. Recently, the development of dual-wavelength instruments has rendered such measurements much more feasible; the basic construction of such an instrument is diagrammed in Figure 10-2. The machine alternates light of two predetermined wavelengths on a cell suspension and measures the difference in absorption between the two. One wavelength is set on the peak of the compound being studied and the other on a wavelength where the compound produces no change. The second wavelength thus provides a reference that substracts away absorption due to other, irrelevant compounds and to light scattering. This strategy is illustrated in Figure 10-3 for the case of cytochrome c, a cell component quite easily measured by this approach. Indeed, owing to the distinctive absorbancy peaks of the cytochromes and oxidative enzymes in general, the technique of dual-wavelength spectrophotometry has been applied frequently to the study of biological oxidations and photosynthesis.

Other modifications of spectrophotometry are often applied to cell suspensions. For example, dual-wavelength measurements taken with

250

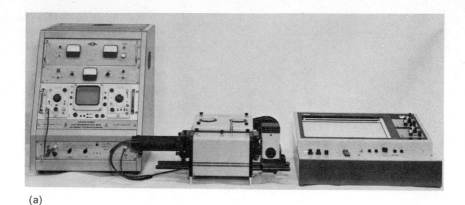

(a)

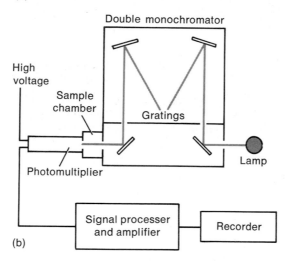

(b)

Figure 10-2 Dual-wavelength spectrophotometer. (a) Photograph of a commercially available instrument. (b) diagram of operation: Light enters the monochromator and falls on two gratings, one set at each wavelength. A spinning wheel "chops" the exiting beam so that light, at first one wavelength and then the other, falls upon the detector (photomultiplier), after having passed through the sample. The electronic part of the apparatus then subtracts the absorbance at one wavelength from that at the other and records the difference. (Courtesy of American Instrument Co., Inc.)

indicator dyes, such as bromothymol blue (discussed earlier), have been used to yield information on internal pH. Such instruments can also be combined with devices for rapid mixing of cells and reactants, enabling one to study rapid reactions, often complete in much less than 1 second. In addition, one is not restricted to absorption spectrophotometry alone, as many cell constituents fluoresce and can be measured by changes in specific fluorescence. The latter approach employs exciting light of one wavelength and measurement of emitted light at another (see Figure 10-4) and is well suited to measurement of many aromatic compounds of biological importance. With this approach, not only can cell constituents be measured by their fluorescence but fluorescent compounds may be added to cells whose degree of fluorescence may then serve as indicator of

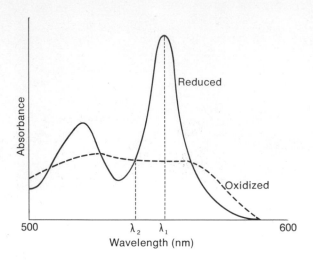

Figure 10-3 Measurement of the degree of reduction of cytochrome c by dual-wavelength spectrophotometry. The instrument is set to record absorbancy at $\lambda_1 - \lambda_2$, so that only changes due to cytochrome c reduction will be recorded. λ_1 is the reduced peak of the cytochrome at 550 nm.

some property of the cellular interior. For instance, compounds exist whose fluorescence is altered by the binding of calcium, so that they have been used as intracellular (or intramitochondrial or chloroplast) calcium indicators; an excellent example is the antibiotic chlorotetracycline. Since calcium plays important roles in many cells (particularly with regard to excitability), such measurements are of the greatest utility.

Finally, fluorescence changes resulting from the addition of organic compounds such as 1-anilinonaphthalene-8-sulfonate (ANS)

have been used as indicators of alteration in the conformation of protein molecules to which they bind. Since conformational changes have been

Figure 10-4 Diagram of spectrofluorometer. (a) represents the exciting beam entering the sample, (b) the emitted beam. The intensity of the emitted beam will be a function of the concentration of the component in the cuvette that gives rise to the fluorescence.

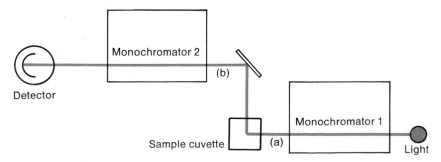

implicated in a number of processes, including transport, energy coupling, and enzyme regulation, such measurements are likely to become of increasing importance.

The direct microscopic observation of cells

Since measurements often produce harmful effects on cells and since, in any case, the techniques employed frequently involve highly complex and specialized apparatus, it is pleasant to note that direct observation often plays a significant role in understanding cellular function. Results of the greatest significance have been obtained by simple observation of unstained cells, and it is worth giving some examples of what may be learned. Clearly, only certain kinds of information may be acquired in this manner and, still more clearly, this kind of tactic in biology depends heavily upon the imagination and insight of the viewer. One might look in vain for direct indication of biochemical changes in cells, although the final effects of biochemical events might well be evident. Consider first what sort of information is available. In the first place, one can determine how many cells are present in a sample and consequently one can tell how rapidly cells are increasing (or decreasing) in number. Moreover, when cells are motile, the degree of motility can often reflect something about the physiological state of a cell population. Likewise, the very shape of cells can reflect something of their physiological state. Mammalian cells grown in culture frequently exhibit the form of a spindle or irregular star (depending on cell type) when in physiologically stable condition, whereas they become round when damaged or placed into a new or unfavorable medium.

Finally, by judicious use of so-called vital stains or by means of phase-contrast microscopy, one can view something of the interior events and structures of cells, and frequently one receives the impression of furious activity as *cytoplasmic streaming* whirls recognizable organelles about in the cytoplasm. Such a vision is obtainable neither by means of exotic biochemical analysis nor by means of the electron microscope, and the view nonetheless contributes to our understanding of cells.

In some instances, quite sophisticated information is obtained by microscopic examination of living cells. The study of junctions between plasma membranes of cells in contact has depended in part upon the technique of fluorescence microscopy. Thus, the fluorescent dye fluoroscein is injected with a micropipet into one of several adjacent cells. The other cells are seen to become fluorescent as the dye diffuses across the cell junctions, an observation that indicates that even relatively large molecules pass across such intercellular junctions. It can also be observed that injection of calcium into the cell inhibits dye migration to adjoining cells and that interference with respiration or ATP synthesis by inclusion of inhibitors or uncouplers also prevents migration. These visual observations have been combined with physical measurements such as estimation of conductivity between cells through the use of microelectrodes thrust into different cells. A detailed picture has emerged that includes a dramatic increase of conductivity when junctional connections are made and the ability to pass molecules with molecular weights as large as those of small proteins. Thus, regions of cell membranes in contact must take on properties quite different from those of the free, unattached, membrane that forms an effective permeability barrier to many molecules with molecular weights less than 100. Again, it must be stressed that these findings have often been obtained with a technique no more complicated than that involving an ordinary light microscope equipped with a light source emitting in the ultraviolet region, thus serving to excite fluorescence of the dye employed in the study.

Cells grown in culture

The development of techniques for growing cells from higher organisms in sterile culture has presented numerous new opportunities for the study of intact cells. Thus, it is possible to disperse cells from a particular organ or plant tissue and obtain at least some cells capable of forming viable *clones* (self-perpetuating lines). The cells may be grown either in suspension or as a single-cell layer (*monolayer*) on the bottom of a container of nutrient medium. The latter form of culture yields a smaller mass of cells than the former but has the advantage of ease of microscopic examination through the bottom of the flask. It should be added that techniques for sterile culture of higher cells are clearly extensions of the more general culture techniques long available for growing unicellular organisms, such as the bacteria and yeasts. Emphasis on growth of cells from multicellular organisms should not obscure the important service to cell physiology provided by the relatively easy availability of large and mainly homogeneous populations of microorganisms.

Cell culture has yielded particularly interesting information about the nutrition of higher cells, information not obtainable in any other way. Indeed, investigations of nutrition necessarily accompanied successful culture, as it soon became evident that mammalian cells required a very large number of nutrients in the growth medium to survive. For example, culture medium for human uterine tumor cells contains no less than 28 different growth factors, including 17 amino acids and sundry vitamins.

The requirement for a complex medium points to an important feature of cells, as found in the intact organism: *Their* medium, the multicellular organism itself, is extremely complex. Cells in their tissue "habitat" exist at a high population density—in a mammalian organ as high as 10^8 per milliliter. Thus, on the average, each cell is likely to be quite close to others and there are excellent opportunities for intercellular transport of nutrients. On the other hand, cells in culture exist at much lower densities, from, say, 10^4 to 10^6 per milliliter, and one consequence of this great dilution with respect to tissue is that small molecules tend to leak out into the culture medium. Thus, it seems quite clear that certain amino acids that cultured cells require can, in fact, be manufactured within the cells but are lost into the medium at a rate more rapid than their synthesis. Thus, in order to maintain adequate intracellular levels of the compounds, they must be added externally.

Cultured cells have been used in a wide variety of physiological and, recently, genetic experiments, and it is important to add a cautionary note about application of such results. The cultured cell is studied in the hope that its properties will enable one to understand cells in their native condition in complex tissue. This hope is often only partially justified, as the process of culturing the cells often leads to catastrophic alteration of properties. Often, tissue-specific features of cells are lost. For instance, cells from endocrine organs often fail to secrete the hormone that they would be expected to elaborate. Sometimes it appears that culturing itself produces such a loss—a sort of generalizing of cell characteristics; in other instances, culture conditions may impose a selection on the cells that survive and produce clones. Since organs are invariably heterogeneous with regard to cell type, it is often impossible to be totally sure about the origins of a particular cell line, an uncertainty that clouds interpretation of results obtained with cultured cells.

Since the important topic of growth is notoriously difficult to study in multicellular tissues, considerable research in this area has centered on cultured cells. It appears that, in general, mammalian cells grow much more rapidly in suspension than in the intact tissue, exhibiting a doubling time of the order of one day, as opposed to months, in tissue. This acceleration implies that a form of control is lost, and considerable effort is presently expended in studying this phenomenon, since control of growth is of considerable practical interest. One element of the control may reside in the mechanism of *contact inhibition*, where cells residing on a monolayer inhibit each other's growth when they grow so dense as to touch one another. Although the mechanism of contact inhibition is not clearly understood, it is of great interest that certain tumor cells fail to exhibit the effect, suggesting that a similar mechanism may occur in tissue and that its failure may have profound and disastrous consequences for the organism.

Studies with beating heart cells

One of the most interesting recent examples of physiological investigation based on simple observation may be seen in the studies by I. Harary and E. C. Slater (see Suggested Reading) on heart cells grown in culture. Cells are isolated from the heart of a newborn rat and grown in a monolayer (see above). These cells retain the property of beating, and the members of a population all beat in synchrony—that is, at the same time. The rate at which they beat reflects the state of metabolism, especially with regard to the generation of energy. For this reason, it has been possible to study energy metabolism in these intact cells without disturbing them beyond the alteration produced by growing them in culture in the first place.

One of the important results that came from these studies was the close dependence of beating rate on the supply of ATP. Addition of *dinitrophenol*, an uncoupler of oxidative ATP synthesis, inhibited beating completely, an effect that was partially reversed upon addition of ATP to the medium. *Iodoacetate*, which is an inhibitor of glycolysis, also under certain conditions produced cessation of beating, suggesting that either oxidative or glycolytic ATP production could support the contractile process. It was also interesting that the glycoside *ouabain*, which inhibits certain energy-requiring transport reactions, stimulated the frequency of beating. Its effect was probably to increase intracellular ATP by inhibiting transport, which leads to its breakdown. Finally, there had previously been some question as to whether muscular contraction actually required ATP itself or perhaps one of the "high-energy intermediates" said to precede it. Since *oligomycin* inhibits ATP formation by acting at the last reaction of ATP synthesis, it would presumably allow one to discriminate between the two possibilities. Indeed, under the proper conditions, oligomycin did inhibit beating, a result that indicated that "high-energy intermediates," whatever their nature, could not replace ATP in supporting contraction.

Thus is it that results of the greatest importance, even in the area of energy metabolism, can come from measurements employing no more exotic apparatus than a stopwatch and perhaps a relatively low-powered microscope.

Suggested Reading

Books

Ackerman, E., *Biophysical Science*, Prentice-Hall, Englewood Cliffs, N.J., 1962, Chs. 26–29.

Bourne, G. H., *Division of Labor in Cells*, 2nd ed., Academic Press, New York, 1970.

Bresnick, E., and A. Schwartz, *Functional Dynamics of the Cell*, Academic Press, New York, 1968.

Ham, A. W., *Histology*, 6th ed., J. B. Lippincott Co., Philadelphia, 1969, Chs. 2–4.

Harris, H., *Cell Fusion*, Harvard University Press, Cambridge, Mass., 1970.

Parker, R. C., *Methods of Tissue Culture*, Harper & Row, New York, 1961.

Paul, J., *Cell and Tissue Culture*, Williams and Wilkins Co., Baltimore, 1965.

Willey, R., *Microtechniques*, The Macmillan Co., New York, 1971.

Articles

Caldwell, P. C., "Intracellular pH," *Intern. Rev. Cytol.*, **5**, 229 (1956).

Chance, B., "Intramitochondrial pH Indication," in *Biochemistry of Mitochondria*, E. C. Slater, Z. Kaniuga, and L. Wojtczak (eds.), Academic Press, New York, 1967.

Chance, B., "Spectrophotometry of intracellular respiratory pigments," *Science*, **120**, 767 (1954).

Cowles, J. C., "Theory of dual wavelength spectrophotometry for turbid samples," *J. Opt. Soc. Am.*, **55**, 690 (1965).

Harary, I., and E. C. Slater, "Studies *in vitro* on single beating heart cells VIII," *Biochim. Biophys. Acta*, **99**, 227 (1965).

Stubblefield, E., "Synchronization methods for mammalian cell cultures," in *Methods in Cell Physiology*, D. M. Prescott (ed.), Academic Press, New York, 1968.

11 The structure of cellular membranes

In Chapter 1 we saw that the structure of cells is very much a matter of membranes. Systems of membranes enclose the cell, and, in eukaryotic cells, isolate the nucleus from the rest of the interior and fill it with complex structures (organelles). Not only do membranes constitute the units from which much of cell structure is built, but they play an essential role in cell activities and, most especially, in the regulation of those activities. Cells not only fail to look like a bag of enzymes, but they do not act as if they were. Many of the most interesting properties of cells, such as irritability and transport, are uniquely associated with membranes, while many of the metabolic reactions of cells take place on membrane surfaces. The compartments formed by membranes within the cell, thus separating potential reactants, have profound significance for the pattern of metabolism. Clearly, the ability of membranes to allow selective passage of some compounds, but not others, and to couple the passage of some with energy-yielding reactions presents significant possibilities for regulation. Finally, by dividing regions of space into compartments that can be made to differ in ion composition, membranes often bear an electrical potential difference between their two sides. We shall see that such membrane potentials are of great importance in cell excitability, translocation of ions, and, very probably, energy coupling in general.

Membrane structure

The membranes of the cell have certain properties in common, whether they constitute the outer boundary of the cell or form the organelles of the interior. All are composed almost entirely of protein and lipid and *protein & lipid* have a remarkably similar geometry, as revealed by electron microscopy. The basic pattern seen in electron micrographs consists of two heavily staining layers with a nonstaining "filling" in between, and with an overall thickness of about 75–100 Å. This configuration is known as the *unit membrane* and forms the basis of all cellular membrane systems.

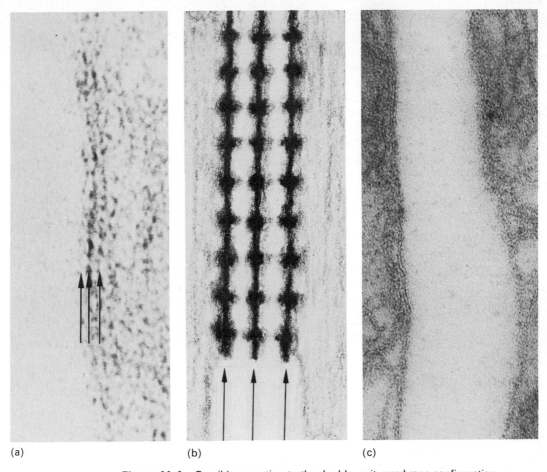

(a)   (b)   (c)

Figure 11-1   Possible exception to the double-unit-membrane configuration.
Recent studies with a number of cellular membranes suggest that it may be necessary
to consider alternatives to the usual view. These electron micrographs are from a paper
by F. L. Crane and J. D. Hall (*Biochem. Biophys. Res. Commun.*, **36**, 174) published
in 1969. They suggest that a single membrane may consist of three layers of protein
with two layers of lipid in between. Although the authors suggest that the common
failure to see all three dense layers may reflect inability of the stain to penetrate the
entire membrane, it must be said that their view is not yet widely accepted. Indeed,
their micrographs may reflect differences between their technique and those yielding
the more usual picture. Their figure is included in this book as an indication that there
is still much to be learned about membrane structure. (a) Section of mitochondrion
fixed in osmium tetroxide and stained with $KMnO_4$. Magnification is $\times$ 640,000.
(b) Enlarged view of (a). Magnification is about $\times$ 2,250,000 (final). (c) A lower-
magnification view of the same mitochondrial membrane. Note that at this level of
magnification, it is difficult to say whether the membrane is double or triple.

For example, the plasma membrane consists of a single unit membrane often with an outer coat of collagen (or other material), while the nuclear envelope is formed from two such unit membranes, closely applied to each other. An example of the unit membrane configuration was shown in Figure 1-4. Although the electron-microscopic view of membrane structure is partially influenced by the methods of preparation used (see Figure 11-1), the classical unit membrane is most often observed. This apparent constancy of membrane structure is interesting in view of the widely different functions of different membranes, where, for example, quite different permeability properties are observed as well as entirely different enzymic contents.

The chemical composition of membranes

The chemical composition of different membranes may be described as a common pattern onto which is imposed variation, depending upon cell type and function. In general, membranes are composed almost entirely of protein and lipid, with much of the lipid in the form of phospho- [*phospholipids*] lipids. Structures of some important membrane lipids are given in Figure 11-2.

Membranes from mammalian tissue contain steroids, including [*mammalian vs. bacterial*] cholesterol, whereas bacterial membranes do not. In addition, myelin

Figure 11-2 Some lipids commonly found in membranes. Note that the first three are phospholipids, the last a steroid.

Phosphatidylethanolamine

Phosphatidylserine

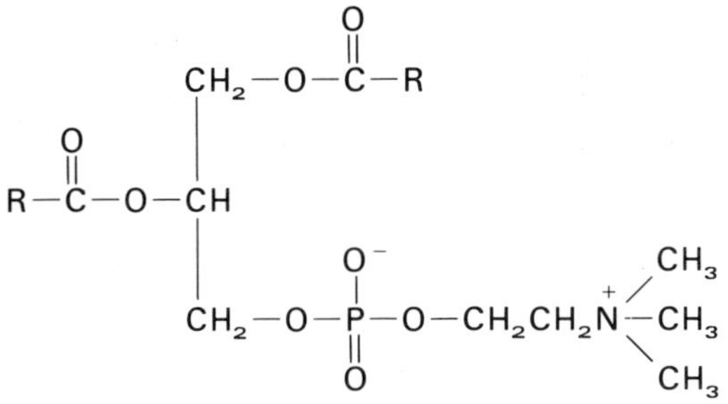

Phosphatidylcholine (lecithin)

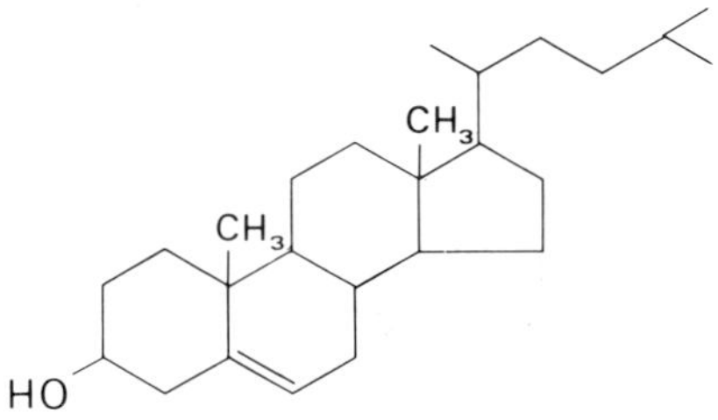

Cholesterol

interesting

membranes from the nervous tissue of animals contain α-hydroxy fatty acids, which are not found elsewhere. It is probable that, as more detailed information about membrane composition becomes available, additional differences will be discovered. When one considers that different membranes play distinct roles in energy metabolism, transport of small molecules, and protein synthesis, as well as other processes, it becomes clear that considerable variation might be expected.

There is a great deal of variation in the protein content of membranes, as well as in the percent composition, which ranges from about 20 per cent in nerve myelin to 80 per cent in erythrocytes. Many of the proteins in membranes are either enzymes or proteins involved in transport, as we shall see, and the type of membrane dictates which proteins will be included. For example, mitochondrial inner membranes contain the enzymes of the respiratory chain, those involved in ATP synthesis, and, apparently, many concerned with transport across the membrane. On the other hand, the outer mitochondrial membrane contains neither ATP-synthesizing enzyme systems nor transport proteins but instead contains a few enzymes, such as monoamine oxidase and kynurenine hydroxylase, which are entirely lacking in the mitochondrial inner membrane and which are thus frequently used as markers for the outer membrane.

insolubility

It has often been suggested that membranes contain a class of proteins that are not enzymes and which exist as important structural elements only. Such *structural proteins* would be expected to bind phospholipids and possibly other membrane proteins, and to be highly insoluble in water (lest they be extracted from membranes by the water surrounding them). A number of proteins with these qualities have been isolated from a variety of membranes, including red-blood-cell plasma membranes and mitochondrial and chloroplast membranes. They may qualify as structural proteins, as they have not been shown to possess enzymic activity, although the failure to demonstrate such activity may only reflect ignorance as to substrate, or denaturation due to isolation procedures, which in such cases are by necessity rather harsh.

Models of membrane structure: the bilayer model

simplicity?

Membranes are of such a thickness (> 100 Å) that they must be composed of layers of relatively few molecules. Indeed, some of the proteins isolated from membranes are quite large and must span at least one half of the total membrane thickness. Thus, in a certain sense, the very scale of membranes suggests that their architecture cannot be excessively complicated, and the wide distribution of a common *unit membrane* pattern has encouraged biologists to propose various models in an endeavor to explain physical and biological features of membranes. Hypotheses of this sort fall into two main categories, which are worth describing, since it

261

becomes increasingly apparent that a realistic understanding of membranes must include aspects of both.

The first approach to the molecular structure of membranes was suggested by Danielli in 1936 and is generally associated with his name, being called the *Danielli* or *bilayer model*. The model is based in part on the high lipid content of membranes, together with an important feature of many lipid molecules, separation of a polar (hydrophilic) end and a nonpolar (hydrophobic) end. For instance, the fatty acid palmitic acid, a common membrane component, contains a fifteen-carbon aliphatic side chain and a carboxyl group (which forms the polar end):

$$CH_3CH_2CH_2CH_2CH_2CH_2CH_2CH_2CH_2CH_2CH_2CH_2CH_2CH_2CH_2COOH$$

Many other components of membrane lipids share this separation of hydrophilic and hydrophobic character, as we shall see, and it is maintained in the many lipids that are esters of such fatty acids, together with polyhydroxy alcohols, such as glycerol. The bilayer model suggests that the central layer of a cellular membrane is a double layer of such lipid molecules arranged in such an orderly fashion that the nonpolar ends are extended toward each other with a uniform array of polar groups toward the outside. The protein of the membrane would then be bound to the polar surfaces of the lipid bilayer. This orientation is illustrated in Figure 11-3.

The bilayer model was proposed on the basis of certain physical measurements using oil–water interfaces, which since 1936 have come to be regarded as irrelevant to membranes. However, as soon as electron microscopy became available and revealed the three layers of the unit membrane, the Danielli point of view took on considerable appeal. For example, the deposition of heavy metals, which are used as electron-dense "stains" in electron microscopy, would be most pronounced in the regions of

Figure 11-3   Bilayer model for unit membranes.

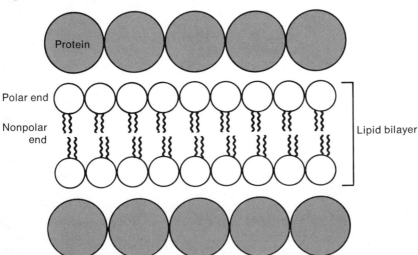

Protein

Polar end

Nonpolar end

Lipid bilayer

the hydrophilic ends of lipids and the outer protein layers, which would yield the characteristic unit membrane picture. The bilayer model, likewise, is in conformity with many of the measurements of physical properties of membranes, including low conductivity, high capacitance, and also several mechanical parameters.

The subunit model

It has also been argued that membranes can profitably be regarded as an array of similar subunits, forming a regular two-dimensional pattern of uniform thickness. This approach, illustrated in Figure 11-4, has grown by analogy with viruses, whose outer layer is often composed of identical protein molecules. The concept of subunit structure as applied to membranes has appeal, as it provides a reasonable mechanism for the assembly of membranes (namely by aggregation of the subunits) as well as a convincing explanation for the constant proportions of certain enzymes located in cellular membranes. Moreover, there is some electron-microscopic indication that certain membranes, such as those of chloroplasts, show a repeating substructure, although the failure to show a similar configuration in all cell membranes represents a serious defect. Positive evidence for the subunit idea would, most convincingly, include the disaggregation of membranes into the identical subunits followed by reassembly into membranes approaching a normal state. Unfortunately, in no case has such disaggregation been demonstrated in an unambiguous manner.

There are, however, reasons to take the subunit model seriously to the extent that it may provide partial explanations for membrane properties when considered together with the bilayer model. For example, mitochondrial and chloroplast membranes may be regarded as possessing a bilayer "backbone" upon which a sort of subunit structure is imposed. In these cases, the subunits would be represented by the "knobs" that appear to contain part of the ATP-synthesizing apparatus and which are

Figure 11-4   Subunit model for membrane structure.

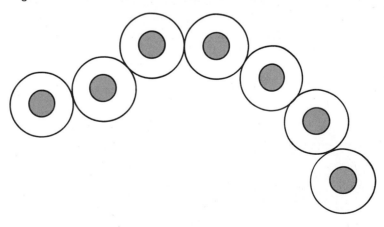

spaced at regular intervals over the inside surface of the inner membranes. It is interesting that a partial demonstration of reassembly from subunits has been carried out with mitochondrial knobs (ATPase), after knobs had been removed from membranes by protein isolation techniques, leaving smooth membranes. Subsequent addition of ATPase to naked membranes yields a membrane bearing knobs on its surface with essentially the same spacing and geometry (Figure 6-2).

Membrane phospholipids

The properties of natural membranes appear more and more to reflect the large amount of phospholipid that they contain. Indeed, in the cell, membranes are the main locations of such compounds, only vanishingly small quantities being found elsewhere. The phospholipids play a major role in forming the characteristic structure of membranes, and attempts to obtain active membrane protein complexes that can carry out such functions as electron transfer (in the case of mitochondrial or chloroplast proteins) generally require that phospholipids be added. There is also considerable variation in the amount of phospholipid in various membranes. The lipid component itself varies from 15 to 20 percent in the case of red blood cells and muscle cells from mammals to as much as 80 percent in that of myelin. Within this fraction, phospholipid accounts for only about 40 percent of the lipids of myelin, while it is practically 100 percent of the lipids of many bacterial plasma membranes.

Phospholipids are most commonly triglycerides in which one of the glycerol hydroxyl groups is esterified to phosphoric acid instead of to a third fatty acid molecule. The presence of a phosphate group on the glycerol "end" of the molecule, in contrast to the highly apolar fatty acids at the other end, makes the phospholipid molecule highly directed as to polarity. This property is of central importance in stabilizing the structure of membranes, as there is a strong tendency for the polar end to associate with proteins and other polar entities (including water) and for the fatty acid end to associate with other lipids. When detergents are used to disrupt membranes, they interfere with the interaction between phospholipids and other molecules. In the simplest case, this can be viewed as resulting from the dipolar nature of many detergents, which act almost as phospholipid analogues and compete for binding at the phospholipid surface.

The simplest phospholipid is *phosphatidic acid*,

$$^-PO_4-CH_2$$
$$CHO-C-(CH_2)_n-CH_3$$
$$\|$$
$$O$$

$$CHO-C-(CH_2)_n-CH_3$$
$$\|$$
$$O$$

which is not common in membranes but which is the basis and the precursor of many other, more common, molecules. A number of phospholipids differ from phosphatidic acid in having alcohols (or amino alcohols) attached to the phosphate by a second ester linkage. Examples include *phosphatidylethanolamine* and *phosphatidylcholine*, which contain, attached to the phosphate, ethanolamine,

$$NH_2CH_2CH_2OH$$

and choline,

$$(CH_3)_3N^+CH_2CH_2OH$$

respectively (see Figure 11-2). In other cases, the molecule esterified to the phosphate is a sugar, such as inositol, a hexose.

The structures of some detergents that disperse membrane structure, owing to their interaction with phospholipids, are illustrated in Figure 11-5, where their dipolar character is evident. A number of these are of great practical importance in the isolation of enzymes and enzyme systems from cellular membranes. In addition, enzymes that break down phospholipids are often used in the same connection. For example, *phospholipase A*, which has been isolated from snake venom, specifically hydrolyzes the

Phospholipase
A & B.

Figure 11-5   Some detergents that disrupt membrane structure. Those with a steroid structure are bile salts, which function as natural detergents in leading to the emulsification of fats in the digestive tract. There is some indication that cholesterol, with its similar structure, serves as a natural intramembrane detergent in promoting fluidity of membranes *in vivo*.

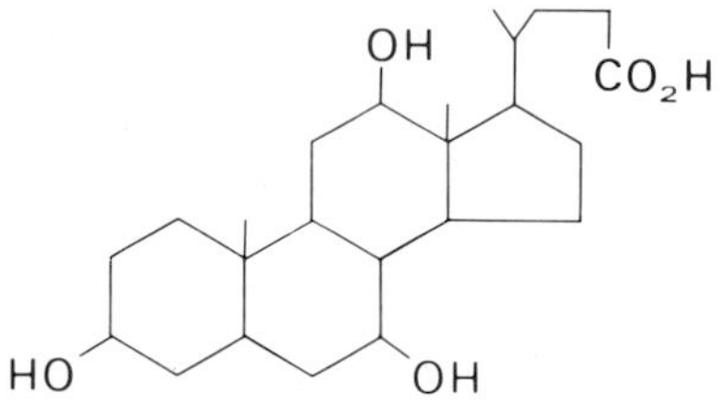

Cholic acid

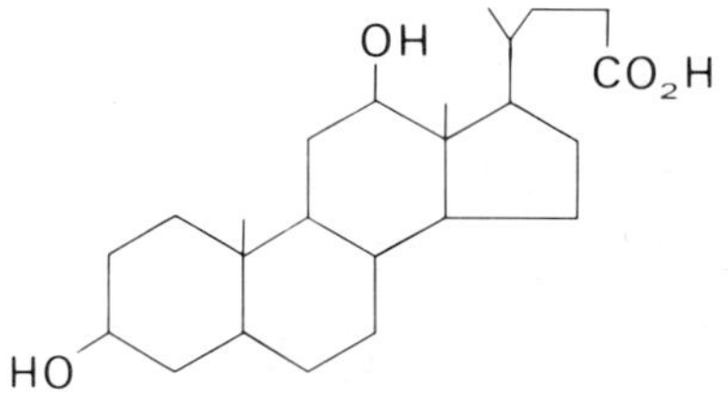

Deoxycholic acid

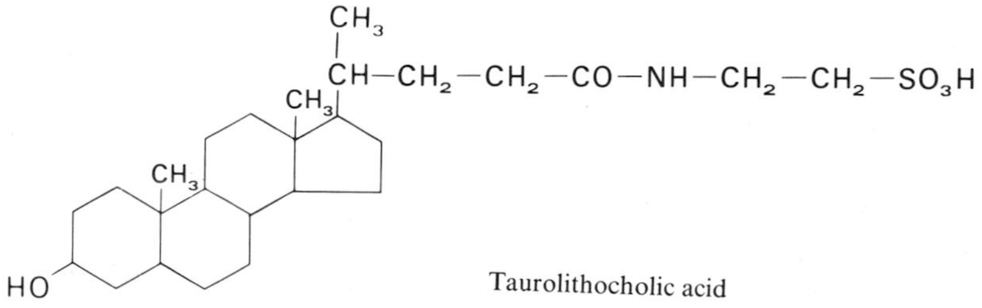

Taurolithocholic acid

$$CH_3(CH_2)_{10}CH_2OSO_3{}^-Na^+$$

Sodium dodecylsulfate

265

fatty acid next to the phosphate group, removing it from the glycerol portion. *Phospholipase B* can effect the removal of both fatty acids.

Studies with artificial membranes

The discovery of artificial systems that duplicate some properties of biological membranes has suggested to some workers that the formation of membranes occurs by way of relatively simple intermolecular interactions. A mixture of purified phospholipids and water forms aggregations (or micelles) that can be of a tubular nature, apparently composed of a double layer of molecules with a separation very roughly that of natural membranes (50–100 Å). Such similarities have suggested that model systems may lead to detailed understanding of natural membranes, but considerable caution should be exercised. For example, the spacing between the layers in micelles is not constant, with great variation associated with the method of preparation and the concentration of dissolved ions. Thus, at least one of the similarities between artificial and biological systems may be due to a fortuitous selection of conditions.

On the other hand, micelles of various compositions are known to exhibit a dynamic shape and size, with the possibility of coalescence to form larger aggregates. In this property they are similar to cell membranes, which also appear able to coalesce and change configuration. There are many instances of fusion of parts of intracellular membranes, of which a good example is the closure of the animal cell membrane after furrowing associated with mitosis. Likewise, it should be recalled (from Chapter 1) that the nuclear envelope is formed from coalescence of endoplasmic reticulum. Indeed, when one views cells through a light microscope, the rapid streaming of cytoplasm and the manner in which organelles are deformed suggests that the cytoplasmic membranes are extremely fluid in nature. The orderly array of cellular membranes in electron micrographs does not reflect the probable dynamic state of the same membranes in the living cell.

When we write of a dynamic state of cellular membranes, we envision the seemingly stable membranes as being continuously formed and broken down, a view that appears to be required by the observation of active cytoplasmic movement in the same eukaryotic cells that possess extensive cytoplasmic membranes (endoplasmic reticulum). Direct evidence for such a dynamic condition comes from the recent demonstration of an isotopic exchange reaction between phospholipids of different cell organelles. Thus, mitochondria with radioactive phospholipid were incubated with unlabeled endoplasmic reticulum, whose phospholipid then became radioactive.

An added indication of the fluidity of membrane structure is the behavior of membranes upon treatment with detergents or mechanical disruption, for example by means of ultrasonic vibrations. Fragmentation of the membrane is followed by the pieces' closing up in the form of vesicles, which can be shown to retain some of their original permeability

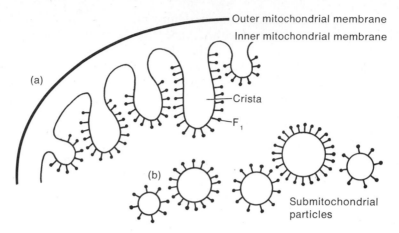

(a)

Outer mitochondrial membrane

Inner mitochondrial membrane

Crista

F_1

(b)

Submitochondrial particles

Figure 11-6   Formation of submitochondrial particles by ultrasonic oscillation. (a) In the intact mitochondria, the configuration of the inner membrane is such that the F_1 (ATPase) knobs point inward. (b) Disruption leads to breaks in the "neck" of cristae. When these breaks reseal, the result is vesicles with the F_1 knobs pointing outward.

Figure 11-7   Electron micrograph of a preparation of a membrane protein, cytochrome b. Note the formation of numerous vesicles. The preparation is negatively stained with phosphotungstic acid; magnification is × 160,000. From S. Yamashita and E. Racker, *J. Biol. Chem.,* **243**, 2446 (1968). (Courtesy of Dr. Efraim Racker.)

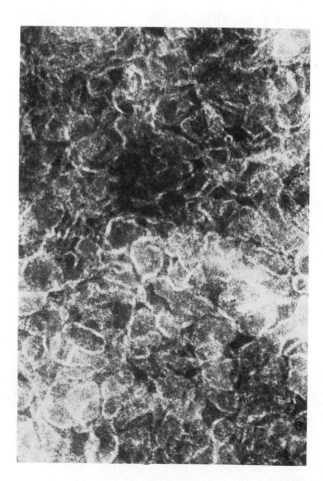

properties. In the case of the mitochondrial inner membrane, the vesicles are called *submitochondrial particles* and appear to be formed from the cristae in such a manner (Figure 11-6) that they are inside out, with the ATPase knobs on the outside instead of within. Similar inverted particles may be obtained from chloroplasts (*subchloroplast particles*). Finally, it is interesting that many purified proteins from membranes, when isolated in the company of phospholipids (which is often unavoidable), form similar closed vesicles. Such structure exhibited in a preparation of mitochondrial cytochrome *b* is shown in Figure 11-7. The tendency of such preparations to form closed vesicles indicates that these structures *stability consideration* are thermodynamically most favorable.

The structure of nerve myelin

Much of our knowledge of membrane properties comes from investigations of relatively few different systems. For example, a great deal of what we know of membrane structure and transport comes from studies with the red-blood-cell membrane with the interior contents gone (ghost). This system is relatively simple to work with, owing to the ease in obtaining very pure suspensions of these cells and the limited metabolic activity that they carry out. However, red cell membranes may not be particularly typical of other cells and, indeed, they do contain a higher percentage of protein than do other membranes studied. Similarly, the study of nerve myelin membranes has been extensive, again, partly because of the relative ease in obtaining material.

The myelin sheath is derived from the plasma membrane of the *Schwann cell*, which exists in close proximity to a nerve axon. The sheath is formed by wrapping many layers of the membrane around the axon, as shown in Figure 11-8. The sheath appears to act as a sort of dielectric insulation, and it is important to note that it is a particularly nonpolar membrane (with a low dielectric constant), owing to the high lipid/protein ratio.

Figure 11-8 Formation of myelin sheath by enfoldment of Schwann cell membranes.

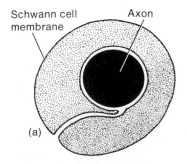
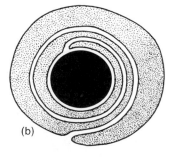

Schwann cell membrane   Axon

(a)   (b)

Suggested Reading

Books

Ansell, G. B., and J. N. Hawthorne, *Phospholipids: Chemistry, Metabolism, and Function*, American Elsevier Publishing Co., New York, 1964.

Chapman, D., *Biological Membranes: Physical Fact and Function*, Academic Press, New York, 1968.

New York Heart Association Symposium, *Membrane Proteins*, Little, Brown and Co., Boston, 1969.

Articles

Drost-Hansen, W., "Role of water structure in various membrane systems," in *Physical Principles of Biological Membranes*, F. Snell, J. Wolken, G. Iverson, and J. Lam (eds.), Gordon and Breach, Science Publishers, New York, 1970.

Hokin, L., and M. Hokin, "The chemistry of cell membrane," *Sci. Am.*, **213**, 78 (Oct., 1965).

Korn, E. D., "Structure of biological membranes," *Science*, **153**, 1491–98 (1966).

Northcote, D. H. (ed.), "Structure and function of membranes," *Brit. Med. Bull.*, **24**, 99–186 (1968).

Rothfield, L., and A. Finkelstein, "Membrane biochemistry," *Ann. Rev. Biochem.*, **37**, 463–96 (1968).

Sjöstrand, F. S., "Molecular structure and function of cellular membranes," in *Regulatory Functions of Biological Membranes*, J. Järnefelt (ed.), Elsevier Publishing Co., Amsterdam, 1968.

Stoeckenius, W., "Electron microscopy of mitochondrial and model membranes," in *Membranes of Mitochondria and Chloroplasts*, E. Racker (ed.), Van Nostrand Reinhold Co., New York, 1970.

Stoeckenius, W., and D. M. Engelman, "Current models for the structure of biological membranes," *J. Cell Biol.*, **42**, 613 (1969).

Vernon, L. P., H. Mollenhauer, and E. R. Shaw, "Structure and function in membranes of photosynthetic systems," in *Regulatory Functions of Biological Membranes*, J. Järnefelt (ed.), Elsevier Publishing Co., Amsterdam, 1968.

12 Transport across membranes

The membranes and membrane systems that divide cells into compartments are of the greatest significance in the regulation of cell activity. Since membranes are often highly exclusive with respect to the chemicals that are able to traverse them, they present many possibilities for regulation by separating potential reactants or by allowing them to come together at closely controlled rates. For example, the plasma membrane is a primary determinant of what compounds contribute to the nutrition of cells: Material that cannot penetrate will not be used. Likewise, mitochondria participate in the energy-yielding oxidation of only those substrates that can penetrate the inner membrane, and the primary product of mitochondrial metabolism, ATP, cannot leave the inner phase of the mitochondrion without the functioning of a specific ATP-carrying system.

The transport properties of membranes are highly specific with regard to type of membrane and phylogenetic origin. The outer mitochondrial membrane is impermeable only to the largest macromolecules and appears to admit all small molecules and ions, while the inner membrane is highly selective. Likewise, the plasma membranes of bacteria (and many other cells) are able to transport such cations as calcium (Ca^{2+}) and potassium (K^+), while those of yeast exclude them completely. *cell wall implication*

In subsequent discussion we shall employ the following definitions. *Permeability* is the property of a membrane allowing passage of a substance across it. The substance must be specified, as it is possible (and likely) that a membrane will pass certain materials and be *impermeable* to others. A substance that passes across a membrane may be called *permeant*. Moreover, to say that a membrane is permeable to a certain molecule does not imply anything about the mechanism by which the passage occurs. Finally, we shall use the terms *transport* and *translocation* in an identical sense, the process of movement of material across the membrane. *Active transport*, as we shall see, is transport that does not take place without expenditure of metabolic energy.

270

Selective permeability: the passage of solutes through membranes

The power of cellular membranes to regulate and specify cell events comes in part from the great selectivity of membranes with respect to different solutes. When such a membrane is interposed between two regions of space, it affects the free diffusion of some molecules to a very slight degree and completely prevents that of others. The manner in which membranes exercise such selectivity is related to the mechanism by which they transport solutes, and one might expect a membrane to exhibit one sort of selectivity if translocation was (for example) by pores and another if it occurred by means of solution of the solute in the lipid phase of the membrane. In fact, membranes act as if several different mechanisms were in effect, including the two mentioned.

In general, large molecules encounter greater resistance in passing through membranes than do smaller ones. Similarly, molecules with a great dipolar character tend to pass through membranes with difficulty when compared to those with less. This is probably due to the formation of a shell of water molecules about the solute molecule, which would result in increasing its effective cross section. In a similar fashion, an asymmetric molecule presents on the average a greater cross section to a membrane and is less able to pass through it than a more symmetrical one of similar molecular weight.

As important as size in determining the permeability of a biological membrane to a given solute is the solubility of the solute in lipid. Lipid forms an important part of cellular membranes, and it is reasonable that a compound soluble in lipid should have an easier road through such membranes. For example, a common observation concerning the study of certain inhibitors of mitochondrial processes is that compounds with long aliphatic side chains inhibit at lower concentrations than corresponding compounds with shorter (or no) side chains. Since the longer side chains confer greater lipid solubility, such compounds are better able to traverse the mitochondrial membrane and produce their effect. The influence of lipid solubility is closely related to the degree of ionization, which generally bears an inverse relationship to it.

In addition to certain general relationships between transport specificity and molecular size, shape, and lipid solubility, there are many examples of more detailed specificity, where a certain molecular structure or the presence of certain groups plays a major role. This sort of discrimination by membranes is strongly reminiscent of enzyme specificity and suggests that membranes may bear sites related to the transport of specific molecules with perhaps a geometric relationship to the structure of the molecules. As an example of such selectivity, one might cite the transport of amino acids, which appear to require such sites. However, there is not a single mechanism for all amino acids, nor is there one for each different amino acid. Rather, there appear to be separate mechanisms (and sites) for the translocation of neutral amino acids (such as glycine), cationic (such as lysine), and anionic amino acids (such as glutamate). Evidence for the three groups comes from the fact that within each group amino acids may be shown to compete for a single site (since the addition of one will suppress the uptake of another), while such inhibition, which is similar to competitive enzyme inhibition, does not occur between members of different groups.

membrane specificity vs. enzyme specificity

    Although the apparent existence of specific sites points to similarities between membrane transport and enzyme activity, it is important to note that membrane specificity appears in many cases to be somewhat less restrictive. In other words, the number of compounds that react at a certain site tends to be larger in the case of membranes, enabling a single site to mediate the entry of several compounds.

The passage of water through cell membranes (osmosis)

Biological membranes are generally quite permeable to water. For example, when red blood cells are placed in solutions of low solute content (hypotonic), they take up water from the medium with consequent swelling. Similarly, cell organelles, such as chloroplasts or mitochondria, swell or contract depending on the solute content of the medium in which they find themselves, indicating that such organelles with closed outer surfaces also allow water to pass readily through their membranes. As a consequence of this property, techniques for the isolation of cells and cell organelles must include the careful selection of a medium with precisely the correct solute concentration to avoid osmotic disruption of membrane structure.

    The passage of water (osmosis) across a membrane into a region of higher solute concentration must be viewed in terms of chemical equilibrium. Consider two compartments separated by a membrane, as shown in Figure 12-1. The membrane, as we said, is highly permeable to water but not to a solute, S. All the S is on the left side only and the membrane forms a barrier to its movement to the right. Water, however, is not so constrained and is free to move across the membrane in either direction. Now, the net movement of material in any situation where equilibrium has not been attained is in the direction from a locally higher concentration toward a lower one. Thus, if the membrane were permeable to S so that it could diffuse freely, S would move toward the right until concentrations became equal on both sides. But S cannot move; only water is able to move across the membrane, and it will also move in the direction of the concentration gradient, from a higher to a lower concentration. Water is at the highest concentration when it is the only component in the compartment (the right side), and it will flow toward the left, where *← water concentration* it is at a lower effective concentration. In principle, the flow would continue until the concentration of water in the two sides became equal— that is, until the concentration of S approached zero. This could occur only when an infinite amount of water had moved toward the left.

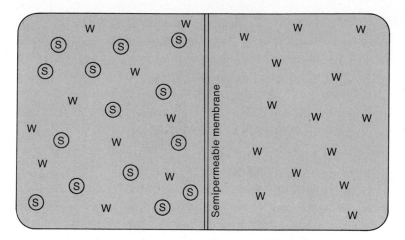

Figure 12-1 Basis for osmosis. On the right-hand side, all molecules are water, which is, therefore, at a higher concentration than on the left side. Water crosses the membrane in the direction of the concentration gradient, according to the normal laws of diffusion. Component S cannot penetrate the membrane, so that only water is able to flow.

Since we are interested in cells, it is clear that the movement of an infinite mass of water across a membrane does not constitute a realistic situation. In fact, a cell possesses a property not identical with the illustration in Figure 12-1—a finite volume. Since its volume is not unlimited, the movement of water toward the interior of a cell would give rise to a pressure, just as pumping fluid into a closed tank would give rise to a pressure exerted against its sides. As water enters the cell (or fluid, the tank) the pressure grows until it is sufficient to oppose the force doing the pumping, and the net inward movement of fluid ceases. At this moment, an equilibrium obtains where the driving force of water entering the cell (a functions of the water concentration gradient) is exactly opposed by

Figure 12-2 Osmotic uptake of water by a cell. If the cell contains a higher concentration of osmotically active molecules than the external medium, an osmotic pressure (from the inside of the cell outward) will be generated. This will be opposed by corresponding pressures from outside the cell, which serve to limit cell enlargement. The latter include the limited elasticity of the plasma membrane as well as the presence of a rigid cell wall in plants and prokaryotes.

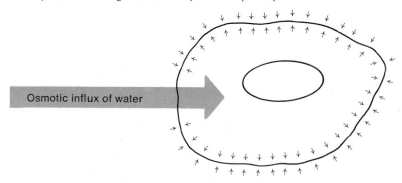

the pressure that it has produced. At a low solute concentration, this *osmotic pressure* is given by

$$\Pi = mRT \tag{12-1}$$

where m is the solute concentration in moles per liter, R is the gas constant, T is the absolute temperature, and Π is given in atmospheres.

A cell plasma membrane encloses a finite volume, but this does not mean that the volume is necessarily constant. Indeed, cells often can be shown to swell or shrink depending upon their osmotic condition, that is, the relative concentrations of solute on the inside and outside. For example, a red blood cell placed in a solution of sucrose, to which its membrane is impermeable, will exhibit an altered volume depending upon the sucrose concentration. If the concentration is changed from 0.3 to 0.15 mole/liter, the cell will undergo an approximate doubling of volume. If the concentration is reduced still further, the cell will continue to swell until the elasticity of the membrane reaches a limit, at which point the cell will burst.

On the other hand, we saw in Chapter 1 that many cells, including bacteria and many plants, are contained in a rigid case of constant volume, so that they are unable to swell to the bursting point. In these instances, high external concentrations of an impermeant solute still lead to shrinking of the cell volume, so that the plasma membrane is pulled away from the cell wall. Increase in solution concentration produces an osmotic pressure, which causes the plasma membrane to become closely applied to the inside of the cell wall. Further increase in concentration only increases the pressure against the cell wall and renders the cell increasingly rigid.

There are two applications of osmotic pressure that are of importance in the study of the cell. In the first place, the relation between osmotic pressure and concentration renders it possible to use Π as a measure of molecular weight, a technique that has frequently been applied to macromolecules, such as proteins, using an osmometer with an artificial membrane. In Equation 12-1, the concentration, m, of solute was given in moles per liter. This equation may be written in terms of grams per liter, c, by dividing by the molecular weight of the solute, MW:

$$\Pi = \frac{cRT}{MW}$$

and

$$MW = \frac{cRT}{\Pi} \tag{12-2}$$

This relation is valid only for low concentrations of solute, so that, in practice, one carries out the measurement at several low concentrations and then extrapolates to $c = 0$.

A second application of osmotic pressure to cell study is the use of cell (or organelle) volume as an indicator of changes in solute concentration. For instance, the active translocation of anions into isolated mitochondria or chloroplasts is accompanied by an osmotic flow of water,

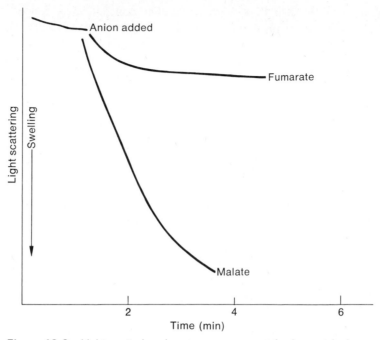

Figure 12-3 Light-scattering changes as a measure of anion uptake by mitochondria. A suspension of mitochondria is incubated with phosphate buffer and a salt of either malate or fumarate anion. In the case of malate, the large change in light scattering indicates that malate is being taken up. Clearly, fumarate does not enter the mitochondria at a comparable rate under the conditions employed in the experiment.

for reasons given above. This flow increases the volume of the organelle, which may be measured by monitoring light scattering produced by the suspension of organelles. In these experiments, an anion giving rise to change in light scattering was judged to penetrate the organelle membrane. A similar approach has been employed in studies with cultured cells; it is interesting that it is quite inapplicable to yeast cells, which appear to maintain a fixed volume.

The driving force for transport

The motion of molecules or ions from one location to another must be regarded as a matter of probability. If the probability that a given molecule will go from A to B is greater than the probability that it will go from B to A, then there will be net movement from A to B, in the more probable direction. When a membrane divides two compartments (Figure 12-1), of which only one contains S, then the probability that S will move from the S-containing region is some finite number, P $(0 < P < 1)$, while the probability that it will go from the empty compartment is obviously 0. Likewise, if a membrane separates two regions of finite, but different

275

concentrations of S, then flow from the higher concentration toward the lower will be more probable, owing to the greater availability of molecules to flow. Another way of saying this is that a concentration gradient can serve as a driving force for diffusional flow of material from one place to another. We saw above that such a gradient produces the osmotic flow of water and that it may be thought of as a force, since, at equilibrium, it may be vectorially opposed by the force exerted by osmotic pressure.

In general, a flow, $\mathbf{J}$, may be expressed as the product of a driving force, $\mathbf{X}$, and a proportionality factor, L. More precisely, $\mathbf{J}$ is often the result of several driving forces and should be expressed as their sum:

flux

$$\mathbf{J} = L_1\mathbf{X}_1 + L_2\mathbf{X}_2 + L_3\mathbf{X}_3 + \cdots \tag{12-3}$$

In the case of simple diffusion, $\mathbf{J}$ represents the transport of material through a unit area orthogonal to the direction of flow. In the simplest case (where there are no gradients of temperature or pressure), the driving force is simply the gradient in concentration dc/dx, where c is the concentration and x the distance in the direction of flow. The proportionality constant is expressed as D, the diffusion coefficient, yielding an expression

$$\mathbf{J} = D\frac{dc}{dx} \tag{12-4}$$

which is valid for diffusion in the steady state (see Chapter 3). Steady state in this case means that the gradient is unchanging with respect to time, a condition that is probably often realized in cells where the gradient may be maintained by metabolic processes or by having a very large external volume. In the more general case of non-steady-state existence of diffusion, the rate is proportional to the second derivative of concentration:

$$\mathbf{J} = D\frac{d^2c}{dx^2} \tag{12-5}$$

These relations remain valid when diffusion takes place in regions of space between which a membrane is interposed. Although the membrane might be said to impose a resistance to flow, this would be expressed by an altered diffusion coefficient, and the rate would still be proportional to the gradient found in any given region. Unfortunately, the precise measurement of a gradient within a membrane is not readily accessible and only an average gradient may be estimated. Similarly, the relevant gradient might be that of a complex between the permeant molecule and a carrier, and this is not easily determined. Such cases, where the driving force is that of diffusion but the actual transport is mediated by a chemical mechanism in the membrane, are called *facilitated diffusion*, and appear often.

facilitated diffusion

Finally, an important property of a concentration gradient serving as a driving force for transport is that transport tends to eliminate the gradient that drives it. As molecules flow down the gradient, the concentration difference is diminished until, at equilibrium, the difference vanishes—that is, gradient is completely collapsed.

Membrane potentials as driving forces

It is often possible, using suitable techniques, to measure an electrical potential across biological membranes. Such transmembrane potentials are closely related to cell excitability when they occur across the plasma membrane (see Chapter 13). The origin of membrane potential will be discussed more extensively below; for the present discussion, it is only necessary to note that it can arise from unequal distribution of ions on either side of a membrane as well as the flow of ions (or electrons) in a directed fashion within the membrane itself.

A membrane potential can serve as a driving force for transport in the case where the material being transported bears an electrical charge, in other words, is an ion. For example, if a membrane (shown in Figure 12-4) bears a potential difference that is positive on the right side, the motion of an anion (negative) from left to right will be favored by the potential. This suggests that the presence of a potential will be reflected in the distribution of permeant ions on either side of the membrane and, in fact, the magnitude of the potential can be measured by measuring such ionic distribution. Thus, if one considers the potential of a closed membrane system, such as a mitochondrion, to be $\Delta\chi$, then it is related to the distribution of a permeant ion, A^+, by the expression

$$\Delta\chi = \frac{RT}{nF} \log_e \frac{(A^+)_{inside}}{(A^+)_{outside}} \tag{12-6}$$

where R is the gas constant, T the absolute temperature, n the number of charges per ion, F the Faraday constant (96,500 coulombs per gram equivalent), and the (A^+)'s the concentrations of ion on either side of the membrane. This is the *Nernst equation* and has been frequently used to give a measure of membrane potentials across cell and organelle membranes. For example, measurement of the distribution of several

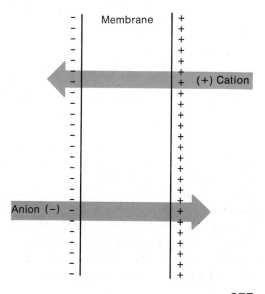

Figure 12-4 Potential across a membrane causing a change in the distribution of ions. Cations will flow toward the relatively negative direction, anions toward the positive one.

uni- and divalent anions across the membranes of isolated mitochondria has yielded values for $\Delta\chi$ of about 30 millivolts, a figure apparently confirmed by recent measurements, where microelectrodes were used to measure the potential directly.

The force acting to drive an ion across a membrane may be thought of as consisting of an electrical component (the membrane potential) and a concentration gradient component. For example, the force acting on a proton (hydrogen ion), X_{H^+}, can be given by the expression

$$X_{H^+} = \Delta\chi + RT \Delta \log_e(H^+) \tag{12-7}$$

where $\Delta\chi$ is the potential difference, R the gas constant, T the absolute temperature, and $\Delta \log_e(H^+)$ the difference between natural logs of the hydrogen ion concentrations on the two sides of the membrane. Utilizing the definition of pH as $-\log_{10}(H^+)$, this may be rewritten:

$$X_{H^+} = \Delta\chi - 2.303 RT \Delta pH \tag{12-8}$$

It is important to add that the forces associated with membrane potential and concentration gradient can be either additive or opposed. For example, it is possible for the movement of an ion in the direction of a favorable concentration gradient to be retarded (or stopped) by a sufficient opposed membrane potential. Movement of a single ion across a membrane must give rise to a movement of charge and, hence, a potential. This potential will, by necessity, be opposed to further movement of the ion, and such movement will be quickly prevented. For this reason, movement of single ions across biological membranes does not occur on a large scale, and when measurable uptake of an ion occurs, almost invariably a second ion is seen to move also in such a way as to maintain electrical neutrality across the membrane. For instance, the frequently massive transport of calcium (Ca^{2+}) across membranes involves the simultaneous transport of an anion, such as phosphate, in the same direction, or the counter-movement of a different cation (such as H^+ or K^+). In either case, except for transient potential differences, electrical neutrality is maintained.

The translocation of solutes against a concentration gradient: active transport

Many membranes are known to translocate molecules or ions from regions at a low concentration to those at a higher concentration. Cells and closed organelles are able to increase the internal concentration of solutes until very high levels are reached and very considerable gradients are established. Cultured mammalian cells are able to concentrate a number of solutes, including amino acids, until the total gradient across the membrane is of the order of 0.1 M. Mitochondria are able to take up calcium and phosphate until the mitochondrial structure is actually deformed by the precipitated calcium phosphate inside, which can equal about 20 percent of the mitochondrial mass. In these cases (and numerous others) transport is opposed to considerable concentration gradients, and clearly a process other than pure diffusion must be available to supply energy. Such a translocation requires metabolic sources of energy and is usually termed *active transport*.

Before we describe active transport, it is necessary to point out that there are many situations in which a "passive" diffusion-linked process might appear active, and one must be extremely careful to exclude them. For example, binding sites within a closed membrane might lead to an apparent concentration of solute by removing it from solution as soon as it passed into the interior (see Figure 12-5). Thus, as long as binding sites were available, the transport would take place against what would look to the external observer like a concentration gradient.

It is also possible to find situations in which a transport process operating against a gradient is coupled to one in the opposing direction that is favored by diffusion. Thus, it has been suggested that the transport of some organic molecules, including sugars and amino acids, might be coupled in some manner to cation transport. The evidence for this includes a significant influence of cation distribution on transport of the molecule in question.

Although transport may occur across cellular membranes against an apparent concentration gradient when the transported species is either bound on one side of the membrane or removed by metabolic reactions, true active transport exhibits two properties that, together, constitute its definition:

1. Active transport must be able to run against a true concentration gradient; that is, the gradient may include only free, unbounded molecules (or ions).
2. Energy for active transport comes from spontaneous (energy-yielding) reactions of the cell. Most often, the energy is available from the hydrolysis of ATP, which is produced by fermentative, oxidative, or photosynthetic processes.

Active transport is undoubtedly a feature of plasma membranes of all living cells and also occurs in the membranes of mitochondria, chloroplasts, and endoplasmic reticulum. Active transport is most dramatic in tissues that are specially modified to take up (or expel) solutes against concentra-

Figure 12-5 Apparently active transport due to internal binding. Binding lowers the internal concentration of free X and so promotes, by mass action, its entry.

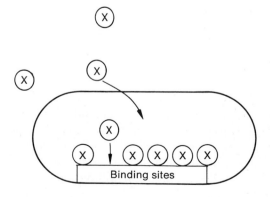

tion gradients, and this includes, as we shall see, nerve and muscle tissue as well as such secretory (or excretory) tissue as the animal kidney, the gills of marine animals, and the avian salt gland.

The source of energy for active transport

Energy for active membrane translocation comes from energy-yielding metabolic reactions of the cell. For this reason, incubation of cells at low temperature, which slows metabolic reactions, may lead to cessation of active transport and, indeed, leaking out of actively concentrated substances. For example, mammalian red blood cells contain a higher concentration of K^+ than is found externally in the blood plasma. When the cells are cooled to 2°C, K^+ is released into the exterior medium, and rewarming to body temperature (37°C) leads to its uptake again.

More specifically, energy able to drive active transport originates in those processes in the chemistry of the cell that are associated with ATP production. Thus, it is not surprising that active transport by aerobic cells is greatly inhibited by dinitrophenol, which was seen earlier to be an uncoupler of ATP synthesis, and by cyanide, which inhibits respiration itself. Similarly, in cells that are unable to trap energy by aerobic processes and that rely on glycolysis, transport against a gradient is inhibited by reagents that interfere with glycolytic ATP production. Finally, transport by photosynthetic cells (and their organelles) is linked to light absorption and the related energy-transferring processes, exactly as one might predict. These observations suggest that inhibition of a transport process by, for example, dinitrophenol, yields a valid test of whether transport is active and implies that ATP, the synthesis of which is prevented by dinitrophenol, is the source of the energy. Furthermore, the fact that dinitrophenol interferes with both oxidative ATP formation and with certain transport events does not necessarily imply that ATP is itself the energy carrier. This is clear when one recalls that the synthesis of ATP appears to occur by way of high-energy intermediates, which are themselves formed as a result of respiratory electron transfer:

$$\text{Respiration} \rightarrow \begin{array}{c} \text{High-energy} \\ \text{intermediates} \end{array} \xrightarrow{\qquad} \text{ATP}$$
$$\text{ADP} + \text{PO}_4{}^{3-}$$

(The arrows, in this instance, denote energy transfer.) Since dinitrophenol is usually thought of as either preventing the formation of the intermediates or leading to their destruction, it is possible that the intermediates are the sources of energy for transport and that there is no absolute requirement for ATP. Note that the role of ATP might even be to lead to the formation of the intermediates by reversal of the last set of reactions. Evidence for transport that does not require ATP per se comes from studies of mitochondrial ion uptake, which requires energy from respiration to cross steep gradients. The inhibitor, oligomycin, which acts between the intermediates and ATP formation, is without effect on ion uptake when the energy is obtained from respiration. Under similar conditions, dinitrophenol eliminates transport completely. On the other hand, uptake may occur without respiration (as in the presence of cyanide), as long as ATP is supplied in the reaction mixture. This ATP-dependent ion transport is, however, eliminated by oligomycin, which suggests that the intermediates, and not ATP itself, are the direct sources of transport power. These results are summarized in Figure 12-6 and indicate that one might expect competition for the intermediates between ATP synthesis and transport, and this is found to be the case. Thus, the addition of Ca^{2+} to a mitochondrial suspension inhibits ATP synthesis markedly, indicating that Ca^{2+} is able to compete quite well for the intermediates. Note that high-energy intermediates might be expected to play an especially significant role in mitochondrial transport, owing to the fact that they are presumably formed in the mitochondrion (see Chapter 6). This is true, and Figure 12-7 shows the effect of added calcium on mitochondrial respiration in the absence of ADP. Without ADP, mitochondria cannot make ATP, and the intermediates build up, a condition that was seen previously to lead to a less than maximal rate of respiration. It should be added that maximal stimulation of respiration upon the addition of a cation such as Ca^{2+} is observed when a permeant anion is included in the reaction mixture. When an anion such as phosphate or acetate is present, it is taken up in company with the Ca^{2+} and aids in preventing buildup of a membrane potential that would oppose further Ca^{2+} transport. It will be discovered below that the mechanism for cation uptake by mitochondria includes an additional means of maintaining electrical neutrality across the membrane. Thus, cation uptake is associated with the outward flux of protons (H^+ or,

Figure 12-6 Energy pathways for mitochondrial ion transport. Transport depends on the presence of "high-energy intermediates" which may be formed as a result of respiratory electron transport or generated from added ATP. It is possible to examine the two sources of "intermediates" separately, as cyanide prevents their formation by one route and the antibiotic oligomycin by the other. When energy is supplied by electron transport, the reactions of ATP synthesis and those leading to ion uptake compete for the "high-energy intermediates" formed. In this figure, the term "high-energy intermediate" is used without prejudice as to the exact nature of the energy-charged state of mitochondria, which, as we have seen in Chapter 6 and shall discuss in this chapter, may reflect either an actual high-energy compound or an alteration in the molecular or charge configuration of the mitochondrial membrane.

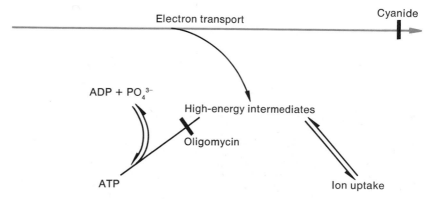

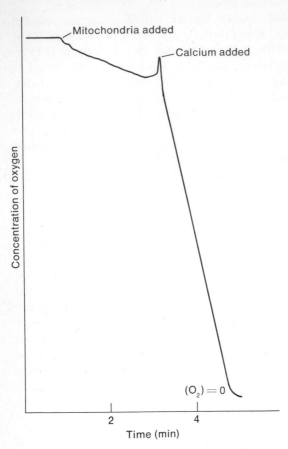

Figure 12-7 Stimulation of mitochondrial respiration by calcium. This is a recording produced by a polarographic electrode sensitive to oxygen (see Chapter 6, especially Figure 6-8). The reaction mixture included phosphate buffer, magnesium chloride, succinate, and water. When calcium (as $CaCl_2$) was added, respiration was greatly increased. Independent measurements employing a radioactive isotope, calcium 45, show that calcium is concentrated within the mitochondrion.

under some conditions, K^+), which assists in equilibrating the charge on either side of the membrane.

It is interesting to note that certain estuarine organisms that are able to carry on transport of ions in response to changes in salinity do so by means of a process that is accompanied by an increase in the respiratory rate of the entire organism. This suggests not only that such transport requires energy (which is self-evident) but that ion-induced respiratory stimulation in mitochondria may reflect a similar process in the intact cell.

It will be seen later that the nature of the high-energy intermediates that we have found to be implicated in mitochondrial transport is, to say the very least, unclear. Not only is the precise chemical nature of such "compounds" still undiscovered, but it has been suggested that the intermediates are not chemical entities at all but rather gradients of electrical charge across the inner mitochondrial membrane.

Transport mechanisms

It should be clear that the remarks made above about membrane specificity all bear a relation to the nature of mechanisms that might be involved. When one discusses the influence of molecular size and shape on transport, it is rather too easy to think of the membrane as bearing pores through which the molecules pass. Similarly, when one speaks of the importance of lipid solubility, the sort of mechanism that comes to mind includes the dissolution of the molecule into the lipid phase of the membrane and subsequent release on the other side. Finally, the discovery of precise chemical specificity in many cases has led to the idea of membrane transport sites analogous to the active site of an enzyme. One is thus free to choose among approaches to transport that seem to bear very little similarity and that, on first examination, appear somewhat mutually exclusive.

Fortunately, one does not have to choose, since real transport is probably a combination of a number of mechanisms. The apparent mutual exclusiveness of the mechanisms is diminished when one notes that membrane transport itself may be divided into a number of phases, including the passage of the molecule into the membrane, the traversing of the membrane, and the exit into the cell or organelle interior. It is quite possible that the different phases of transport may each contribute to the specificity of the process as a whole and that one or another of them may play a greater role depending on the nature of the solute molecule involved.

Several additional transport mechanisms have been suggested to account for specific properties of membranes. For example, studies of disaccharide transport by bacteria have led to the idea that transport may be effected by an enzyme termed a *permease* (see Chapter 12). The permease is believed to be a protein because its synthesis is prevented by known inhibitors of protein production, such as chloramphenicol. Furthermore, it is an induced enzyme—it is synthesized in response to the presence of its substrate. It is likely that a bacterial permease is one instance of a more general phenomenon, transport mediated by a protein, and it is probably unfortunate that the term permease has been adopted, since it tends to set it apart and imply a distinction that may not exist.

The precise role of a protein in a transport event is not yet clear, but there appear to be a number of possibilities. It may be that conformational changes in a protein molecule are involved in the actual translocation. One might imagine a protein molecule situated in a cellular membrane and oriented in such a way that a conformational change (alteration in tertiary structure) would reorient the active site so that a small molecule bound to it would be brought close to the "inner" side of the membrane (see Figure 12-8). Similar mechanisms have been suggested to account for active transport coupled to the hydrolysis of bound ATP, whereby the binding of ATP would lead to such structural changes as to effect net transport of a different molecule.

An additional manner whereby a protein situated in a membrane could lead to translocation of a chemical group is illustrated in Figure 12-9. This mechanism, which was suggested by Mitchell and Moyle (see Suggested Reading), requires an enzyme situated in a membrane with steric restrictions as to which reactants may approach it from each side. In the hypothetical example, the environment in the membrane around the enzyme would be such as to permit X—A to approach it from the

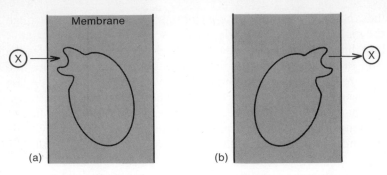

Figure 12-8 How a conformational change of a transport protein can give rise to transport across a cellular membrane. The actual alteration in conformation (shown in a highly diagrammatic fashion) might be induced by the binding of X to the active site of the protein or it might be the result of some other event, such as ATP hydrolysis or binding of some other molecule to the protein.

left and Y—A from the right. The sequence of transport might include X—A donating the A group to the enzyme, followed by the diffusion of X to the left (the only direction in which it is free to move). Y would then approach the active site and receive the A group. Y—A, being free to move only toward the right, would leave the site, giving rise to a net translocation of A from left to right. This proposal has been very influential in pointing out that the only requirement for translocation by an enzyme is the sort of anisotropy suggested, and has led to some very specific suggestions as to the relationship between energy production and transport. It is important to realize that, although the proposal seeks to describe group translocation, it may be of great generality, since the

Figure 12-9 Possible mechanism for group transport. An enzyme is shown to be located at a "pore" in the membrane. The membrane is so structured that the enzyme is accessible to X on the left and Y on the right. X is able to donate group A to the enzyme only from the left, while Y can pick it up only on the right, leading to a net translocation of A from left to right.

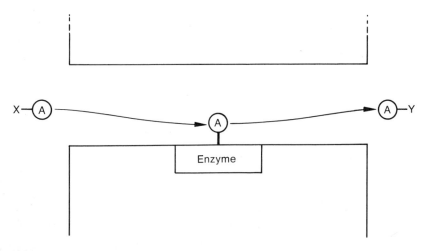

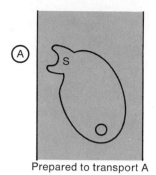
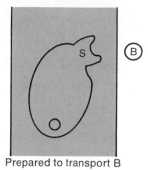

Prepared to transport A   Prepared to transport B

Figure 12-10   Simple model for exchange transport. Site S is able to transport A when it is oriented toward the left, B when toward the right. Transport is imagined to take place through a conformational change in the carrier protein, so that transport of either species will leave the carrier positioned at the side of the membrane toward which transport has occurred. When B is absent, and A is being transported, the carrier returns to the left (so that it can pick up another molecule of A) only by some relatively slow random process. When B is present, the conformational change involved in B transport will rapidly return the carrier toward the left and therefore stimulate A transport. The presence of A will stimulate B transport in a like manner.

group A could represent a whole molecule, such as a sugar or amino acid, bound to the carrier molecules, X and Y.

Additional examples of transport mechanisms include those involving the coupled transport of more than one solute, either in the same or in opposite directions. Uptake of divalent cations (such as Ca^{2+} or Sr^{2+}) by mitochondria appears in many cases to be linked with the uptake of certain anions (such as phosphate or acetate). We saw that such linked translocation provides for maintaining a balance of charge across the membrane, but, in fact, similar linkage can be demonstrated where one (or both) of the solutes translocated is uncharged. Thus, the linkage appears to reflect a basic feature of the transport mechanism; a simple example of such a proposed mechanism is illustrated in Figure 12-10. In it, the transport of A to the right is coupled to that of a similar molecule, B, to the left. We imagine the transport to be mediated by a protein bearing an active site free to move, perhaps because of conformational changes in the protein. When A is translocated, it first binds to the site, S, and is carried on it to the right, where it is released. The site then eventually returns to the left side and picks up another A molecule. On the other hand, if there is a high concentration of B on the right side, and if B has a similar ability to bind to the site, a net flow of B to the left will tend to promote the orientation of the site in the left direction. Thus it will tend to stimulate A transport by ensuring that the site will be, on the average, more able to receive A. To the external observer, the two fluxes will support each other and appear coupled.

Finally, examples of linked transport of the sort described fall into three categories, which are summarized in Figure 12-11. It will be noted that, in the cases of both *symport* and *antiport* translocation, the flows

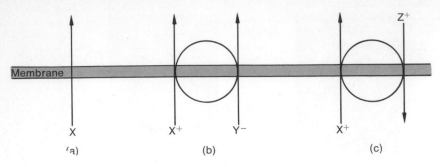

Figure 12-11 Classes of linked transport across membranes: (a) uniport, (b) symport, and (c) antiport. See text for details and examples.

are vectorially directed so that charge distribution across the membrane is not altered. An example of *antiport* translocation has been given above in the case of exchange between Ca^{2+} and H^+, and an additional instance, the Na^+–K^+ ATPase of red blood cells, is discussed below. An interesting example of a *symport* relationship is found in the uptake of substrates by mitochondria. Krebs cycle intermediates, such as citrate, α-ketoglutarate, or malate, which are oxidized by mitochondria, are anions at neutral pH and are taken up in the presence of univalent cations (such as K^+). Incubation of isolated mitochondria in K^+-free medium results in a lower rate of substrate uptake and, hence, in diminished oxidation.

The preceding examples represent only a few of the mechanisms that have been proposed to account for transport phenomena. They are included here to give an idea of the type of thinking that is brought to bear on the subject and are not intended to be inclusive. For a more complete discussion, see the book by Christensen listed in Suggested Reading. If any conclusion is possible about transport mechanisms, it would be that there is a rich selection of proposals available to explain a given transport event, and that studies of specificity and competition are apt to be useful in choosing among them.

Transport-linked ATPase

Although high-energy intermediates of ATP synthesis are involved as energy sources for transport, at least in some instances, we have seen that ATP may be used to generate them. Thus it should not come as a surprise that the addition of an actively translocated solute to a suspension of cells or suitable organelles in the presence of ATP should lead to the hydrolysis of some of that ATP. This hydrolysis might well be termed a sort of ATPase activity, with the special property that it is stimulated by the molecular (or ionic) species translocated. For example, sodium and potassium give rise to ATPase activity in intact red blood cells. The ATP hydrolysis appears to be closely linked to the transport of the ions, since both the ATPase and the transport are influenced to a similar degree by the transport inhibitor, ouabain, which is a steroid. In this instance, the

ATP used in transport is a product of metabolism and we have seen that cooling red blood cells to 2°C prevented transport by eliminating the energy supply. Most ATP synthesis in red blood cells appears to come from substrate-level phosphorylation in glycolysis and the glycolytic inhibitor, iodoacetate,

$$I-\overset{\displaystyle H}{\underset{\displaystyle H}{\overset{|}{\underset{|}{C}}}}-COO^-$$

inhibits the transport process.

Red-blood-cell ATPase is an example of an antiport, since uptake of K^+ and extrusion of Na^+ are linked. Hydrolysis of one molecule of ATP yields inward transport of two K^+ and outward movement of three Na^+ ions, and it has been shown that the enzyme, located in the plasma membrane, is accessible to ATP on the inside only.

A third cation is involved in the reaction: Mg^{2+} is required for the reaction, measured either as transport or ATP hydrolysis, and the addition of a chelator of divalent cations, EDTA (ethylenediamine tetraacetate), inhibits.

The protein responsible for Na^+–K^+ ATPase activity has been isolated from red-blood-cell membranes by treatment with a mild detergent and has been partially purified. It has a molecular weight of approximately 6.7×10^5 daltons. Similar ATPases occur in nerve tissue, in the electric organ from the electric eel, in secretory tissue, and, indeed, are probably general constituents of cellular plasma membranes. Other proteins have been recently isolated from plasma membranes of a variety of cells which appear to contribute to the transport of lactose, leucine, sulfate, calcium, and phosphate. The calcium-binding protein of animal cells is interesting in that its formation is induced by vitamin D, which is known to stimulate calcium uptake in the intact organism.

Ion-conducting antibiotics

A number of antibiotics that have been long known to interfere with cellular energy coupling are able to conduct cations across cellular membranes. These compounds are often cyclic in structure (forming a closed ring) and are made up, at least in part, of amino acids. In contrast to cell proteins, the amino acids comprising these antibiotics include those of the D (or unnatural) configuration, and the cyclic structures also include other (nonamino) organic acids.

These antibiotics were first studied extensively because of their ability to uncouple oxidative or photosynthetic phosphorylation. It has been recently concluded that the uncoupling reflects the ability of these compounds to permit rapid uptake of univalent cations across organelle membranes, producing stimulation of respiration and elimination of ATP synthesis in a manner analogous to that by calcium (see Figure 12-7).

Normally, mitochondrial membranes are quite impermeable to univalent cations and the antibiotics act by forming a complex with the ion so that it can penetrate. The cyclic character of these compounds enables them to accept the ion in their core, which is hydrophylic, and surround it with an outer shell, which is hydrophobic and permits its rapid passage through the membrane.

The structure of one such antibiotic, *valinomycin*, is given in Figure 12-12, and the reader should note its circular nature and the availability of both relatively polar (carbonyl) and nonpolar (alkyl) groups. Presumably, the polar groups become closely associated with the transported ion.

Ion-conducting antibiotics have been shown to transport cations across mitochondrial, chloroplast, and plasma membranes and to affect the conductivity of artificial membranes as well. The last observation indicates that the action of these compounds is simply to make univalent cations more freely soluble in membranes and does not imply anything about the transport mechanisms in specific biological membranes.

Figure 12-12   Structures of two ion-transporting antibiotics.

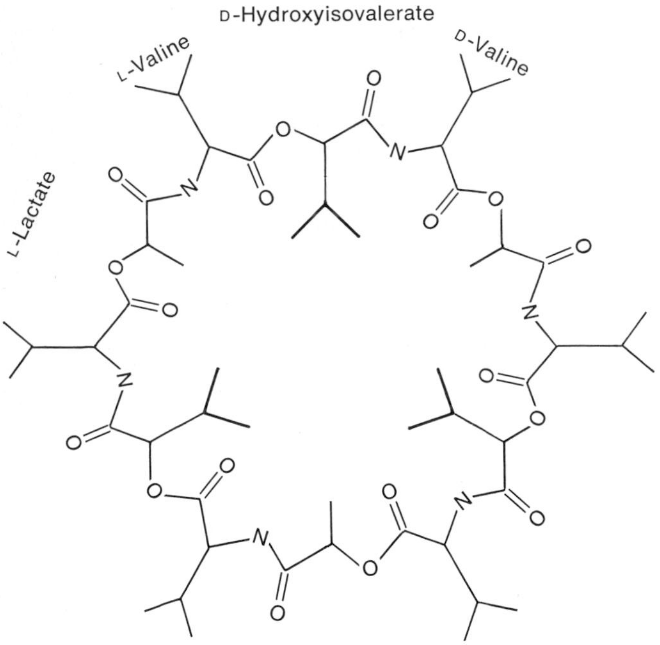

Valinomycin

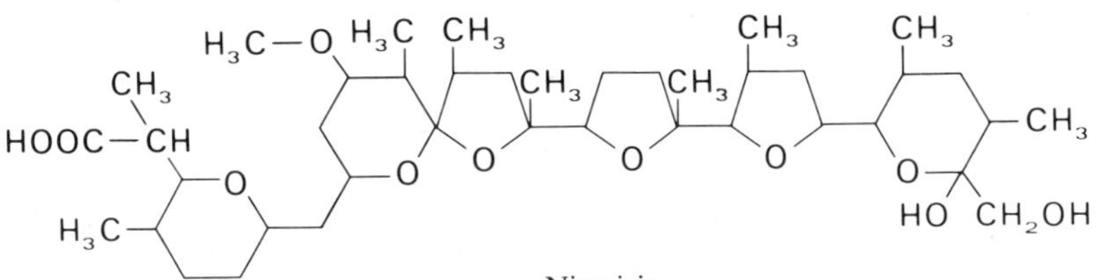

Nigericin

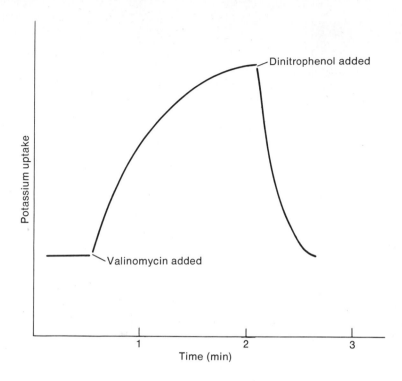

Figure 12-13 Valinomycin-induced uptake of potassium by rat liver mitochondria. Mitochondrial suspension is incubated in a medium containing phosphate buffer, succinate (as energy source), and potassium chloride. Addition of valinomycin leads to rapid uptake (as measured by removal of potassium from the medium). Subsequent addition of the uncoupling reagent, dinitrophenol, causes still more rapid release, indicating that the uptake and retention of potassium requires energy— that is, that it is a form of active transport. Potassium is measured using a specific potassium electrode, which functions in much the same manner as a glass hydrogen ion electrode. (Unpublished experiment; courtesy of Mr. Robert H. Turner.)

Table 12-1   Compounds that transport ions across natural and artificial membranes

Compounds	Ions transported
Gramicidin A	H^+, Na^+, Li^+, K^+, Rb^+, Cs^+
Valinomycin	K^+, Rb^+
Nigericin	K^+–H^+ exchange
Uncouplers	
(e.g., dinitrophenol)	H^+

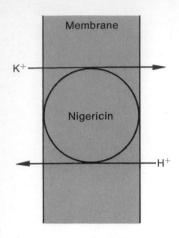

Figure 12-14 Nigericin antiport: a potassium–proton exchange. The antibiotic transports K^+ by surrounding the ion and, thus, rendering it lipid-soluble. The proton is transported on the return trip of the antibiotic, being carried at its carboxyl group.

Different members of this class of antibiotics exhibit different specificity for cations (see Table 12-1), and the variations probably reflect the relationship between the geometry of the antibiotic and the volume of the ion. This relationship is a strict one, as the discrimination between K^+ and Na^+ by valinomycin is in favor of K^+ by a factor of 10,000. It should be added that *gramicidin* possesses a linear (noncyclic) structure but is nonetheless believed to wrap itself around the cation in a fashion similar to the others. Finally, *nigericin* (Figure 12-12) is of interest because it catalyzes an exchange between K^+ and H^+, so that its action does not produce a *net* charge difference across a membrane (see Figure 12-14). The availability of ion-conducting antibiotics has been of great value in the study of membranes and, especially, in the examination of ion distribution and membrane potential.

Finally, the uncouplers of oxidative and photosynthetic phosphorylation, such as dinitrophenol, which we met earlier appear to represent a special case of ion-conducting compounds. These compounds (and many are known) are invariably both weak acids and lipid-soluble, and are able to conduct protons (H^+) across natural and artificial membranes by

Figure 12-15 Conduction of protons across a membrane by a weak acid (a phenol). Note that the phenol will carry protons in either direction, as dictated by the concentration gradient. Transport will stop when the concentrations are equal on both sides of the membrane.

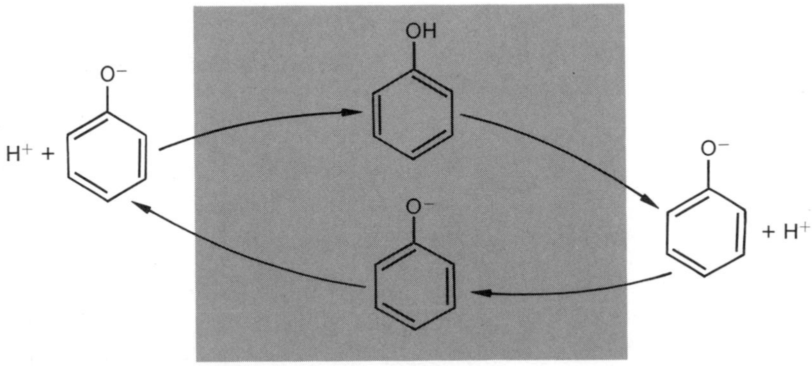

a mechanism like the one illustrated in Figure 12-15. It is probable that the influence of uncouplers on energy production is related to their ability to conduct protons; the nature of the possible connection is discussed below.

The effect of narcotics on transport

We have seen that some antibiotics are able to conduct ions across membranes. Other compounds of medical interest produce effects on membranes that modify their ability to carry out translocation of various ions and molecules. For example, butanol, an anesthetic, inhibits the passage of glycerol across red-blood-cell membranes of humans, rodents, and birds, while stimulating the process in some other organisms, including pigs, bats, and reptiles. Urethan, also an anesthetic, inhibits a number of transport events in various types of cells. Several local anesthetics, including butacaine and nupercaine, as well as the tranquillizer chlorpromazine, all alter the mitochondrial membrane in such a way as to increase the exchange between divalent cations (Ca^{2+} and Mn^{2+}) and hydrogen ion. The result of this interaction is that, in the presence of the narcotic, large amounts of protons are extruded when the divalent cations are actively taken up. The connection between this effect and the influences of these drugs on nerves is not yet certain. It is clear, however, that there is a correlation between the ability of such drugs to be effective and their solubility in lipids. This correlation is not strange, since the binding of them to the phospholipid of the membrane would tend to concentrate the drugs in the cellular membranes, where they presumably produce their effects.

The reversibility of ion-linked ATPase

We have seen that the passage of ions across a membrane against a concentration gradient may be linked to the breakdown of ATP, a process that might be described as an ion-stimulated ATPase. It is interesting to imagine possible consequences of reversibility of the reactions involved here. Since the reverse of ATPase activity is ATP synthesis, and since it is not difficult to imagine ions moving in the direction of a concentration gradient, there is no reason why, in principle, such ion movement should not be coupled to the net *formation* of ATP.

Such an idea has been vividly proposed by Mitchell (see Suggested Reading) to account for mitochondrial ATP synthesis coupled to electron transport. This suggestion has had the effect of throwing the whole idea of high-energy *chemical* intermediates into question and has emphasized very strongly the vectorial nature of many energy-linked processes. Mitchell has noted that ATP synthesis in mitochondria and chloroplasts appears to require an intact organelle with the membranes in good condition. Any treatment of these organelles that disrupts the membrane structure diminishes the ability to make ATP, although the electron transfer reactions themselves may be unimpaired. Attempts to prepare soluble phosphorylating preparations appear to have been quite unsuccessful. Mitchell suggests that the inability of workers to isolate the high-energy intermediates of ATP synthesis may be because they do not exist, and that the synthesis is in reality performed by a membrane-linked ATPase and driven by a sort of ion transport.

This view of ATP synthesis, which is called by Mitchell *chemiosmotic coupling*, is summarized in its simplest form in Figure 12-16. It depends on several postulates, all consistent with present knowledge of energy metabolism:

1. Both the oxidoreductions of the respiratory chain and the ATP-synthesizing reactions are vectorial—that is, they occur in the membrane of mitochondrion or chloroplast in an oriented fashion. The membranes in which they occur (in the mitochondrion—the inner membrane, and in the chloroplast—the thylakoid membrane) are known as the *coupling*

Figure 12-16   Synthesis of ATP coupled to a hydrogen ion gradient, as suggested by Mitchell (see Suggested Reading). According to this hypothesis (presented in its most simplified version), the enzymes of electron transport are so situated in the coupling membrane that net hydrogen ion flow (see Figure 12-18) occurs in the outward direction, thereby generating a membrane potential and a proton gradient. That gradient could drive an ATPase in the direction of ATP synthesis, provided the enzyme is located in the membrane in such a way as to be inaccessible to water as such, and accessible to OH^- from the outside only. Moreover, the enzyme would be accessible to ATP, ADP, PO_4, and H^+ from the inside only. Thus, the gradient of protons established by the respiratory chain would lead to a very low internal proton concentration, which would tend to draw protons away from the ATPase, driving that reaction in the direction of synthesis. Recent forms of this hypothesis are considerably more complex, including the effect of a membrane potential on ATP synthesis, as well as the introduction of something rather like chemical intermediates. The central feature of the mechanism remains the coupling of the two opposing vectors (respiration and ATPase), leading to net synthesis of ATP. It should be added that some features of the ATPase have been verified. For instance, ADP and ATP must approach it from the inside, and special transport mechanisms exist to get them there. The efflux of H^+ associated with respiration has also been verified directly (as we shall see in this chapter).

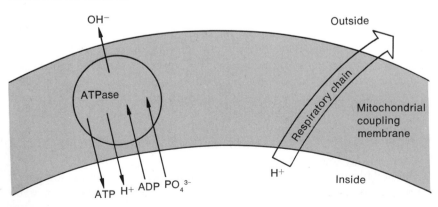

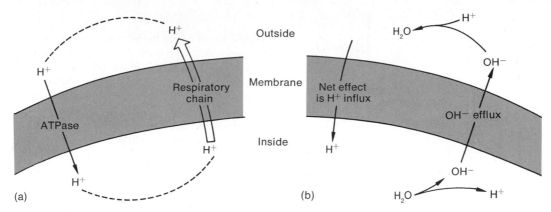

Figure 12-17 Proton circuit according to the chemiosmotic hypothesis. (a) Protons are "pumped" out by the action of the respiratory chain and return via the reversible ATPase (leading to ATP synthesis). (b) An illustration of why the OH^- efflux associated with the ATPase is effectively identical to a H^+ flow in the opposite direction. As hydroxy ion moves outward, dissociation of water occurs to replace it. This leads to an accumulation of H^+ on the inside. At the same time, arrival of hydroxy ion outside leads to the formation of water with a loss of free H^+. Thus, OH^- movement to the left leads to exactly the same effect as if H^+ had been transported toward the right.

membranes and are permeable to water and impermeable to protons and other ions. Protons and certain other ions can be transported across the membrane by means of the symport and antiport mechanisms outlined above. In what follows, we shall largely consider the situation in mitochondria, although similar treatment may be made with respect to chloroplasts or bacterial plasma membranes.

2. The respiratory chain is oriented in the coupling membrane so that substrates must approach from the inside and so that the passage of electrons along the chain produces a net flow of negative charge (in the form of the electrons themselves) in an inward direction and a net flow of protons in an outward direction. This arrangement is illustrated in Figure 12-18. The counterflow of electrons and protons is envisioned by Mitchell as resulting from the alternation in the respiratory chain of redox carriers that carry electrons only (for example, the cytochromes) and carriers that carry both electrons and protons (for example, the flavoproteins). The action of the respiratory chain thus gives rise to both a pH gradient (through flow of protons) and a membrane potential (through flow of electrons) and it will be recalled from Equations 12-7 and 12-8 that the two, taken together, comprise the force able to drive a proton across a membrane. It will be noted that the two flows render the mitochondrion interior basic and the outer surface of the coupling membrane positive with respect to the inner one.

3. ATP is postulated to be synthesized by reversal of the action of an ATPase enzyme located in the coupling membrane. The equilibrium of the ATPase reaction

$$ATP + H_2O \rightleftharpoons ADP + PO_4$$

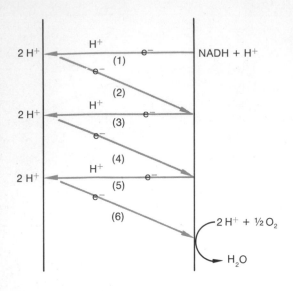

Figure 12-18 How the respiratory (or photosynthetic) electron transfer chain could be folded to give rise to a net H^+ gradient. Mitchell envisions that the carriers (flavin enzymes, cytochromes, etc.) alternate between those that carry both electrons and protons (e.g., flavin enzymes and ubiquinone) and those that carry electrons only (e.g., cytochromes). Thus, each pair of carriers would transport one pair of protons. For this reason, one ATP molecule would be produced for every pair as the protons returned via the ATPase (see Figure 12-17). Each pair would thus constitute a site of oxidative (or photosynthetic) phosphorylation, an arrangement that corresponds well to the known levels of ATP synthesis with different substrates.

is far in the direction of ADP + PO_4, but its reversal (ATP synthesis) could be effected by making the concentration of water in the region of the enzyme very low. One way that the concentration of water might be lowered is by removal of protons toward the inside and the removal of OH^- toward the exterior. Since motion of the anion OH^- toward the outside is vectorially identical to movement of protons in the opposite direction, the synthesis of ATP by the ATPase could be associated with the net transport of a proton in the inward direction. Likewise, hydrolysis of ATP would then be associated with proton transport outwards. According to these ideas, the coupling of oxidation to ATP synthesis could result from coupling the net outward proton flux, due to the respiratory chain, with the inward flux, associated with the ATP synthetic reaction (see Figure 12-17). The result would be a closed loop of proton transport, passing through the redox system and the ATPase. It is important to keep in mind that, just as the force acting to move protons across a membrane has a ΔpH and a potential component (Equation 12-8), so does the coupling of oxidation and ATPase include both terms. We have considered only the coupling between proton fluxes (that is, only the pH term), but it should be clear that the proton flux through the ATPase system and hence ATP synthesis will be driven by both the pH difference and the membrane potential and, indeed, under some conditions, the potential appears to be the predominant influence.

It should be stressed that the essential feature of the preceding mechanism is the spatial orientation of the respiratory chain in the coupling membrane, giving rise to vectorial movement of protons and charge and to the spatial orientation of the ATPase, such that its action in the synthetic direction is associated with a vectorially opposed proton movement.

Without considering the theory in much greater detail, it is possible to note that none of the postulates are inconsistent with what is known about mitochondrial and chloroplast membranes, and also that the theory explains as much of what is known about the properties of ATP synthesis as does the purely chemical theory. To give one example, the chemical theory states that the action of an uncoupling agent, such as dinitrophenol, is to effect the hydrolysis of a high-energy intermediate, whereas the chemiosmotic view is that it serves as a hydrogen ion carrier within the membrane. In the latter case, the uncoupler, being dissolved in the membrane lipid, carries H^+ through the membrane and prevents an H^+ gradient from being formed as a result of electron transport. This suggests that uncouplers should be lipid-soluble and able to carry protons, that is, weak acids, and we have seen that both conditions are generally met. Uncouplers should probably be viewed as short-circuiting the proton loop between the respiratory chain and the ATPase (see Figure 12-15).

The chemiosmotic point of view also leads to a coherent explanation of active transport of ions, namely, that they are propelled by the same charge gradient that acts on protons. Thus, a charge gradient is formed as a result of the operation of the respiratory chain, and this gradient would favor uptake of cations such as Ca^{2+} or Sr^{2+}. Movement of cations down the gradient (that is, inward) would tend to collapse the gradient, that is, equalize the charge across the membrane. If the gradient were collapsed by ion movement, it would obviously become unavailable for ATP synthesis by the membrane ATPase. Addition of an ion-conducting antibiotic, such as valinomycin, allows univalent cations (such as K^+) to be taken up into mitochondria with the expenditure of energy. This is most easily explained if the antibiotic simply renders the ions more

Figure 12-19 Hydrogen ion efflux associated with the action of the respiratory chain. (a) Mitochondria isolated from mouse liver are incubated under argon until completely anaerobic. Then oxygen, dissolved in a salt solution, is added with a syringe. The pH of the suspension is monitored with a glass electrode and extremely sensitive pH meter. A burst of H^+ is produced, which is proportional to the amount of oxygen added. When the efflux is over (oxygen is exhausted), the H^+ leaks slowly back into the mitochondria. (b) This return is greatly accelerated by the addition of the uncoupler carbonyl cyanide m-chlorophenylhydrazone (CCCP).

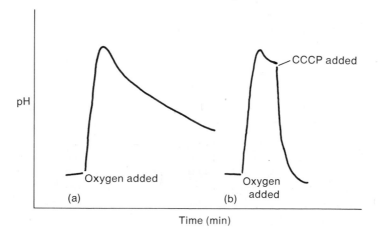

easily permeant and the ions are then propelled in a purely electrostatic fashion by the charge gradient.

The chemiosmotic point of view is sufficiently detailed to make predictions that may be experimentally tested. For example, the action of the respiratory chain should give rise to proton extrusion from mitochondria, and this prediction has been fulfilled. Figure 12-19 shows the burst of H^+ when a small amount of oxygen is added to anaerobic mitochondria in the presence of a substrate. The decay of the curve is due to a leak of protons back into the mitochondria and it is greatly accelerated by the addition of an uncoupler, as would be expected, if the uncouplers act by carrying protons across the membrane. Likewise, the action of the ATPase should be associated with proton flow through the membrane, and this has been observed as well.

Chemiosmotic coupling in chloroplasts

We have said that the chemiosmotic view is also applicable to photosynthetic energy coupling, and, in fact, some of the most interesting evidence for its validity has come from studies using chloroplasts. We saw in an earlier chapter that proton flow was closely associated with photosynthetic phosphorylation. It appears that the photosynthetic electron transfer chain may be oriented in the thalykoid membrane, so that there, as in mitochondria, electron flow would produce a proton gradient ΔpH and a membrane potential. An important distinction between chloroplast membranes and mitochondria is that the action of the respiratory chain leads to expulsion of protons whereas that of the photosynthetic chain leads to proton uptake. In other words, the vectorial character of the two membrane systems is exactly opposite.

An important experimental result, which suggests that a proton gradient is able to drive ATP synthesis in chloroplasts, is the observation that a rapid shift in the external pH of chloroplasts in the dark can give rise to ATP formation. Chloroplast fragments (which still contain closed thalykoid membranes) are incubated at an acid pH (about 4) and then rapidly brought to pH 8.5 in the presence of ADP and phosphate. This jump in pH produces a temporary pH gradient across the membrane, which appears to give rise to ATP synthesis. Since the pH gradient is formed externally by changing the pH, it does not depend on the electron chain for its occurrence, and the ATP formed in this fashion is insensitive to addition of electron chain inhibitors (see Figure 7-10).

Finally, the chemiosmotic view is instructive in understanding the inhibitory effect of ammonium ion upon photosynthetic ATP synthesis. It appears that ammonium ion crosses the membrane in the uncharged form of ammonia after liberating a proton outside the chloroplast (Figure 12-20). It picks up a proton again inside to form ammonium, with the net effect of having short-circuited the proton flow through the chloroplast ATPase, thus eliminating ATP synthesis.

A great deal more might be said in support of the chemiosmotic view

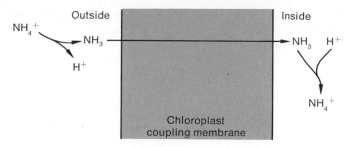

Figure 12-20 Uncoupling of photosynthetic ATP synthesis by ammonia. It appears that ammonium is transported in the uncharged (ammonia) form. For this reason, its uptake leads to the appearance of a proton on the outside and the disappearance of one on the inside. Thus, there is a net outward flux of protons, which has the effect of short-circuiting the ATPase. (The reader will note that the direction of proton flow associated with photosynthesis is opposite to that associated with respiration.) It is also clear that when ammonia uncouples, it effectively replaces a proton gradient with an ammonium gradient; the former could be coupled to ATP synthesis, the latter could not.

(the reader is referred to Mitchell's papers listed in Suggested Reading). On the other hand, available information does not permit acceptance of the theory in its entirety until aspects of it have been subjected to much more extensive examination. We have considered this theory in such detail, partly because it is subtle enough to preclude cursory treatment, and partly because its central feature, the vectorial nature of energy metabolism, appears to expand our understanding of many features of cell function. The chemiosmotic view appears relevant not only to oxidative and photosynthetic phosphorylation but, more generally, to situations where ions traverse membranes, whether they be plasma membranes of bacteria or the excitable membranes of nerves (discussed in the next chapter). Finally, the chemiosmotic theory has, with startling clarity, brought to the attention of cell biologists two important ideas: (1) Membranes are intrinsically vectorial in nature, and (2) transport and respiration (or photosynthesis) are closely interconnected.

The sidedness of energy-coupling membranes

The central feature of the chemiosmotic hypothesis is that membranes possess a *sidedness*—that is, a vectorial quality—so that the membrane potential produced by electron transport shows a characteristic polarity. Measurement of the proton movement associated with the functioning of the electron transport chain indicates that the sidedness of mitochondria and bacteria is similar, with protons passing outward, in both cases, as a result of respiration. It will be recalled that the bacterial plasma membrane is the locus of the bacterial respiratory chain, corresponding to the inner membrane of mitochondria.

Submitochondrial particles, which are formed when intact mitochondria are subjected to ultrasonic oscillation, appear to be inside out with respect to mitochondria. This may result from their originating from cristae, as shown in Figure 11-6. Such mitochondrial fragments exhibit proton uptake (instead of expulsion) on respiration. Similarly, chloroplasts expel protons on action of the photosynthetic chain, making them similar to submitochondrial particles (and unlike mitochondria) in their polarity. The polarity of membranes from mitochondria and chloroplasts is subject to additional tests, as the ATPase knobs (see Figure 11-6) are on the inside of mitochondrial inner membranes, outside in chloroplasts, and outside in submitochondrial particles.

Cell compartmentation and regulation

In Chapter 1 is was clear that cells are divided into a number of compartments by their internal membranes, whereas in this chapter we have seen that the same membranes have considerable specificity as to the passage of chemicals. This suggests that compartmentation could play a significant role in the regulation of cell activity, since the passage of molecules and ions between different parts of the cell could be strictly regulated. Furthermore, much translocation across membranes is dependent on energy production by the cell, so that energy metabolism, which is itself under tight control, influences transport in a fundamental way.

Experimentally, the question of compartmentation often arises when it is discovered that a cell organelle is not equally permeable to all materials or when a chemical compound appears to exist in distinct populations within the cell, with barriers to mixing between them. Similarly, changes in the reactivity of cells on disruption of structure often point to the existence of compartments, the barriers between which prevent the free reaction of the product of one enzymic reaction with another enzyme.

For example, the important compound oxaloacetate appears to exist in two compartments within mitochondria, being able to produce inhibition of succinate oxidation (see Chapter 5) only while in one of them. Other cases exist where a reaction occurs only when ATP is available, in situations that suggest that the role of ATP is to allow active transport to eliminate a membrane-imposed barrier to the complete reaction. Many cases of apparent compartmentation in cells could be cited, but instead we shall consider one central one in some detail—barriers to the free migration of pyridine nucleotides.

We have seen that an important aspect of the role of pyridine nucleotides (NAD^+ and $NADP^+$) in respiration is their ability to move about and thus serve as mobile electron carriers. We have seen, for example, that the acceptance of electrons in glycolysis may be coupled to the donation of them in the aerobic oxidation of NADH by the respiratory chain. This requires, at first sight, that NADH should be able to pass from the soluble cytoplasm, where the glycolytic enzymes are located, to the mitochondrion, where the oxidation by molecular oxygen occurs. A possible alternative would be the aerobic oxidation of NADH by enzymes situated in the extramitochondrial cytoplasm, but such reactions do not, in fact, appear to occur. It is therefore interesting to learn that isolated, intact mitochondria are quite without the ability to oxidize external NADH but that the oxidation occurs rapidly when mitochondria are damaged.

It is clear, on the other hand, that mitochondria oxidize internal NADH, since it is formed in the course of oxidation of a number of substrates whose oxidation is NAD-linked and which pass freely through the mitochondrial membrane. Thus, it is necessary to search for a mechanism that will explain the known role of NADH in mediating electron transfer between the reactions of glycolysis and the respiratory chain. Similar problems exist with respect to the role of NADPH, but one might suppose that the two difficulties are related, since a mechanism allowing NADH to ignore the barrier, together with the transhydrogenase reaction described earlier, would settle things for NADPH as well.

Transcompartment substrate cycles

It is clear that the mitochondrial barrier to NADH entry may not be avoided by any sort of special NADH transport mechanism, since the transport has been impossible to observe in undamaged mitochondria. Therefore, the passage of electrons from the extramitochondrial to the

Figure 12-21 Glycerol phosphate cycle.

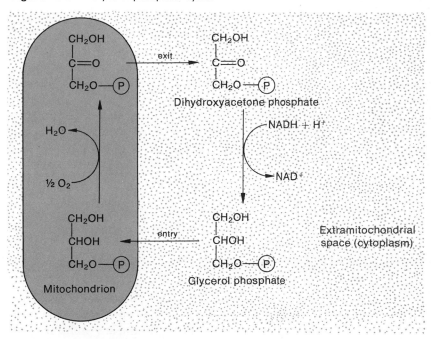

299

intramitochondrial compartments should be thought of as the transfer of electrons from outer to inner pyridine nucleotides (the reduction of internal NAD^+ by external NADH). Clearly, something must cross the mitochondrial membrane, and, since isolated mitochondria do not oxidize external NADH, its oxidation is likely to involve some component of cytoplasm that may be lost during the isolation procedure for mitochondria.

A method whereby electrons may be passed from external to mitochondrial pyridine nucleotides has been described and is called the *glycerol phosphate cycle*, summarized in Figure 12-21. The cycle depends on the presence of two enzymes for the oxidation of glycerol phosphate— a mitochondrial one, which is a flavoprotein passing electrons to oxygen via the cytochrome system, and an extramitochondrial pathway, where the electron acceptor is NAD^+. External NADH first reduces dihydroxyacetone phosphate. The product of this reaction is NAD^+ and glycerol phosphate. The mitochondrial membrane is permeable to glycerol phosphate, which enters the mitochondrion and is there reoxidized by the respiratory chain to form dihydroxyactone phosphate, which is free to diffuse to the outside and react again with NADH. The two reactions may be written in such a way that they add up to the net oxidation of NADH by mitochondria:

Extramitochondrial

$$NADH + H^+ + \text{dihydroxyacetone phosphate} \longrightarrow NAD^+ + \text{glycerol phosphate}$$

Mitochondrial

$$\text{Glycerol phosphate} + \tfrac{1}{2} O_2 \longrightarrow \text{Dihydroxyacetone phosphate} + H_2O$$

Sum:

$$NADH + \tfrac{1}{2} O_2 + H^+ \longrightarrow NAD^+ + H_2O$$

It is probable that this mechanism accounts for a large part of NADH oxidation in mitochondria from a number of tissues. Additional cycles of a similar nature have been suggested using other NAD^+-linked substrates, but in all cases positive evidence is meager.

Finally, it is worth remarking that the glycerol phosphate cycle is itself subject to regulation in the cell insofar as it requires the presence of one of the intermediates of glycolysis for activity. This ensures that, under conditions where glycolysis is not continuously providing for the cytoplasmic reduction of NAD^+, mitochondrial oxidation of NADH will not occur, an arrangement that tends to conserve any NADH in the extramitochondrial space when its rate of production is low.

Epithelial membranes

Having discussed the theory of membrane transport in some detail, let us turn now to an actual system where transport is measured. Our knowledge of transmembrane translocation is frequently influenced by the accident of choice of experimental material. For example, a great deal of information has come from the study of red-blood-cell membranes for the compelling reason that they are relatively simple to obtain and that they possess little metabolism to interfere with the measurements of transport events. The only difficulty with the red cell system is that it may be quite atypical, in that the composition of red cells is hardly representative of membranes in general, considering its unusual protein/lipid ratio (discussed earlier). Nonetheless, one must seek enlightenment where it can be obtained and the red cell has been most productive in that regard.

Another biological system that has been very productive in the understanding of membranes has been the epithelial surface, which includes a number of quite disparate regions of animals where solute transport takes place on a large scale. For example, the epithelium of intestine is the locus for nutrient uptake for many animals, and it is reasonable that the study of transport should concentrate on such a system. Another sort of epithelium that has been of much interest has been frog skin, which is able to transport a number of molecules and ions. Frog skin has the advantage of being simple to study in that it may be removed in large areas and experiments may be carried out on a relatively gross scale. The opinion of frogs about this usefulness goes unrecorded.

Although various epithelial systems are of great convenience in experimental manipulation, they also present some special difficulties in the interpretation of results. The main difficulty is their complexity, for, whereas one would like to look at a single uptake event, many epithelial membranes are a number of cells thick and are composed of different cell types. The apparent uptake through such a "membrane" may, in fact, include initial uptake, intracellular movement, transcellular translocation, and a final transport event on the other side of the whole system. In such a complex situation one cannot even be sure that the permeability of a cell is a constant function. Since cells are oriented in the tissue, it is quite possible for cells to be permeable to an ion on one side and not on the other, leading to the situation in which the ion can be taken up by the epithelium but not released on the other side.

Let us first consider the manner in which an epithelial system can be studied. In the case of a system with flat topology, such as frog skin or a "slab" of intestinal epithelium, the region is excised and mounted with some sort of a ring arrangement so that it separates two solutions, which can be then manipulated and studied chemically and electrically. This arrangement is illustrated in Figure 12-22. Other epithelial systems are not topological planes and must be studied differently. For instance, intestinal epithelium is often studied as a topological ring (see Figure 12-23), while other tissues, such as gallbladder, urinary bladder, gastric lining, and kidney tubules, are each studied in characteristic ways. Nor should we omit the fact that frequently the most effective experimentation on an epithelial system is with the intact animal. Much knowledge of nutritional uptake by the intestine has come from studying the ingestion of radioactive sugars and other compounds by normal animals. A particularly useful technique on the *in vivo* scale has been the *intestinal*

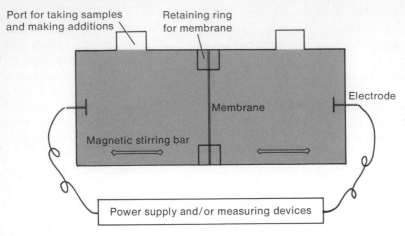

Port for taking samples and making additions

Retaining ring for membrane

Membrane

Electrode

Magnetic stirring bar

Power supply and/or measuring devices

Figure 12-22   System for studying epithelial membranes.

loop, where a loop of intestine is excised, the cut ends rejoined, and the loop mounted in the form of a letter U—with both ends in contact with the exterior through the body wall. This system preserves as much as desired of the normal innervation and blood supply to the loop so that the transport can be measured in a simple manner with healthy epithelium.

To return to the frog skin, it is of interest that the system has been studied for over a century. One of the first accurate measurements of a biological potential was that taken by Raymond, who found about 100 millivolts across the frog skin. It was discovered (about 70 years ago) that this potential resulted from the asymmetric transport of sodium (or lithium) with a higher flux in the inward direction. This net flux can occur against a considerable gradient and is coupled to the metabolism occurring in the cells. The rates of sodium flux for a number of different systems are compared in Table 12-2.

It is clear that the transduction of metabolic energy to the measured sodium flux occurs by means of a Na^+–K^+ ATPase system of the sort described earlier. This is supported by the finding that the flux of sodium and the consequent potential require the presence of at least a low concentration of potassium on the inward side of the skin (that is, the side toward which the net flux of sodium occurs).

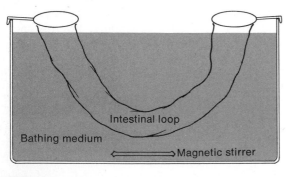

Intestinal loop

Bathing medium

Magnetic stirrer

Figure 12-23   *In vitro* intestinal loop. Note that the contents of the outer medium and that within the lumen of the intestine can be varied independently.

Table 12-2   Sodium flux in some epithelium systems

System	Flux (pmoles/sec/cm²)
Red blood cell	0.1
Nerve (squid)	33
Bladder (toad)	140
Skin (frog)	834
Small intestine (rat)	1,330

To illustrate the wide range of measurements that can be performed with the frog skin epithelium system (owing, in part, to its macroscopic size), we turn to the question of the makeup of the net flux through the membrane system. In the first place, it is possible to measure net flux as the electrical current passing through the whole system. At the same time, the voltage may be measured independently and, indeed, can be manipulated externally by passing current through the skin through the insertion of electrodes into the fluid chambers. These arrangements are illustrated in Figure 12-22. The composition of the current in terms of ionic species can be followed by the use of radioactive tracers, such as sodium-22, chlorine-36, and others. It is even possible to measure the flux of sodium in both directions at once by employing a double-label technique. In this case, sodium-22 is introduced into one chamber, while sodium-24 is placed in the other. After the system has been allowed to carry out transport under the conditions of interest, the content of both isotopes in both chambers can be measured. This is possible because, of the two isotopes, sodium-24 emits beta particles (electrons), while sodium-22 emits positrons of much lower energy, which then give rise to some gamma particles by means of electron capture. Since detection of beta and gamma particles allows for their discrimination, it is simple to separate the two fluxes. It should be added that this sort of measurement is valid only for fluxes taking place during relatively short periods of time, as only initially do the contents of the two chambers contain pure isotopes. In other words, the occurrence of transport has the effect of mixing the isotopes.

Many experiments have been performed by this method. For one thing, it is possible to cancel the potential across the skin by opposing it externally so that flux can be measured isotropically under conditions where the ions are not traversing a potential gradient. This is an important experiment, because it shows that the sodium ions are not somehow moving "down" a potential slope, produced by the distribution of some other, unspecified ion. Indeed, sodium motion occurs even in the absence of potential, so that one can conclude that the movement of sodium is itself active and, under natural conditions, can give rise to the measured potential difference. Figure 12-24 illustrates results of a sodium-flux experiment, where part *a* shows the influence of altering the external sodium (and chloride) concentration. It should be added here that failure to provide a permeant anion (such as Cl^- but not SO_4^{2-}) leads to

303

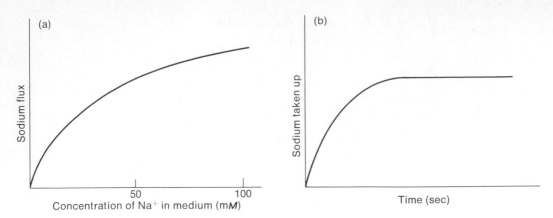

Figure 12-24 Sodium flux across an epithelial membrane. (a) The effect of increasing the NaCl concentration. Note that the transport system finally becomes saturated; finally, in other words, increasing the Na concentration produces relatively little effect. (b) The time course of uptake.

diminished transport of the cation, sodium. It also develops that the amount of chloride translocated (when present) is less than sodium. This may reflect some counterflow of protons (as well as added sodium or potassium ions) in the opposite direction, which work to maintain equivalence of charge. Part *b* of the figure shows the kinetics of sodium influx. Finally, it should be said that, in such experiments, if cyanide is added to the system, the flux is inhibited and the trans-skin potential relaxes to zero. This finding is in conformity with the requirement of ATP of a sodium–potassium ATPase system that appears to underlie these phenomena, since the effect of cyanide would be to poison the oxidative processes in the cell mitochondria, thus eliminating most of their ATP synthesis.

More exhaustive analysis of the frog skin (and other epithelial) systems becomes very complex, owing to the complicated nature of the system being studied. For one thing, there appears to be pools of sodium that either do not equilibrate with the remainder of the sodium in the membrane or, worse, do it slowly. Thus, any model that endeavors to describe the system must include such sequestering of ions as well as the numerous interactions, some little understood, among the individual cells of the skin.

Our emphasis on the relationship between ion flux and potential difference should not mislead us regarding the function of the flux in the intact organism. Clearly, the frog is not equipped with a voltmeter nor are his enemies. The occurrence of the potential difference is a reflection of the movement of ions, and it is important to ask what role the ionic movement may have in the first place. Clearly, in many instances, the flux itself is of importance. For example, kidney cells and, to an extent, the skin of amphibians carry out ion transport as part of the organism's regulation of internal salt balance. It is also clear that sodium transport underlies, in many cases, the transport of an organic molecule such as a sugar or an amino acid. For example, the epithelium of rabbit intestine

304

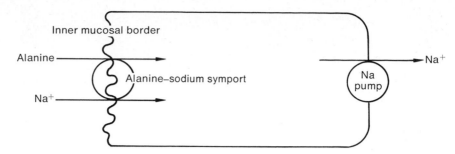

Figure 12-25 Alanine transport by mucosal cell. Alanine is absorbed from the lumen of the intestine by a carrier mechanism that is coupled to sodium uptake. This "symport" is passive, in that transport is driven by a concentration gradient (i.e., low sodium inside the cell). This low sodium concentration within the cell is maintained by an energy-requiring sodium pump at the other end of the cell.

carries out the active uptake of the amino acid alanine, and this process requires sodium transport as well if it is to occur at a maximum rate. In this case, the linkage between transport of the amino acid and the sodium appears to reflect the affinity of a single carrier molecule situated in the plasma membrane for both categories of ion. The carrier appears to transport either ion at a greater rate when both are present. In this case, the uptake is not directly active and the ions enter by a diffusional process. On the other hand, uptake requires that the intracellular concentration of sodium be maintained at a low level so that the gradient will favor its entry. This is managed by an active extrusion of sodium at the other "end" of the cell, an efflux that does require metabolic energy. These relationships are illustrated in Figure 12-25.

Pinocytosis

When we refer to the transport of molecules across a membrane, we usually refer to a process wherein the molecule being translocated is small in comparison to the macromolecules comprising the membrane. Thus, movement of a large protein through a membrane would necessarily involve major disruption of the membrane structure. It is, however, known that cells do take in macromolecules much too large to traverse the membrane in the usual sense. For example, cells suspended in a medium containing the enzyme ribonuclease undergo hydrolysis of their intracellular RNA, indicating that the enzyme has somehow entered. In fact, this process occurs by means of an engulfment procedure by which the plasma membrane sends out folds that surround some of the external region (see Figure 12-26). The folds fuse, so that what was exterior becomes a vacuole within the cell. Subsequent breakdown of the vacuole or its fusion with a membranous structure (such as lysosomes or the endoplasmic reticulum) leads to the inclusion of its contents in the cytoplasm proper. When the engulfment is of a purely liquid sample of the environment, the process is called *pinocytosis* (cell drinking); when there are particles

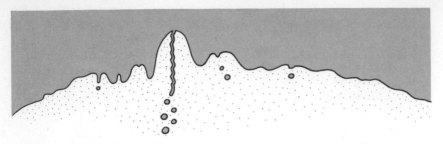

Figure 12-26 Surface region of an animal cell during pinocytosis. Note the formation of pinocytotic vesicles at the end of a channel, which, in turn, originates in a projection of the cell surface.

present, it is called *phagocytosis* (cell eating). It should be clear that there is a continuum of size leading from soluble molecules to particles so that the two names do not reflect a real mechanistic difference.

Pinocytosis (or phagocytosis) occurs in many free-living cells, including amoebas, as well as many cells of multicellular organisms. For example, there are specialized cells of human beings, such as the Kupffer or phagocytic cells of the liver and leukocytes, where this sort of activity is especially pronounced. In general, the extent of phagocytosis can be estimated by suspending cells in a medium containing radioactively labeled protein and by measuring the rate of uptake of the label. Conversely, phagocytosis can be measured with the electron microscope by adding to the cells a suspension of colloidal gold or iron or a very large protein, such as ferretin, and then examining them with the electron microscope.

Pinocytosis does not occur continuously but is stimulated by a variety of agents that may be present in the cellular medium. For example, proteins and amino acids stimulate cells to commence pinocytosis as do a number of inorganic and organic cations. Anionic substances do not work and such negatively charged macromolecules as nucleic acids are without effect. In addition, proteins only induce pinocytosis when they are at a pH such that they bear a net positive charge. It appears that the stimulation of pinocytosis requires binding of the positive ion to the cell surface prior to the initiation of plasma membrane activity. When pinocytosis occurs, it is interesting that the potential across the plasma membrane undergoes a drastic collapse, presumably as external charge is brought within the cell by the infolding process.

It should be noted that the occurrence of pinocytosis emphasizes the importance of the dynamic nature of membranes. Clearly, the self-sealing character of membranes noted earlier is of significance here. A little reflection will also suggest that the plasma membrane must be continuously regenerated in the region where pinocytosis occurs, since material is constantly being taken away toward the center of the cell in the form of the pinocytotic vesicles. When the latter vesicles break down, their components appear to re-enter the pool of membrane precursors and are then available for regeneration of the plasma membrane. It has been estimated that the intracellular membrane component pool comprises about 10 times as much protein and phospholipid as that actually in the form of cellular membrane at a given time.

In a sense, pinocytosis can be regarded as an active, but not selective, transport mechanism. The event does require metabolic energy and, at least in principle, can lead to movement of a molecular species against its concentration gradient. Obviously, the pinocytotic vesicle does not select which of the molecules in the medium it will include, so that the uptake reflects, very strictly, the percentage composition of the medium. In any case, pinocytosis is inhibited by agents that interfere with cellular-energy production, such as carbon monoxide and cyanide. These inhibit mitochondrial oxidation and, therefore, ATP synthesis. It is interesting that cells of the *sarcoma* (cancer of mesodermal tissue) carry out pinocytosis that is not inhibited by these metabolic poisons, since these cells are presumed to be less dependent on oxidative phosphorylation for their energy (see Chapter 17).

Suggested Reading

Books

Bolis, L., and B. A. Pethica, *Membrane Models and the Formation of Biological Membranes*, John Wiley & Sons, New York, 1968.

Cereijido, M., and C. A. Rotunno, *Introduction to the Study of Biological Membranes*, Gordon and Breach, Science Publishers, New York, 1970.

Christensen, H. N., *Biological Transport*, W. A. Benjamin, New York, 1962.

Crone, C., and N. A. Lassen, *Capillary Permeability*, Academic Press, New York, 1970.

Harris, E. J., *Transport and Accumulation in Biological Systems*, Butterworth & Co., London, 1956.

Johnson, F. H., H. Eyring, and M. J. Polissar, *The Kinetic Basis of Molecular Biology*, John Wiley & Sons, New York, 1954, Ch. 11.

Lehninger, A. L., *The Mitochondrion*, W. A. Benjamin, New York, 1964, Chs. 7–8.

Passow, H., and R. Stampfli, *Laboratory Techniques in Membrane Biophysics*, Springer-Verlag, Berlin, 1969.

Robertson, R. N., *Protons, Electrons, Phosphorylation and Active Transport*, Cambridge University Press, Cambridge, 1968.

Schoffeniels, E., *Cellular Aspects of Membrane Permeability*, Pergamon Press, Oxford, 1967.

Stein, W. D., *Movement of Molecules Across Cell Membranes*, Academic Press, New York, 1967.

Tosteson, D. C. (ed.), *The Molecular Basis of Membrane Function*, Prentice-Hall, Englewood Cliffs, N.J., 1969.

Articles

Albers, R. S., "Biochemical aspects of active transport," *Ann. Rev. Biochem.*, **36**, 727–56 (1967).

Caldwell, P., "Energy relationships and the active transport of ions," in *Current Topics in Bioenergetics*, R. Sanadi (ed.), vol. 3, Academic Press, New York, 1969.

Fox, C. F., "The structure of cell membranes," *Sci. Am.*, **226**, 30 (Feb., 1972).

Green, D. E., and D. E. Young, "Energy transduction in membrane systems," *Am. Scientist*, **59**, 92 (1971).

Greville, G. D., "A scrutiny of Mitchell's chemiosmotic hypothesis of respiratory chain and photosynthetic phosphorylation," in *Current Topics in Bioenergetics*, R. Sanadi (ed.), vol. 3, Academic Press, New York, 1969.

Katchalsky, A., and R. Spangler, "Dynamics of membrane processes," *Quart. Rev. Biophysics*, **1**, 127 (1968).

Klingenberg, M., "Mitochondrial metabolite transport," *Federation of European Biochemical Societies Letters*, **6**, 145 (1970).

MacRobbie, E. A. C., "The active transport of ions in plant cells," *Quart. Rev. Biophysics*, **3**, 251 (1970).

Mitchell, P., "Chemiosmotic coupling in oxidative and photosynthetic phosphorylation," *Biol. Rev.*, **41**, 445 (1966).

Mitchell, P., "Translocations through natural membranes," in *Advances in Enzymology*, F. F. Nord (ed.), vol. 29, John Wiley & Sons (Interscience Division), New York, 1967, p. 33.

Pardee, A., "Membrane transport proteins," *Science*, **162**, 632 (1968).

Pressman, B. C., "Energy-linked transport in mitochondria," in *Membranes of Mitochondria and Chloroplasts*, E. Racker (ed.), Van Nostrand Reinhold Co., New York, 1970.

Schmidt-Nielsen, B., "Comparative aspects of transport of hypertonic, hypotonic solutions by epithelial membranes, Introduction," *Fed. Proc.*, **30**, 3 (1971).

Siekevitz, P., "On the meaning of intracellular structure for metabolic regulation," in CIBA Foundation Symposium, *Regulation of Cell Metabolism*, G. E. W. Wolstenholme and C. M. O'Connor (eds.), Little, Brown and Co., Boston, 1959.

Skou, J., "Enzymatic basis for active transport of Na^+ and K^+ across the cell membrane," *Physiol. Rev.*, **45**, 596 (1965).

308

13 Excitability

A central feature of living matter is *excitability*, by which is meant the characteristic ability of living things to respond to a stimulus from without. A *stimulus*, which is a change in the environment, evokes a *response*, which is a characteristic change on the part of the organism. An example of a generalized sort of excitable response might be the change in plasma membrane potential when the immediate environment of the cell surface is perturbed. The contraction of a muscle fiber upon electrical stimulation provides an additional example, as does phototaxis, described in an earlier chapter. In all cases, cells exhibit an extremely rapid response (of the order of milliseconds) to a stimulus that is not necessarily of the same kind as the response. For instance, a mechanical perturbation may elicit an electrical response and vice versa. There are examples of a relatively slow response to the environment, as in the case of induced synthesis of an enzyme, where the stimulus is the presence of its substrate in the growth medium. The completion time for such a response might be of the order of a cell generation (minutes to hours instead of milliseconds), and we shall reserve the term *excitability* for the more rapid events.

Excitability appears to be closely associated with the distribution of charge across cellular membranes, most particularly the plasma membrane. We shall see shortly that an excitable response involves a rapid alteration in the plasma membrane potential in such a manner that the disturbed potential can spread across the surface of a cell from the original site of stimulus. This transferal of an excitable response over a cell surface may be viewed as a mechanism for communication, such that the entire cell may respond in a somehow coordinated fashion to a spatially isolated stimulus. One may regard nerve cells, with their long, impulse-conducting *axons*, as specially modified so that this property of intracellular communication can be of benefit to multicellular organisms. While excitability has been most studied in the specialized case of nerve cells, it is a general property of cells and has been demonstrated in a wide phylogenetic selection of cell types.

The resting potential

Plasma membranes exhibit a characteristic potential difference between their two sides of between 50 and 100 millivolts (10^{-3} volt), the outside being positive with respect to the interior. As long as the cell metabolism is maintained, the potential remains, although it is abolished by cyanide and other poisons that eliminate energy metabolism in aerobic cells. It may be measured either by the insertion of a microelectrode into the cell (which must be done with minimal damage) or estimated by considering the distribution of permeant ions according to the Nernst equation (Equation 12-6). Not only can the membrane potential be estimated by the distribution of ions, but it is the direct consequence of the enforced distribution of cations, according to the following general properties:

1. The plasma membrane is relatively permeable to K^+ and Cl^-, much less permeable to Na^+, and impermeable to other anions.

2. There is an energy-requiring sodium–potassium "pump," which transports Na^+ outward and K^+ inward. The pump is of the antiport variety in that the two flows are tightly linked. If K^+ is absent from the external medium, Na^+ cannot be extruded. The transport system requires energy, either in the form of ATP or substrate oxidation. Thus, in the presence of cyanide, which prevents oxidative ATP production, added ATP has been shown to support transport.

3. Since the active transport of Na^+ and K^+ occurs by an antiport mechanism, there is equal movement of charge in both directions, and the pumping will not alter the membrane potential directly. However, the leak of K^+ back out of the cell is much more rapid than the return of Na^+ to the interior, owing to the different permeabilities in the cases of the two cations. Since the net leak of positive charge is greater in an outward direction, a positive charge is built up on the outer surface until it is sufficiently great to oppose further leak of K^+; at this point, a *steady state* obtains, where the linked active transport of Na^+ and K^+ is exactly equal to their return. At this point, Cl^- will be found to be partitioned across the membrane according to the Nernst relationship.

It will be noted that the antiport pump of the plasma membrane is similar (or identical) to the Na^+–K^+ ATPase described earlier for the red-blood-cell plasma membrane. It should also be recognized that the resting potential of cells has an analog in the transmembrane potential of such organelles as chloroplasts and mitochondria, although the origin of the potential may be quite different in each case.

The action potential

The *action potential* is a transient change in the membrane potential following some sort of stimulation by electrical, mechanical, or chemical means. It occurs in all excitable plant and animal membranes and has been most extensively studied in nerve and muscle. The potential change

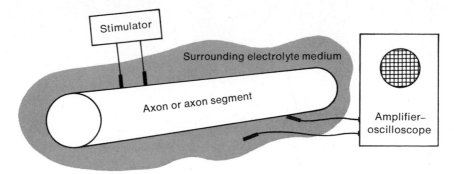

Figure 13-1 Arrangement for the measurement of action potential in nerve axon.

does not occur simultaneously over an entire membrane but is localized. The localized potential alteration, however, moves as a wave over the membrane surface, so that in the case of an elongated nerve cell it passes along the fiber at constant rate and constitutes the nerve impulse. The action potential should be considered as a means of rapid communication between different regions of a cell, and the most obvious question about its mechanism relates to its ability to radiate over the cell surface.

Measurement of the action potential can best be accomplished, in the case of single cells, by placing a recording electrode on the cell surface, completing the circuit through the recording device via an electrode in the surrounding fluid, but not touching the cell. A second pair of spatially removed electrodes is used to stimulate (see Figure 13-1). Several properties of the action potential may be illustrated by this sort of procedure:

1. There is a characteristic *threshold* of stimulus. Thus, if electrical stimulation is used, impulses of a current below a specific level will produce no action potential. Once the threshold is exceeded, greater stimulus to a single cell soon reaches the point where it does not produce a greater action potential. The potential is thus for all practical purposes an *all-or-none* event somewhat analogous to a binary digit in a digital computer, where a particular number is either 1 or 0.

2. The form of an action potential is illustrated in Figure 13-2. The duration of the signal is of the order of 1 millisecond, so that the recording of the peak must employ a rapid form of measurement, usually an oscilloscope. Since to record the potential two electrodes must necessarily be employed, the action potential is most frequently recorded as a *biphasic* signal (Figure 13-2). This occurs when the altered potential passes first one electrode and then the other. If, for example, a positive shift in potential moves along a nerve fiber, the first electrode to measure it will record a positive signal with reference to the second, whereas when it reaches the second electrode, the second electrode will be more positive than the first, so that the difference fed to the oscilloscope will appear negative. Finally, the maximum amplitude of action potential that is recorded is of the order of 20 millivolts, although the exact value may vary considerably, depending on experimental conditions. Whereas the outer surface of the membrane is positive in the case of the resting potential, it is negative in

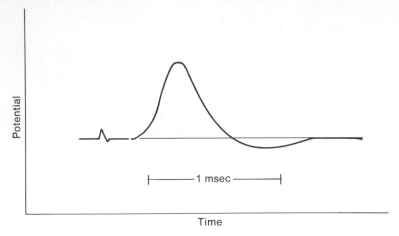

Potential

Time

——— 1 msec ———

Figure 13-2   Form of a biphasic action potential. The small spike preceding the main peak is an artifact due to electrical stimulation.

that of the action potential. This reversal of polarity is of great significance in understanding the underlying basis of excitability.

3. After passage of an action potential over a particular region of cell surface, a period of about 1 millisecond follows when stimulation fails to trigger a second action potential. This period is known as the *refractory period* and represents a recovery time for the membrane. It has the effect of limiting the frequency of pulses and hence the rate of information transfer across the cell surface. In the case of a nerve cell, one may say that the *information channel* exhibits a finite limit of information transfer rate, so that increased information flow can only be achieved by using additional channels.

4. The velocity at which an action potential moves along a membrane may be measured by separating two recording electrodes by a known distance and subtracting the time required for a peak to reach the near one from the time required to reach the far one (see Figure 13-3). Velocities thus measured range from 5 centimeters per second in nerves from especially phlegmatic invertebrates to about 100 meters per second in some vertebrates.

Figure 13-3   Measurement of the velocity of nerve transmission. The use of an oscilloscope with an accurately timed sweep enables one to time the passage of the impulse as it moves from one electrode to the next.

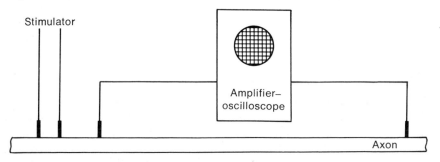

Stimulator

Amplifier–oscilloscope

Axon

Even the most rapid velocities recorded are many orders of magnitude too low to be explicable on the basis of any form of simple electrical conduction (as in a wire). Moreover, as a purely electrical signal passes along a conducting material, it becomes diminished in amplitude as the resistance of the material takes its toll. It is a property of action potentials that the amplitude is *independent of distance* from the point of stimulation, an observation that suggests that some form of active renewal of the signal occurs as it moves.

The ionic basis of the action potential

The resting potential of a membrane was explained in terms of an active Na^+–K^+ pump together with different rates of leakage of the two ions, Na^+ being by far the slowest. The reversal of charge polarity in a nerve cell reflects a sudden alteration in permeability of the cell to Na^+, which might be thought of as the rapid (ca. 0.1 millisecond) opening of a *Na^+ gate*. Sodium floods into the cell, and inward movement of the positive charge finally renders the outer surface of the membrane negative. When inward Na^+ flux is completed (when the peak in action potential is attained) the Na^+ gate closes and a K^+ gate opens, so that K^+ leaks out, restoring the original polarization of the membrane. Since both Na^+ and K^+ have been displaced in opposite directions away from their resting distribution, the final phase of the action potential response must be the operation of the Na^+–K^+ pump to exchange the ions to their original levels. This last phase of excitability is the only one that directly requires metabolic energy in the form of ATP. The first two phases use energy previously stored in the two ionic gradients of Na^+ and K^+.

Figure 13-4 Sequence of events in a nerve action potential. Thus, sodium influx precedes efflux of potassium. Finally, the two ions are re-exchanged by the action of an ATP-linked pump. Until the exchange is performed, additional action potentials cannot be transmitted (refractory period).

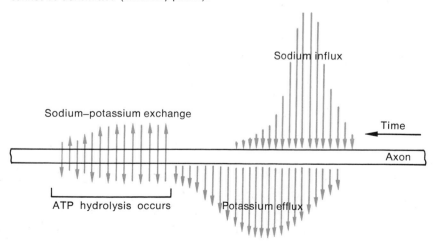

Opening of the Na^+ gate is triggered by a local potential change such as that produced by electrical stimulation of a nerve. Once the gate opens, the charge dislocation is magnified by the inward flux of Na^+, so that greater opening occurs. Thus, the Na^+ gate is, in a sense, autocatalytic, and it is not surprising that a stimulus great enough to produce any effect at all (threshold stimulus) leads to a maximum (all-or-none) effect. The same property of the Na^+ gate accounts for the spread of the action potential; inward flux of Na^+ produces a charge dislocation in neighboring parts of the membrane, which opens Na^+ gates there. The flux of Na^+ in the neighboring region leads to charge dislocation and, hence, gate opening in regions still more removed. The action potential will be propagated in both directions from the point of stimulation, but behind the "wave front" formed by the potential peak, there follows a region where the refractory period is in effect, so that no back propagation can occur (see Figure 13-4).

There is considerable direct experimental evidence for the version of action potential advanced here. For example, ionic flux has been measured across the membranes of nerve by means of radioactive isotopes of Na and K with results that are in conformity with the picture given. It has been possible, moreover, to remove most of the cytoplasm from large nerve cells and to replace it with salt solutions. Effects of altering salt balance across the membrane occur as would be predicted, and it is important that the membrane remain quite intact; normal transmission

Table 13-1   Examples of chemical transmitter substances

Compound	Structure	Distribution (as transmitter)
Acetylcholine	$(CH_3)_3$ N+ CH_2 CH_2 O $COCH_3$	Nerve–muscle junction (as well as other synapses)
Norepinephrine	(structure)	Postganglionic fibers of sympathetic nervous system
Glutamate	COO^- CH_2 CH_2 $CHNH_2$ COO^-	Some crustacean systems
γ-Aminobutyrate	COO^- $CH_2CH_2CH_2NH_2$	Some crustacean systems

can be observed even with cytoplasm removed, but only as long as there exist no short circuits in the form of holes in the membrane.

In the preceding section, we have presented a highly generalized pattern of excitability. The association between action potential and ionic flux appears to be a universal feature of excitation, although the exact nature of the flux may differ in different cells. For example, in some algae, the role of Na^+ appears to be fulfilled by an anion, Cl^-, while divalent cations, such as Ca^{2+}, are involved in some invertebrate cells. Moreover, we have not touched on the transmission of action potential between different cells, of which the best example is the animal synapse. Such transmission appears to be mediated either by one of a class of compounds called *chemical transmitters* or, in a few cases, by a direct mechanism involving charge movement. Some examples of chemical transmitters are shown in Table 13-1.

Finally, it should be added that the precise mechanism of "opening" the ionic gates of excitable membranes has yet to be described and remains one of the most compelling questions in cell study. A number of recent studies of this question have concentrated on using artificial membranes, in many cases "doped" with ion-conducting antibiotics (see Chapter 12), and potential changes have been produced in such membranes with many of the properties of a natural action potential.

The electroplax

An interesting extreme in the evolution of excitable membranes is the *electric organ* of certain species of fish (for example, the catfish and the eel). The organ, which is homologous with muscle and develops from the same embryonic primordia, is composed of individual units called electroplaxes, each of which is highly membranous and which can, on stimulation, develop an action potential of the order of 100 millivolts. The electroplaxes, however, are stacked in such a manner that their potential differences add vectorially, and the organ as a whole can generate a potential difference of several hundred volts.

The ability of these fish to deliver a jolt of such magnitude enables them to protect themselves effectively against predation as well as to stun their own prey. Since the stunning effect of high voltages can probably be attributed to a generalized action at the organism's excitable membranes, it is especially interesting that the electric fish itself is not stunned by its own discharge. It has been shown recently that other electric fishes rely on their electrical discharge for information about their (often muddy) surroundings. These fish emit periodic discharges of relatively low voltage, which produce an electrical field about the animal, and the presence of objects, within the field, with dielectric properties that differ from those of water distort the field. The fish appears able to monitor his own field and so can detect foreign objects.

The electroplax has been a useful tissue for the study of membrane processes in general. For example, the addition of ion-conducting

315

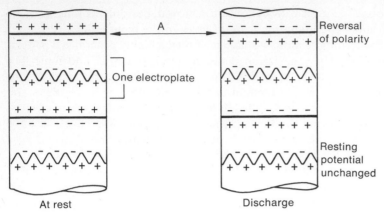

Figure 13-5   Electric organ of certain tropical fishes. In the electric eel, *Electrophorous electricus*, the electric organ includes membranes derived from both neural and muscle tissue. In this diagram, the straight lines denote neural membranes, the wavy lines, membranes of nonneural origin. Discharge occurs when the polarity of neural membrane (arrow A) undergoes a change in direction. Since no opposing change occurs in the next membrane, the effect of many of the plates discharging together is additive, in the sense of batteries arranged in series. In this fashion, the individual membrane-potential changes of 70 millivolts are summed up to yield a jolt of as much as 500 volts for the entire organ.

antibiotics to an isolated electroplax produces potential changes that reflect the ion specificity of the antibiotic added. The addition of gramicidin produces a transient potential change, whereas the addition of valinomycin does not. This appears due to valinomycin's greater specificity, which allows it to conduct potassium across a membrane but not to admit sodium. Gramicidin conducts both ions, so that it may be suggested that, in this system, at resting potential potassium is close to equilibrium across the membrane, while sodium is not. Thus, making the membrane freely permeable to potassium by the addition of valinomycin is without effect, while opening the membrane to sodium with gramicidin collapses the resting potential, producing the change noted.

Suggested Reading

Books

Bourne, G. H., *The Structure and Function of the Nervous System*, vols. 1–3, Academic Press, New York, 1969.

Brazier, M. A. B., *The Electrical Activity of the Nervous System*, 3rd ed., Williams and Wilkins Co., Baltimore, 1968.

Eccles, J. C., *The Physiology of Synapses*, Springer-Verlag, Berlin, 1964.

Katz, B., *Nerve Muscle and Synapse*, McGraw-Hill Book Co., New York, 1966.

Lajtha, A. (ed.), *Handbook of Neurochemistry*, Plenum Press, New York, 1969.

Wiersma, C. A. G. (ed.), *Invertebrate Nervous Systems: Their Significance for Mammalian Physiology*, University of Chicago Press, Chicago, 1967.

Articles

Adrian, R. H., "The ionic basis of excitability," in *The Molecular Basis of Membrane Function*, D. C. Tosteson (ed.), Prentice-Hall, Englewood Cliffs, N.J., 1969, p. 249.

Baker, P. F., "The nerve axon," *Sci. Am.*, **214**, 74 (Mar., 1966).

Bennett, M., "Comparative physiology: electric organs," *Ann. Rev. Physiol.*, **32**, 471 (1970).

Changeux, J. P., and J. Thiery, "On the excitability and cooperativity of biological membranes," in *Regulatory Functions of Biological Membranes*, J. Jarnefelt (ed.), Elsevier Publishing Co., Amsterdam, 1968.

DeRobertis, E. D. P., "Molecular biology of synaptic receptors," *Science*, **171**, 963 (1971).

Hurlbut, W. P., "Ion movements in nerve," in *Membranes and Ion Transport*, E. E. Bittar (ed.), vol. 2, John Wiley & Sons, New York, 1970, p. 95.

Jain, M. K., R. Marks, and E. H. Cordes, "Kinetic model of conduction changes across excitable membranes," *Proc. Nat. Acad. Sci. U.S.*, **67**, 806 (1970).

Martin, A. R., and J. L. Veale, "The nervous system at the cellular level," *Ann. Rev. Physiol.*, **29**, 401 (1967).

Rashbass, C., "Sense organs—transducers of the environment," in *Progress in Biophysics*, W. Fuller, C. Rashbass, L. Bragg, and A. C. T. North (eds.), W. A. Benjamin, New York, 1969.

14 Mechanical work by cells

The power of motion is a general property of living matter. Cells are able to move their parts and, often, themselves by a variety of means, and thereby perform mechanical work. We have described the fact of irritability (or excitability) in terms of stimulus and response, and the response of a cell often involves motion. This is in keeping with the significant ability of many cells and aggregates of cells (multicellular organisms) to move in space in such a manner as to minimize certain disadvantages connected with one particular region. This aspect of physiology is often included in the study of behavior and is exemplified by the movement of a cell away from a noxious stimulus or toward a relatively beneficent portion of space (one containing food). Such directed mechanical response is clearly of the greatest importance in considering the control of cell activities, and it must be said that relatively little is known about the details of its mechanism and, further, that experiments directed to such investigation have proved not at all easy to design.

Cell motion falls into two categories—motion of the whole cell and motion within the boundaries of a cell. The former takes place by means of flagella and cilia (short, numerous flagella) as well as by the little-understood process of gliding motility found in blue-green algae and Myxobacteria. The muscle cells of complex organisms move as a unit by means of intracellular muscle filaments involving mechanisms to be described later.

On the other hand, intracellular motion includes such varied processes as protoplasmic streaming, events connected with mitosis, contraction of such organelles as mitochondria, and pinocytosis. It is well known that, in mitosis, chromosomes are drawn apart by a process associated with the spindle apparatus and that some form of mechanical work is associated with the separation of a dividing cell into its two progeny. The limiting membrane of cells is likewise capable of movement, as seen in the active inpocketing associated with the engulfment of external fluid (pinocytosis). Special organelles with the power of motion, often bearing a close relation to the cell membrane, include trichocysts and contractile vacuoles, found in algae and protozoa. Mitochondria have been shown to undergo cycles of contraction and swelling related to events associated with energy transfer. A contractile protein rather like that of muscle has been isolated from mitochondria and is believed by some to provide the basis for mitochondrial contraction, although this is not universally accepted. Similar cycles of volume change have been observed in isolated chloroplasts.

Although the various types of cell motion are associated with different parts of cells and different cells, it appears likely that most, or all, share a common underlying molecular basis. Thus, wherever cell movement has been studied on the molecular level, it has turned out to involve a class of contractile proteins of which muscle actin and myosin are the best-known examples. Moreover, the energy source for movement is the hydrolysis of ATP to yield ADP and inorganic phosphate. In this connection, generally one or more of the contractile proteins exhibits ATPase activity and therefore clearly represent the locus where the chemical form of energy (the terminal phosphate bond in ATP) is converted into the mechanical work of contraction.

General features of muscular contraction

Since most of what is known about cellular movement has been learned from studies of muscle, we shall describe muscular contraction in some detail and only then branch out to other systems. Muscle tissue is highly adapted for transducing chemical-bond energy into mechanical work, and it is worth examining its structure and some of the gross features of the contractile process before examining contraction on a molecular level. In the following discussion, we shall concentrate on the properties of striated muscle from vertebrates, although most general conclusions will be applicable to other sorts of muscle as well.

The organization of a striated muscle cell is illustrated in Figure 14-1. The cell is long and cylindrical, the long axis being the direction of contraction. The cell, in addition to exhibiting many of the ordinary organelles of the eukaryotic cell, including the nucleus, ribosomes, and mitochondria, contains numerous *fibrils*, which represent the contractile apparatus of the cell and which are made of repeating units, the *sarcomeres*. Finally, the sarcomeres possess a characteristic pattern of banding, consisting of a broad *A-band*, a narrower *I-band*, and, within the I-band, a dark *Z-line*.

When muscle cells contract, there can be a shortening of about 30 percent, and the length of an individual sarcomere can contract from about 3 to about 2 microns. To a first approximation, contracting muscle does not undergo a change in volume, so that the shortening is accompanied by thickening. Moreover, it is possible for muscle to exert a force without actually contracting in length (as in the case where muscle is mechanically kept at constant length while stimulated to contract). Such a situation is termed *isometric contraction* in contrast to *isotonic contraction*, where shortening occurs. Maximum force can be exerted in the isometric case. Finally, muscular contraction is accompanied by the liberation of heat, so that available chemical energy gives rise to both work and heat. The

319

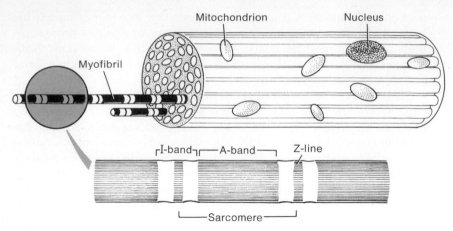

Figure 14-1 Construction of skeletal muscle. An intact muscle consists of numerous, long cells (also called muscle fibers). Within the cell are numerous myofibrils, which are the contractile organelles. These show a characteristic repeating structure, which is shown in more detail below. The striated appearance of muscle tissue results from the myofibril's band configuration. Each repeating unit is referred to as a sarcomere, which is about 2.5 microns in length.

heat represents a nonproductive dissipation of energy, so that the process is increasingly inefficient as the heat of contraction increases. It is interesting that isometric contraction is accompanied by much less heat loss than isotonic; in other words, heat loss is largely associated with the process of shortening rather than the generation of force.

We shall see that the energy for contraction is obtained by the hydrolysis of the terminal phosphate bond of ATP, and that it has been possible to correlate the amount of ATP hydrolyzed to ADP with the amount of work done. Thus, under optimal conditions, about 6.6 kilocalories of work is obtained per ATP molecule hydrolyzed. Since the free-energy change for ATP hydrolysis is of the order of 10 kilocalories, one can conclude that the conversion of chemical-bond energy to work is quite efficient, and, indeed, much more efficient than machines usually devised by man.

The initiation of contraction

In an intact animal, skeletal muscle contracts when stimulated by a motor (that is, muscle-stimulating) nerve. In contrast, other kinds of muscle, including heart and the smooth muscle of the gut, contract rhythmically even when all nerve connections are removed, so that they appear to be self-stimulating. The motor nerve is in contact with the plasma membrane of a skeletal muscle cell in a restricted region called the *myoneural junction*, and, when the nerve is stimulated, it releases the transmitter substance, acetylcholine, which produces a local potential change at the muscle-cell surface. If the change in potential exceeds the threshold for the muscle cell, an action potential travels over the cell membrane, and the cell contracts. The properties of the action potential are similar to those described in Chapter 13 for the case of nerves. Likewise, the muscle cell exhibits a resting potential of about 70 millivolts with the exterior positive, and, on stimulation, the action potential is of opposite polarity. The stimulus for muscle excitation can be either stimulation of an attached motor nerve or, experimentally, direct electrical, chemical, or mechanical stimulation. The action potential travels over the surface of a muscle cell at a rate of about 1 meter per second, which is considerably slower than most vertebrate nerves.

The excitability of a muscle cell is, as we have seen, closely connected with the passage of ions across the plasma membrane, while contraction itself is a property of the fibrils and more particularly the contractile proteins therein. It appears that communication between the primary excitation event and contraction is by way of ion movement, specifically movement of Ca^{2+}. Although the detailed mechanism is not understood, it is likely that an action potential at the plasma membrane is linked to the release of Ca^{2+} in the immediate region of the contractile proteins by means of a system of tubules (the *T system*), which is continuous with both the plasma membrane and the endoplasmic reticulum, also called, in the case of muscle, the *sarcoplasmic reticulum*. The continuity of the plasma membrane and this system of intracellular tubules is of the greatest importance in the initiation of the contraction, for it provides an instantaneous way for the depolarization of the membrane to influence ionic balance in the immediate region of the muscle fibers. This is essential for an explanation of contraction, since it has been shown that all fibers of vertebrate skeletal muscle cell can contract synchronously, so that the fibers in the center of the cell and those near the plasma membrane must be fired off at once. The time required for calcium (or ATP) to diffuse from the plasma membrane to the interior would be intolerably long for this to occur without intervention of the tubule system. Calcium ion then would serve as the trigger for contraction of the fibril proteins, and it is interesting that, under some conditions, addition of Ca^{2+} does lead to contraction in *in vitro* systems derived from muscle.

Studies with purified muscle proteins indicate that the picture must be more complex than the statement above suggests. For one thing, the binding of ATP to actomyosin (as measured by viscosity change) leads to its "contraction" and, under suitable conditions, this can occur in the absence of calcium. Indeed, it is important to separate in one's mind the process of ATP binding, which appears to trigger conformational changes in model systems, and ATP hydrolysis, which requires Ca^{2+} and which may be involved in returning the muscle protein to its initial state. Thus, in such purified systems, there are two reactions that need be considered, both of which are spontaneous (with negative standard free-energy changes) and both of which appear to be related to events in intact muscle:

(I) $Actomyosin_r + ATP \rightleftharpoons Actomyosin-ATP_c$

(II) $Actomyosin-ATP_c + H_2O \overset{Ca^{2+}}{\rightleftharpoons} Actomyosin_r + ADP + P_i$

321

In these reactions, r stands for "relaxed" (that is, high-viscosity) actomyosin, and c denotes the "contracted" (low-viscosity) state. In this system, it seems that the role of calcium is restricted to the return to the relaxed state, the reaction that leads also to ATP hydrolysis.

While these studies with model systems appear to be instructive in trying to understand the contractile process, they cannot be regarded as perfect representations of the situation in living muscle, and the exact relationship between the two sorts of systems remains tantalizingly obscure. For one thing, the role of calcium appears to be quite different in the two systems. We saw that calcium was required, not for contraction in the model system but rather for the relaxation process. In muscle cells, on the other hand, calcium plays a role in initiating contraction, and relaxation, in part, seems to involve removal of calcium into vesicles of the sarcoplasmic reticulum.

Considerable thought has gone into trying to reconcile the results from the two sorts of experiment, but without notable success. In considering the possible role of calcium and ATP in the contraction–relaxation cycle, it is also important not to ignore the presence of mitochondria, which are numerous in muscle and which are found in close proximity to the fibers where contraction occurs. For example, most of the free calcium in heart muscle is contained in the mitochondria, so that the permeability of the mitochondrial membranes to calcium must significantly regulate the concentrations of calcium in the extramitochondrial space. This possibility suggests the sort of mechanism that might account for the initiation of contraction and relaxation in intact muscle and which might also be in conformity with studies on purified proteins. Thus, we can consider the movement of calcium in several phases in the muscle:

1. Stimulation results in the release of calcium from the sarcoplasmic reticulum. Calcium enters the free cytoplasmic space (space 2 in Figure 14-2).

2. Calcium is actively taken up by mitochondria with the expenditure of energy. As we saw in Chapter 6, this amounts to a transitory uncoupling of the mitochondrial respiratory process. One effect of mitochondrial calcium transport is an increase in the transport of ATP or ADP across the membrane, which is often measured as an inward flux of ATP, whose negative charge causes it to be carried in with positive calcium ion. Under the conditions that occur in the muscle cell, with a high intramitochondrial ATP concentration and inorganic phosphate accompanying calcium entry, the effect is more likely a release of ATP from the mitochondrion, and it is hypothesized that this ATP binds to the nearby actomyosin to trigger contraction.

3. Relaxation probably can be decomposed into two events. First, a decline in the ATP concentration in mitochondria, owing both to the efflux of ATP and the utilization of energy to support calcium uptake, leaves the mitochondrion in an energy-poor state. Calcium is known to leak rapidly from such mitochondria and is hence available to stimulate the ATPase reaction at the myofibril (reaction II, above).

4. The second, and final, phase of relaxation is then the active uptake of calcium from space 2 (Figure 14-2) into the vesicles of the sarcoplasmic

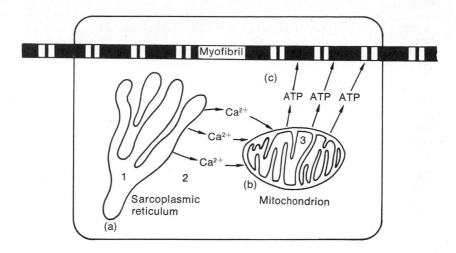

Figure 14-2 Membrane events associated with the triggering of muscle contraction. This diagram shows the three topological spaces that appear to be significant in contraction and relaxation. 1 represents the space within the sarcoplasmic reticulum, 2 the free cytoplasm, and 3 the intramitochondrial space. (a) On stimulation, calcium is released from the sarcoplasmic reticulum. (b) It is taken up by the mitochondria by energy-linked transport. This leads to release of ATP (c), which is then available to stimulate and provide energy for contraction at the myofibril. For additional details, as well as a description of relaxation in these terms, see text.

reticulum, an uptake that is coupled to ATP hydrolysis and which leaves the reticulum charged with calcium and ready to initiate another cycle of activity.

It should be added that the details of this sequence of events are presently under intensive study and the overall picture should be regarded as a sort of working hypothesis. These events are consistent with much of what is known about muscular contraction as well as with the known properties of muscle mitochondria. A central feature of this approach is the conclusion that the mitochondria must be regarded as more than ATP factories and that their ability to transport calcium is not an artifact, but an essential part of cellular-energy transformations.

Earlier studies of the process of relaxation led to the conclusion that there was a chemical *relaxing factor*, which was required for completion of the cycle, and a great deal of effort went into its isolation and analysis. This factor turned out to possess ATPase activity and was sedimented at high velocities in the ultracentrifuge. Electron microscopy of the relaxing factor revealed that it consisted of closed membranous vesicles and it now seems certain that the factor is, in fact, fragments of the sarcoplasmic reticulum. The ATPase activity was a reflection of the coupled transport of calcium, and the ability of these vesicles to promote relaxation represents their uptake of calcium, which is then sequestered in their interior. The ability of fragments of the sarcoplasmic reticulum to form closed vesicles is reminiscent of the formation of submitochondrial particles and subchloroplast particles from those organelles.

The sliding-filament explanation of contraction

The characteristic banded appearance of striated muscle appears to reflect periodicity on the molecular level and has led to an ingenious theory of muscular contraction, first proposed by Huxley (see Suggested Reading). They suggest that the bands (A-band and I-band) arise from the overlap of many parallel filaments of two sorts, as shown in Figure 14-3. The thin filaments, composed of the protein *actin*, are attached to a plate, the Z-line, while the thick filaments, which are largely composed of the protein *myosin*, are arranged between. In the region where the two classes of filament overlap, each thick one is surrounded by six thin filaments.

Contraction of the muscle cell appears to occur by means of the two classes of filaments sliding along each other, so that the degree of overlap is increased. Thus, contraction can be considered in terms of the intramolecular forces between the two classes of protein molecules, which are arranged in close array and parallel to each other. Indeed, cross-linkage between the two molecules has been demonstrated directly by electron microscopy and the basis of such cross-linkage is discussed below. The sliding-filament model is consistent with virtually all information available about muscular contraction and about the detailed morphology of muscle.

For example, if muscle is stretched, it loses the ability to exert maximum force. According to the sliding-filament model, this result is a direct consequence of the geometry of the filaments. Since the force of muscle is generated by interaction between the two classes of filament, it will depend on the degree of overlap between them. In stretched muscle the overlap is diminished and so the force is lessened.

According to this model, the actual contractile event is associated with the interaction between the two protein classes and must be cyclic in nature. Thus, one increment of shortening must include the following:

1. Establishment of a cross-link between the two classes of protein,
2. Breakage of the link on binding of ATP,
3. Hydrolysis of ATP (the myosin ATPase reaction), and
4. Creation of a new link, but with the two molecules in a new orientation with respect to each other.

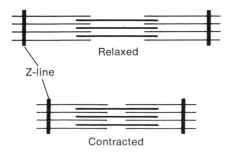

Relaxed

Z-line

Contracted

Figure 14-3 Sliding-filament model for muscular contraction. When myofibrils are examined with the electron microscope, the relaxed and contracted states differ in the relative dimensions of the bands in such a fashion as to support this model. In the diagram, the thin filaments attached to the Z-line are composed of the protein actin; the thick filaments are composed of myosin.

It is important to note that the formation of cross-linkage between molecules does not require ATP binding but that breakage does. This feature is in conformity with the well-known phenomenon of *rigor mortis*, which is an irreversible contraction of muscle upon the death of an animal. This can be construed as the result of spontaneous formation of cross-linkages under conditions where ATP is no longer available for their breakage. The role of ATP in link breakage is also consistent with the rough correlation between ATP expended and length of contraction. Holding muscle in a contracted state without further shortening requires relatively little energy, presumably because relatively little breakage and re-formation are taking place.

Contractile proteins of muscle

A number of proteins have been isolated from muscle and are deeply implicated in the contractile process. For example, *actin* is the major component of the thin filaments. It has been isolated by salt extraction of muscle and when so obtained is a globular protein with a molecular weight of about 50,000 to 70,000 daltons. When globular actin (G-actin) is incubated with ATP in the presence of Mg^{2+}, it polymerizes to form a much larger aggregate, fibrous actin (F-actin), which probably more closely resembles the native actin of the muscle filaments. Actin is also interesting in that it contains the highly unusual amino acid ε-(*N-methyl*)-*lysine*,

$$\begin{array}{ccccc} CH_3 & & & & NH_2 \\ | & & & & | \\ HN-CH_2-CH_2-CH_2-CH_2-C-COOH \\ & & & & | \\ & & & & H \end{array}$$

which also occurs in some other proteins associated with cell motility (see below). G-actin also possesses binding sites for ATP and for calcium ion, binding one of each per protein molecule. Closely associated with actin in the thin filaments is a second protein, *tropomyosin*, whose function is unknown and which is largely (>90 percent) in the form of the alpha helix. In turn, closely associated with tropomyosin is another protein, *troponin*, with a molecular weight of about 86,000 daltons. This protein complexes strongly with tropomyosin and appears to function in the regulation of contraction through the binding of calcium (see Table 14-1).

Tropomyosin occurs in two forms, A and B. Of these, A is fibrous (insoluble in water) and occurs in the muscle of certain mollusks that have the interesting property of locking in the contracted state. It has been supposed that tropomyosin A in some way provides the "catch" mechanism. Tropomyosin B is water-soluble and is found in all muscle that has been studied. When tropomyosin B and fibrous actin are mixed, they form a complex that is thought to be the state in which the two molecular species occur in the contractile apparatus.

Table 14-1   Protein composition
              of fibers from rabbit
              skeletal muscle

Component	% of total protein
Myosin	54
Actin	25
Tropomyosin	11
Other proteins	10

The thick filaments are composed mostly of *myosin*, which is a large protein with a molecular weight of about 500,000 daltons. The molecule is greatly elongated, with a length of about 1,600 Å and with about 60 percent alpha helix, and is water-insoluble. Electron micrographs show myosin to have a globular end and a filamentous end, as shown in Figure 14-4. The globular end appears to be the site of greatest affinity for actin and so probably participates in cross-linkage between the two kinds of filaments. Myosin also possesses ATPase activity, and the locus of activity resides on the globular end. This is known largely from studies where myosin was cleaved with a proteolytic enzyme to yield *heavy meromyosin* (and ATPase activity), which includes the globular end, and *light meromyosin*, which is filamentous.

The ATPase activity catalyzed by myosin has a number of interesting properties. For one thing, it exhibits two pH maxima, one at pH 9.5 and one at about pH 6. The enzyme is most active in hydrolyzing the terminal phosphate of ATP but reacts also with the other nucleotide triphosphates, CTP, GTP, and ITP. The protein carries a number of titratable sulfhydryl groups on its surface, of which some are essential for ATPase activity.

In the sliding-filament approach, contractility is a property of neither actin nor myosin alone but rather of their interaction. Likewise, neither protein is contractile alone, but it is possible to combine them *in vitro* so that some of the contractile behavior of muscle may be demonstrated. A

Figure 14-4   Myosin molecule. The tail is superhelical in the sense that each of the two units that are wound together is itself composed of protein in the alpha helical configuration. Binding to actin, as well as ATPase activity, is associated with the globular heads. The point at which the proteolytic enzyme trypsin cleaves myosin into light and heavy meromyosin is shown with an arrow.

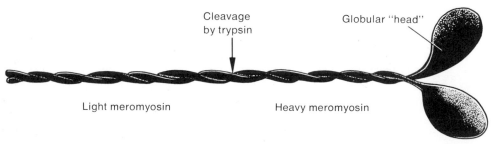

Cleavage
by trypsin

Globular "head"

Light meromyosin   Heavy meromyosin

complex of actin and myosin may be prepared by mixing the two proteins and is called *actomyosin*. Actomyosin, like myosin alone, catalyzes an ATPase reaction,

$$ATP + H_2O \rightleftharpoons ADP + PO_4^{3-}$$

which requires Mg^{2+}. In this requirement, actomyosin ATPase differs from myosin ATPase, which is actually inhibited by Mg^{2+}, so that the connection between the two activities is unclear.

The actomyosin complex is the most useful of the model systems alluded to earlier and studies with it have been influential in providing evidence for the sliding-filament explanation for contractility. We noted that binding of ATP to actomyosin leads to a change in viscosity, and the mechanism by which this occurs lends support to the idea that there is a useful connection between the properties of the purified protein complex and intact muscle. The characteristic viscosity decrease is illustrated in Figure 14-5, and it is clear that the effect is readily reversible. Thus, when the ATPase action of the complex has hydrolyzed the added ATP to form ADP and phosphate, the viscosity resumes its prior value. It is now clear that the viscosity change in the presence of bound ATP is the result of dissociation of the actomyosin complex into the components, actin and myosin. This dissociation appears to be identical to the breakage of bonds between the two strands in the sliding-filament model, so that the viscosity changes are probably a valid measure of an event associated with the contraction of intact muscle, at least as far as the sliding-filament model is itself an accurate representation of reality.

The validity of studying systems of purified muscle protein is supported also by experiments with actomyosin that has been precipitated into an insoluble sheet that contracts (and performs work) on addition of ATP.

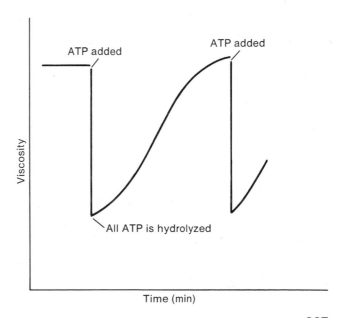

Figure 14-5 Viscosity change undergone by actomyosin on addition of ATP. Since actomyosin exhibits ATPase activity, the amount of added ATP decreases with time. When it approaches zero, the viscosity returns to its original state. Subsequent addition of more ATP allows a second cycle of viscosity change.

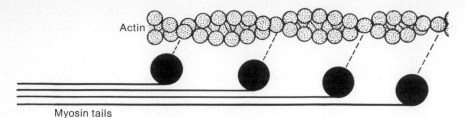

Actin

Myosin tails

Figure 14-6 Schematic view of how myosin heads may serve as a point of attachment between actin and myosin. (Recall that the binding site for actin resides on the head in the heavy meromyosin fraction.) The process of sliding (during contraction) is envisioned as requiring a sequence of breakage and reformation of bonds between the heads and the actin filament. In simplistic terms, the heads may provide a sort of a ratchet mechanism.

When this occurs, the ATP is hydrolyzed to ADP and inorganic phosphate, so that the ATPase reaction is coupled to mechanical work. A number of properties of the actomyosin system differ from contraction by real muscle, but actomyosin nonetheless serves as a useful model.

The energy source for contraction

The discovery of ATPase activity in muscle protein might be expected to remove any ambiguity as to the source of energy for contraction. Quite on the contrary, the energy source for *in vivo* motion has proved somewhat elusive, with a number of molecules being suggested as possible candidates from time to time. In the first place, contraction may take place under either aerobic or anaerobic conditions. In the absence of oxygen, energy production proceeds by way of glycolysis, leading to the production of lactate and ATP. When glycolysis is blocked by an inhibitor, such as iodoacetate, contraction can still take place, but with a loss of intracellular energy. Under such conditions, it is of obvious interest to see what molecules decline in concentration, as one would expect any one that was "energy-rich" to become exhausted.

Under such conditions of limited energy input and repeated contraction, one observes that the ATP content of muscle cells remains constant, while that of creatine phosphate declines. The reason for this turns out to be not a primary role for creatine phosphate in contraction but, rather, a reaction whereby creatine phosphate transfers phosphate to ADP. This reaction is catalyzed by the enzyme *creatine phosphokinase* and has an equilibrium far in the direction of creatine and ATP:

$$^-O-\overset{\overset{O}{\|}}{\underset{\underset{O^-}{|}}{P}}-NH-\overset{\overset{CH_3}{|}}{\underset{\underset{NH_2}{|}}{\overset{+}{C}}}-N-CH_2-COOH + ADP \;\rightleftharpoons\; H_2N-\overset{\overset{CH_3}{|}}{\underset{\underset{NH_2}{|}}{\overset{+}{C}}}-N-CH_2-COOH + ATP$$

Creatine phosphate   Creatine

ATP is, in fact, the primary energy source for muscular contraction, but as soon as any is hydrolyzed under the catalysis of actomyosin, the thermodynamically favorable phosphate transfer takes place to maintain the ATP level in the cell. Evidence for the direct involvement of ATP comes from a number of arguments, but primarily from the fact that, when creatine phosphokinase is inhibited by dinitrofluorobenzene, contraction is entirely at the expense of ATP hydrolysis. A role for ATP as a primary energy source is entirely consistent with its known major function in other energy-requiring processes, as well as the striking juxtaposition of mitochondria (sites of major ATP formation) to the contractile fibers of muscle cells (where it is used).

Creatine phosphate is, however, obviously not without function in the muscle, where it provides an important reservoir of energy. When the creatine phosphokinase reaction is inhibited by dinitrofluorobenzene, recovery from each contraction event is slower, since net ATP synthesis from ADP and phosphate is required (instead of the rapid transfer of phosphate from creatine phosphate). Obviously, the supplies of either ATP or creatine phosphate in muscle tissue are finite and the ultimate source of energy for movement must be some form of primary energy conservation, such as oxidative phosphorylation or anaerobic glycolysis (organisms with muscle not being noted for carrying on photosynthesis). In general, red muscle tissue, such as the dark meat of birds and most mammalian muscles used for running, gains most of its energy from mitochondrial oxidative phosphorylation with acetoacetate and fatty acids used as substrates in the case of noncontracting muscle, and glucose used as well when activity is high. On the other hand, white muscle (such as the best part of a turkey) relies more heavily on anaerobic glycolysis for energy, which is thus much less efficiently obtained. It should be added that the visual distinction between white and dark meat reflects the concentration of mitochondrial oxidative enzymes, such as cytochromes, as well as that of the oxygen-binding pigment myoglobin, both of which are higher in the case of dark meat. Myoglobin acts to ensure that the oxygen

Figure 14-7 Flow chart illustrating energy source of contraction. Note the role of creatine phosphate as energy buffer. The reader will recall that the myokinase reaction leads to the conversion of ADP to ATP + AMP: 2 ADP $\rightleftharpoons$ ATP + AMP

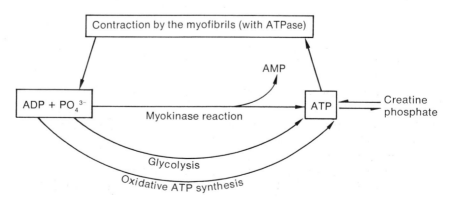

content of muscle tissue be stabilized, obviously a much more important matter in cases where energy production relies on oxygen for the carrying out of oxidative phosphorylation.

Oxygen debt

It is well known that, following muscular exercise, rapid breathing is continued for some time, even though the need for ATP for contraction would be expected to have diminished. This increased ventilation is produced by a number of factors, including a lowered pH of the blood due to the presence of increased lactic acid. The lactic acid is formed by the reactions of glycolysis during the period when oxidation via the cytochrome system and the Krebs cycle is unable to keep up with the demand; the effect of the continued increased oxygen consumption is to oxidize some of the lactic acid to CO_2 and water, with concomitant ATP synthesis. This ATP is utilized, in turn, to convert the remaining lactic acid to glycogen, a reaction that occurs by means of the pathway for gluconeogenesis (Chapter 8), enzymes for which are located largely in the liver. Thus, in vertebrates, the metabolic events associated with muscular contraction are not confined to muscle tissue alone and occur, via the bloodstream, in remote sites in the body.

Diseases of muscle

It is a truism that the study of diseases depends heavily on so-called "pure" research into the normal functioning of biological systems. Indeed, the use of public funds for unapplied biological research is usually justified on the grounds that practical (medical) benefits will eventually accrue. It is not commonly recognized, however, that the reverse line of application is also possible. In fact, the study of the cell has often benefited from highly practical investigations of human disease, which often represents an instructive permutation of the normal situation. For example, there exist a number of degenerative diseases of muscle tissue, grouped together under the general heading *myopathy* (or muscular dystrophy). The study of these conditions has not led to significant advances in their clinical management, although there appear to be hopeful signs that improved understanding is not far off. Such diseases have, however, provided a strong stimulus to understand the operation of the normal muscle fiber, and we shall now describe some features of these investigations.

We shall be primarily concerned here with *congenital myopathies*, which are inborn and apparently genetic. There are similar conditions resulting from injury to the nerves that activate skeletal muscle, as well as those having a nutritional basis. In the latter connection, it was found a number of years ago that when rats were maintained on a diet low in vitamin E (α-tocopherol)

$$\text{H}_3\text{C} \quad \overset{\text{CH}_3}{\underset{\text{H}_3\text{C} \quad \text{H} \quad \text{H}}{\bigcirc}} \quad \overset{\text{CH}_3}{\underset{\text{H}}{\text{O}}} \overset{}{\underset{\text{H}}{\overset{}{|}}} \text{CH}_2\text{CH}_2\text{CH}_2\overset{\text{CH}_3}{\overset{|}{\text{C}}}\text{HCH}_2\text{CH}_2\text{CH}_2\overset{\text{CH}_3}{\overset{|}{\text{C}}}\text{HCH}_2\text{CH}_2\text{CH}_2\overset{\text{CH}_3}{\overset{|}{\text{C}}}\text{H} \underset{\text{CH}_3}{}$$

they developed a progressive degeneration of muscle quite similar to that observed with human congenital muscular dystrophy. This led to attempts to reverse the course of the human disease by medication with vitamin E—unfortunately without success. Subsequent experiments centered on other quinone-like compounds, including ubiquinone, but these were not beneficial either. Studies with vitamin E-deficient rats also indicated that the vitamin played some role in reproduction, as the deficient animals became sterile. This result encouraged the use of vitamin E in all sorts of human sterility (and other sexual problems), an approach that was likewise of questionable use. Indeed, examination of the catalogs of less reputable drug firms reveals that vitamin E is still being sold with the claim of effectiveness in cases of sterility, although, as we have said, the evidence in support of this theory has come from studies with rodents. No well-defined vitamin E-deficiency disease has been described in human beings, and this fact should be taken as an indication that the transfer of results from one species to another should never be done in a careless fashion.

To return to muscular dystrophy, the apparent genetic nature of the disease has led to the hope that it would turn out to be a simple metabolic disease, that is, one resulting from the lack of one cellular enzyme (as in the case of *phenylketonuria*, where cells lack *phenylalanine hydroxylase*). An indication that this might be a fruitful approach appeared in the form of the observation that patients with dystrophy usually excreted abnormal amounts of creatine, a compound important in muscle energy balance, in their urine. This, in turn, has led to the findings that the enzyme *creatine phosphokinase* is low in dystrophic muscle and that it occurs with an altered pattern of isoenzymes (see Chapter 4) and with a probable alteration in primary structure. These results have encouraged some workers in the field to regard dystrophies as a genetic deficiency of this single enzyme.

However, additional study of dystrophic muscle reveals a much more complicated story. For example, many other enzymes of muscle are reduced in concentration, and it is likely that many of these may have simply leaked out of muscle tissue owing to general disruption of the muscle structure. Many of the enzymes that are low in dystrophic muscle are elevated in the serum, so that this is a likely explanation. It is probably significant that most enzymes that appear to leave the muscle are those that are found in the nonmembranous part of the cell. The enzymes of the mitochondrion, such as *succinate dehydrogenase* or *cytochrome oxidase*, are found at normal levels, while such "soluble cytoplasmic" enzymes as *aldolase* and *phosphoglucomutase*, which are involved in carbohydrate utilization, are considerably lower.

When dystrophic muscle is examined with an electron microscope, it appears that the normal muscle fibers have degenerated and been progressively replaced with connective tissue and fat. It also appears that the capacity of muscle fibers to regenerate is lost, a property that separates dystrophies from other diseases of muscle. Changes in the contractile properties of muscle tissue can probably be attributed entirely to the degeneration of muscle fibers and not to alteration in the contractile proteins of the surviving ones. Despite this, muscular dystrophy has served as a stimulus for extensive investigations of muscle proteins in the hope that some clues might turn up.

Finally, another dimension of the complexity of the disease becomes apparent from the observation that cells from mice and chickens with genetic muscular dystrophy respire more rapidly than do normal controls. It appears that the increased cell respiration leads to an increase in ATP synthesis and is not, therefore, some sort of uncoupling of oxidative phosphorylation. Moreover, the increase in respiration has been noted in liver cell mitochondria from dystrophic mice and so probably cannot be attributed to some degenerative change in muscle fibers. Thus, there is some evidence that the disease is a general one of the organism, not affecting muscle alone. Whatever changes are behind the increase in respiration (and there is some indication that they may reflect the permeability of cellular membranes) seem to lead to a much more harmful effect in muscle than in other tissues and to lead to the clinical observation that the disease is primarily one of muscle.

Thus, we find that a single category of muscle disease has led to studies of ultrastructure, genetics, isoenzymes, and cell respiration, and it seems likely that these studies will prove of great importance in understanding the functioning of normal muscle. One should therefore not think of pure cell research as providing information that flows in a unidirectional direction toward medical applications. The interaction is much more nearly a two-way one, so that a student of the cell is well-advised to move as freely as possible, and with as little self-consciousness as possible, between the applied and fundamental realms of investigation.

Other contractile processes: flagella

There appears to be a unity of mechanism for contractile events in cells where the remarks made above about muscle apply to the other contractile systems. The reason that most of our present knowledge about the area comes from studies with muscle stems largely from such practical matters as the medical implications of muscle physiology and the fact that muscle protein may be obtained in very large quantities, permitting a variety of experiments that would be quite impossible with the minute quantities of flagellar or spindle-fiber protein found in a cell. It does appear, however, that in those cases studied the power of motion is associated with a protein, and, where it is possible to say anything about the physical characteristics of that protein, it looks rather similar to the actomyosin system. Thus, the eukaryotic flagellum contains protein with demonstrable ATPase activity, and the activity requires either Mg^{2+} or Ca^{2+}. When animal sperm cells are extracted with glycerol (which removes ATP), they lose their motility, which can be subsequently restored by the addition of ATP. It is likely that flagella do not carry out metabolism leading to ATP production, but that the ATP hydrolyzed in the course of movement diffuses into the flagellum from the cell interior.

The reader will remember from Chapter 1 that cilia and flagella exhibit a characteristic tubular structure with nine double tubules arranged in a circle and enclosing two central ones. Ever since this 9 + 2 arrangement has been known, workers in the field have endeavored to relate this structure to the process of ciliary movement. For example, it has been repeatedly pointed out that ciliary beat could be based on the contraction of several tubules on one side of the array, drawing the tip of the cilium in that direction. If the tubules were indeed the contractile portion of the whole unit, their hollow centers would serve as channels for the diffusion of ATP to be hydrolyzed in providing energy. It should be mentioned that the movement to be accounted for in any theory of cilia is quite complicated and variable in different organisms. For example, the cilia of protozoa beat with a slightly circular component, those of the epithelium of the respiratory tract of human beings beat in a single plane, while the flagella of animal sperm beat with a wave-like undulation. The rate of beating is likewise quite variable, extending from about 5 to 30 beats per second.

Another approach that has been taken in trying to understand the motion of eukaryotic flagella has been to suppose that the tubules were, in effect, hydraulic chambers that enabled the whole structure to beat by the pumping of fluid, selectively, into certain of the tubules. This approach, which has been neither proved nor rejected, is rendered less credible by the lack of known mechanisms for rapid-enough pumping of fluid into the tubes, together with the known ATPase activity in the structures.

Recent studies have centered on the analogy between ciliary and muscular movement and on the likelihood that the fundamental mechanisms are the same and involve a contractile protein that is also an ATPase. One is encouraged here by the observation that the loss of motility in sperm is associated with the loss of ATP from the flagella. Furthermore, as we said, extraction of sperm flagella with glycerol, which removes ATP, leads to a loss of motion, which can be restored on re-addition of ATP. The location of the ATPase appears to be in the peripheral ring of tubules and, more specifically, at the asymmetric arms that extend from one of each pair (see Figure 14-8). This finding is in conformity with the view that the contraction of the outer tubules provides the motive force.

Still more recently, investigators have turned to an effort to resolve the various protein components of cilia, in the expectation that they will be able to effect the reconstitution of functional cilia. The studies have progressed to the point where the disruption of cilia leads to the separation of two fractions, one containing the circular array of outer tubules but lacking the asymmetric arms and devoid of ATPase activity. On adding the other fraction (containing ATPase) back to the outer tubules, in the

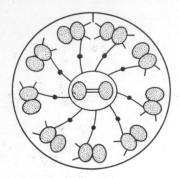

Figure 14-8 Characteristic structure of eukaryotic flagella.

presence of a divalent cation, the ATPase activity becomes associated with the tubules again and the asymmetric arms reappear. Besides confirming the identity of the ATPase with the asymmetric arms of the outer tubules, these studies have permitted a beginning to be made in the characterization of the ATPase protein, which is obviously desirable if one is to examine the similarity of muscular and flagellar contraction.

The studies described above have only been possible through the availability of rather large amounts of isolated and relatively pure flagella or cilia. Similarly, an understanding of the role of the basal granule has come, in part, from the availability of preparations of isolated granules from the protozoan *Tetrahymena*. It will be recalled from Chapter 1 that basal granules are closely similar to centrioles in possessing characteristic ninefold symmetry and that they are believed to serve as the locus for the

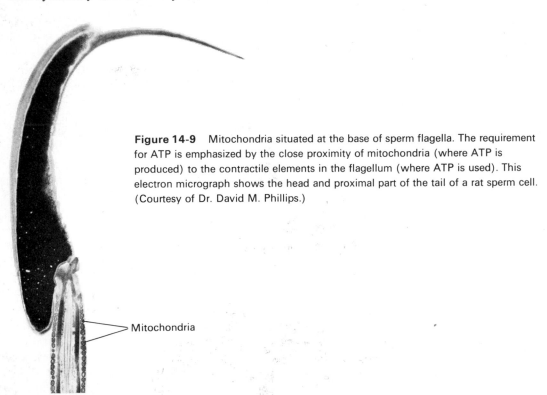

Figure 14-9 Mitochondria situated at the base of sperm flagella. The requirement for ATP is emphasized by the close proximity of mitochondria (where ATP is produced) to the contractile elements in the flagellum (where ATP is used). This electron micrograph shows the head and proximal part of the tail of a rat sperm cell. (Courtesy of Dr. David M. Phillips.)

Mitochondria

Figure 14-10 Relation of basal granule to flagellum. Note that the outer tubules extend into the basal granules, while the central ones do not.

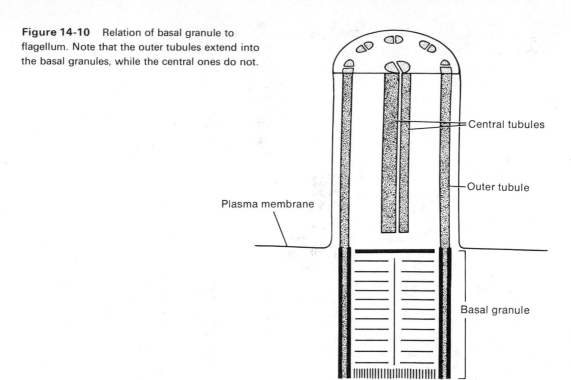

synthesis of the flagellar structure. The relationship between basal granules and flagella is illustrated in Figure 14-10, and supports this idea. We also noted in Chapter 1 that there was genetic evidence that the basal granules in at least some organisms are endowed with genetic autonomy. In other words, they appear to be self-replicating and would, therefore, be expected to contain their own genetic material. This expectation is in conformity with the analysis of the composition of protozoan basal granules (kinetosomes), shown in Table 14-2. These figures were obtained with granules that were isolated from cells, by means of digestion of the remainder of the solid cell structure with a detergent, followed by centrifugation of the insoluble basal granules. Since it is possible, and even likely, that detergent extraction of the granules serves to remove some protein, phospholipids,

Table 14-2   Composition of basal granules from the protozoan Tetrahymena

Component	Dry weight (%)
Protein	50
Carbohydrate	6
Lipid	5
DNA	3
RNA	2
Other	34

and other material, the compositions shown must be regarded as highly approximate. There is no doubt, however, that the granules contain RNA, and most importantly, DNA, so that they are probably equipped to maintain at least a measure of autonomy. It must be added that cyto-chemical studies also have shown the occurrence of DNA in the basal granules as they exist in the intact cell, a demonstration that depends on the specific binding of the dye, acridine orange, which exhibits a specific absorption (and fluorescence) spectrum when bound there.

It now seems clear that basal granules serve as the loci of cilium and flagellum synthesis. The origin of the basal granules is much less obvious, since there are some rather perplexing features to be explained. For one thing, no one has observed basal granules (or centrioles) in the act of replicating, whether by fission or budding. Furthermore, there are proto-zoans that exist in an amoeboid state followed by a flagellated state. When in the amoeboid condition, the organism does not contain any basal granules; they appear to be formed *de novo* when the cells begin to make flagella. These and other electron-microscopic observations have led some workers to suggest that basal granules arise from specially modified mitochondria. It has also been suggested that the centrioles located near the nucleus somehow give rise to the peripheral basal granules, but the exact relationships are clouded, in part, by the question of whether the two are, in fact, distinct organelles.

The question of the replication of the centriole located in the region of the nucleus is, likewise, a source of some uncertainty. In general, it appears that, when cells replicate, a new centriole (*procentriole*) is seen to materialize near the pre-existing one and oriented at a right angle to its axis. A procentriole is apparently a much shorter version of the centriole per se with the same cross-sectional geometry. The molecular basis for its synthesis or for the right-angle orientation with respect to the centriole is not known.

Bacterial flagella

In contrast to the situation in eukaryotic cells, bacterial flagella do not show the 9 + 2 arrangement of tubules and, indeed, are much smaller than "true" flagella or cilia. They are about 0.15 micron in diameter and are, therefore, below the limit of resolution in the light microscope. When bacterial flagella are viewed in the electron microscope, they appear to possess a helical structure with a pitch of the helix that is characteristic of a particular species.

Bacterial flagella are essentially composed entirely of protein which, in contrast to eukaryotic flagella, does not exhibit ATPase activity. Likewise, the bacterial flagellar proteins contain no bound ATP, so that it is likely that ATP hydrolysis associated with flagellar movement occurs within the cytoplasm of the cell. In any case, since the bacterial flagella are not enclosed by a membrane, it is not clear how ATP could reach the more distal parts of the organelle without diffusing away.

Isolated prokaryotic flagella can be isolated in pure form and can be treated with extreme pH to dissociate into subunits that are identical and are called *flagellin*. These are globular proteins with molecular weights of about 40,000 daltons and a diameter of about 50 Å. This protein usually exhibits a characteristic amino acid composition, lacking such common amino acids as tryptophan, proline, hydroxyproline, and histidine, and containing methyllysine (also found in muscle actin). Under suitable conditions, it is possible to demonstrate the reaggregation of the protein subunits to form long strands, quite like the native flagella.

Cells that have been shorn of flagella by rapid stirring in a blender remain viable (but obviously not motile). These are able, however, to regenerate new flagella, and the process is inhibited by inhibitors of protein synthesis, such as chloramphenicol, which probably acts at the 50-S subunit of the prokaryotic ribosome. The mechanism of bacterial flagellar motion is not known, but several features of the process are interesting. For one thing, the motion is much less complex than is the case with eukaryotic cells. Bacterial flagella are situated either in a polar pattern (with a few flagella at the poles of the cells) or peritrichously (distributed widely over the cell surface). In the case of polar flagella, the flagellum rotates around the cell in a path describing a cone, so that the whole bacterium rotates in the opposite direction (see Figure 14-11). Since the bacterium is generally spiral itself, its rotation drives it through the fluid medium. Some bacteria can achieve a rate of 50 microns/second, which should be compared with their length of no more than a few microns. It is interesting that some bacteria, such as *Rhodospirillum rubrum*, that have polar flagella at each end, can reverse direction very rapidly, flipping their flagella in the appropriate direction in apparent coordination with each other. This appears to imply some form of communication between the

Figure 14-11 Motion of a bacterium endowed with polar flagella. The concerted motion of the flagella in one direction drives the bacterium in counterrotation. Since it is spiral in shape, it moves forward in the medium.

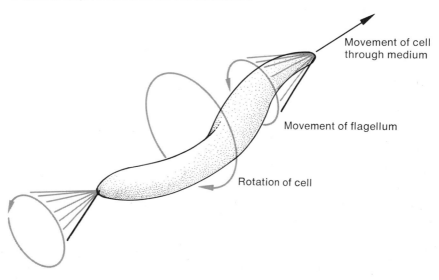

Movement of cell through medium

Movement of flagellum

Rotation of cell

two ends of the cell, but it could also result from the passive movement of the flagellum at one end in response to the change in direction produced by the other.

Finally, it should be added that little progress has been made in discovering the mechanism of prokaryotic flagellar action. Presumably, ATP is the source of energy for the movement, since treatments that interfere with ATP synthesis prevent flagellar motion. The manner in which ATP hydrolysis can be coupled to the rotation of the flagellum is, however, quite unknown, although the transformation of chemical bond energy to mechanical work is probably accomplished in the basal granules, within the cytoplasm. These, as might be expected, are much smaller than the granules from eukaryotic cells and do not possess the characteristic centriole type of structure.

Cytoplasmic motion

Intracellular motion appears to be a universal property of living cells. In some instances, such as amoeboid movement, the intracellular movement of protoplasm is related to the motion of the cell as a whole, while, in many other cases, the intracellular movement occurs when the cells are immobile.

On the whole, cytoplasmic movement has been rather difficult to study owing to its relative inaccessibility to direct measurement. For this reason, many of the results that are available have come from studies of an all-or-none nature, that is, the discovery of conditions that either completely prevent movement, as viewed through a microscope, or that restore it when it is, somehow, suppressed. Accurate measurements have been mostly limited to studies with acellular slime molds whose plasmodia, which are multinucleate, are centimeters across and carry on cytoplasmic movement on a scale large enough to be measurable. Unfortunately, it seems apparent that the extrapolation from slime molds to cells, which are a few microns in diameter, cannot be made with any safety, and universal theories of movement should probably not be formed at present.

Indeed, it is likely that no single theory will be suitable to account for all forms of intracellular movement, and we shall mention some of the hypotheses that seem to provide partial explanations below. There are some generalizations, however, that appear valid, and probably the central one is the observation that the hydrolysis of ATP is more than likely the energy source in all forms of cytoplasmic movement that have been studied. Thus, it has been reported that cytoplasmic streaming in algae is stimulated upon the addition of ATP to the sea water in which they are maintained. Likewise, ATP appears to stimulate movement of cytoplasm in slime-mold plasmodia, and it is interesting that, in this instance, inhibition of oxidative ATP synthesis in mitochondria does not prevent movement, while inhibition of glycolysis, the reactions of which are found in the nonmembranous cytoplasm, prevents it. Thus, it would appear that

ATP from mitochondria is unavailable for streaming in these organisms or that glycolytic ATP production is sufficient.

Since slime molds represent one of the few opportunities to obtain quantitative data on cytoplasmic motion, it is worth mentioning the sort of experimental arrangement that has been used with success. A plasmodium of the slime-mold species *Physarum polycephalum* exhibits a sort of cytoplasmic streaming in which the direction of flow changes cyclically with a period of about 1 minute. Thus, the body of the plasmodium extends filaments that are withdrawn and, then, re-extended, in a rhythmic manner. Indeed, the flow of cytoplasm occurs not only in the filaments but throughout the entirety of the plasmodium, which can be seen in some cases to pulsate. Measurement of the movement of cytoplasm is accomplished by placing the plasmodium in a double chamber, which divides it into two halves connected by a thin strand of protoplasm (see Figure 14-12). Movement is measured as a pressure difference between the two half-chambers. In this way, both the volume of protoplasm that moves per unit time and the force with which it moves can be accurately measured. Since ATP stimulates movement, its addition to one side of the chamber promotes a net movement toward the other side.

A number of mechanisms have been proposed to account for protoplasmic movement, and it is probable that a mixture of them will turn out to conform to reality. Analogies with muscular contraction have led to the development of the idea that the transduction of chemical energy (from ATP) to the mechanical energy of protoplasmic movement involves a contractile protein with ATPase activity. This hypothesis has been supported by the isolation of an ATPase from *Physarum*, which exhibits a number of the required properties similar to those noted in the case of muscle actomyosin, including Ca^{2+}-induced stimulation of the ATPase activity as well as viscosity changes on addition of ATP. Electron-microscopic examination of the superficial cytoplasm of *Physarum* reveals fibers large enough to be seen with the light microscope, which may represent the contractile ATPase and which have therefore been postulated as the sites of motive force. Although there is, as we have seen, considerable presumptive evidence linking ATPase to movement, a direct and unambiguous demonstration of the connection has yet to be found.

It is also probable that a simple contractile mechanism cannot be invoked to account for other forms of cytoplasmic movement, such as

Figure 14-12 Use of a double chamber to measure protoplasmic movement of a slime-mold plasmodium.

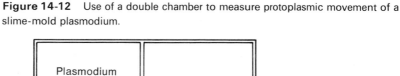

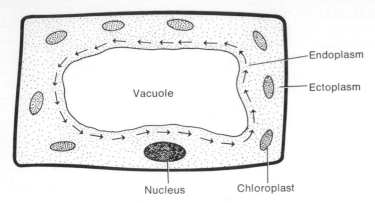

Figure 14-13   Localization of cytoplasmic motion of a plant cell in the endoplasm.

the rotary streaming that occurs in many plant cells (there called *cyclosis*). In this case, streaming often occurs in a zone of cytoplasm (the *endoplasm*) between a central vacuole and peripheral *ectoplasm*, relations between which are diagrammed in Figure 14-13. The rotational movement of cytoplasm can be quite rapid (50 microns/minute) and appears to require active ATP synthesis for its support. The boundary between ectoplasm and endoplasm is sharply discontinuous, and the force that drives movement of the endoplasm must be applied at that interface. The mechanism by which this is done is not known, but analogies with the sliding-filament theory of muscular contraction may be productive, as such an arrangement could produce shear forces across an interface, such as the endoplasm–ectoplasm surface, provided the forces between the filaments were exerted across that interface.

There is considerable evidence that rotary streaming does not require the influence of a particular region of the cell. For example, experimental elimination of the central vacuole of plant cells does not eliminate streaming. Neither does surgical removal of the nucleus. Indeed, streaming can be demonstrated in fragments of cells containing small amounts of both endoplasm and ectoplasm. Significantly, when the fragments contained only endoplasm, no movement was observed. Finally, a number of electron-microscopic observations have shown that the fixation of cells in which streaming had been taking place yields a pattern of microtubules (see Chapter 1) oriented in the direction of streaming. These tubules have been postulated as being identical with the sources of the force that drives the movement of cytoplasm.

Amoeboid movement

Examination of a living amoeba with a light microscope reveals a form of movement depending on a change in the shape of the cell. Cytoplasm is seen to flow in the direction of net movement, so that an extension of

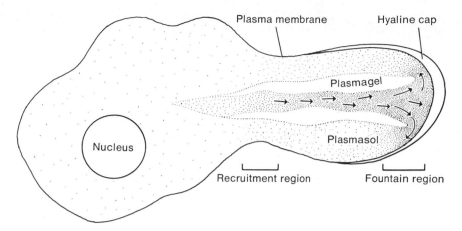

Figure 14-14   Features of pseudopod extension by an amoeba. See text for details.

the cell is poured out in the form of a *pseudopod*, into which, finally, the remainder of the cell is retracted. Wide variation on this pattern is observed and the pseudopods can be very large, containing a major part of the total cytoplasm, or they can be thin filaments, as found in the foraminifera or in some mammalian cells grown in culture.

The mechanical features of amoeboid motion have been studied largely (and appropriately) in species of *Amoeba* and related genera, and are outlined in Figure 14-14. The distal part of the pseudopod is devoid of particles and is called the *hyaline cap*. Behind this cap, a stream of cytoplasm moves toward it, spreading out in the fashion of a fountain just before reaching the cap. Often, this flow of cytoplasm appears to pass through the center of a tube of more solid cytoplasm (*plasmagel*), which contributes to the shape of the pseudopod and provides a channel for the bulk movement of fluid cytoplasm (*plasmasol*). The plasmasol that goes to make up the fountain is "recruited" at the posterior end of the pseudopod, where plasmagel is converted to the more fluid plasmasol form.

The manner in which the movement occurs is, as in the case of cytoplasmic movement, presently unclear. It has been suggested that regional contraction might be responsible for the net flow described. On the other hand, a shear force analogous to the sliding-filament model has been suggested, with the interaction occurring between the central flux of plasmagel and the plasmasol tube surrounding it. What is certain, however, is that the energy for this movement, like the other forms of locomotion that we have discussed, is the hydrolysis of ATP. Extraction of amoebas with glycerol (a treatment that, as we have seen, removes ATP) leads to a cessation of movement, which is restored by the addition of ATP + Mg^{2+}. Injection of ATP into the posterior region of amoebas also leads to a stimulation of motion. Finally, ATPase has been isolated from such organisms and has been postulated to play a role in their movement in a manner analogous to its role in other forms of cellular movement.

Suggested Reading

Books

Bendall, J. R., *Muscles, Molecules and Movement*, Heinemann, London (1969).

Bourne, G. H. (ed.), *The Structure and Function of Muscle*, vols. 1–3, Academic Press, New York, 1960.

Wilkie, R. D., *Muscle*, St. Martin's Press, New York, 1968.

Articles

Gibbons, I. R., "The biochemistry of motility," *Ann. Rev. Biochem.*, **37**, 521–46 (1968).

Holwill, M., "Contractile mechanisms in cilia and flagella," in *Current Topics in Bioenergetics*, D. R. Sanadi (ed.), vol. 2, Academic Press, New York, 1967, pp. 288–334.

Huxley, H. E., "The mechanism of muscular contraction," *Science*, **164**, 1356–66 (1969).

Porter, K. R., and C. Franzini-Armstrong, "The sarcoplasmic reticulum," *Sci. Am.*, **212**, 72 (March, 1965).

Sandow, A., "Skeletal muscle," *Ann. Rev. Physiol.*, **32**, 87 (1970).

Stracher, A., and P. Dreizen, "Structure and function of the contractile protein myosin," in *Current Topics in Bioenergetics*, D. R. Sanadi (ed.), vol. 1, Academic Press, New York, 1966, p. 154.

Zierler, K. L., "Diseases of muscle," in *Biochemical Disorders in Human Disease*, 3rd ed., R. Thompson and I. Wootton (eds.), Academic Press, New York, 1970, p. 489.

Part IV

Cell growth
and genetic
regulation

15 Cell growth and division

Cell growth is the manufacture of cells by cells and represents a summation of all the biosynthetic and energy-capturing reactions discussed thus far. For our purposes, growth is defined as the *increase of cell mass with time*. Any control that is exerted on the reactions that comprise growth may be expected to influence the growth rate itself. Similarly, normal growth is unlikely to be maintained under conditions where important phases of synthesis or energy conversion are interfered with. Cell growth is often thought of as identical to cell division, although, in fact, the two are quite distinct. For example, a cell population can be shown to grow (increase in mass) under various conditions without cell division taking place. Furthermore, all our measurements of growth can be shown to be quite unrelated to the manner in which cells divide. Thus, if cells divided into thirds instead of halves, the growth rate, as well as our methods of describing it, would be unaltered.

Growth rates of cell populations

An accurate description of growth will be seen to involve a certain amount of mathematics which, however, only reflects an extremely simple feature of growth, the fact that cell material is its own factory. This may be expressed in a number of ways, including the statement that cells are autocatalytic (they make cells), but, most significantly, this implies that the rate of increase in cell material is proportional to the amount of cell material present to begin with. In other words, if one has twice the cell material, one observes twice the rate of increase in cell material, because twice the number of factories turn it out.

Now, it is possible to speak of the "amount" of cell material, but, to be a little more specific about units, it is useful to see how growth might actually be measured and represented. The units commonly used in studies of cell growth include the number of cells and the mass of cell material. The two may be interconverted readily as long as the average mass of a cell in a population remains constant, and this is generally the case, except under certain conditions to be described later. Expression of the mass of a population of cells growing in suspension might well be expressed in terms of milligrams per milliliter.

Actual measurement of the number of cells in a population includes such techniques as visual counting of cells in the known volume of a cytometer chamber, photometric measurement of the light scattering by a turbid suspension of cells, and estimation by a method called a *viable cell count*. This last, which is usually applied to bacterial populations, involves plating known volumes of known dilutions of the original sample evenly on solid medium in a petri dish, and counting the colonies formed. One assumes that each colony originated with a single cell, and this assumption is sufficiently true to allow the method to be used extensively. In contrast with other techniques, the viable cell count gives only the number of cells that are in good enough condition to divide and give rise to colonies, so that results may be lower than cell counts by optical methods. It must be stressed that all these techniques are suited to the estimation of the growth of cells in liquid culture only. Growth of cells in a multicellular organism or in a colony of single cells is extremely difficult to estimate and under such conditions appears quite complex.

In the following discussion we shall ignore the method by which cell growth is followed and shall use cell number and cell mass interchangeably, denoting both as N. First, let us consider a population of cells growing in liquid culture. These might be any sort of cell, including bacterial or mammalian, and our only requirements are that there be no unusual conditions influencing growth and that there be a constant volume. Since we found that the rate of growth of a population of cells was proportional to the amount present to begin with, we can say that

Growth rate $= k \times$ (the amount of cells)

where k is a proportionality constant. This may be written more simply as

$$\frac{dN}{dt} = kN \tag{15-1}$$

where dN/dt is the rate of change of the amount of cells, N, with respect to time—the growth rate. This equation may be rearranged to give

$$\frac{dN}{N} = k \, dt$$

This may be integrated

$$\int \frac{dN}{N} = k \int dt$$

which gives[1]

$$\log_e N = kt + C \tag{15-2}$$

---

[1] If this step is not clear, the reader may either look it up in a table of integrals or accept it on faith.

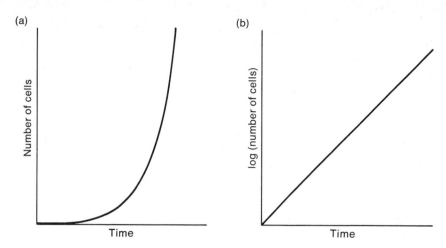

Figure 15-1 Growth of a cell population. (a) The number (or mass) of cells is expressed as a function of time. The curve is an exponential one. (b) The log of the number of cells is plotted against a linear time axis (a semilogarithmic graph).

where C is an integration constant, and the logarithm is taken to the base e ($e = 2.718\ldots$). It must be said that the base of the log is not of great importance in this equation, since the constants may be adjusted to take care of different bases. Indeed, logs to the base 10 or 2 are commonly used in growth studies, although the base e is the only one that springs directly from the integration. Equation 15-2 may be expressed as the log of the number of cells being proportional to time. The constant C simply sets the base line for the increase and can be moved about. If Equation 15-2 is a valid representation of growth (which it is), one could predict that plotting the log of N versus time would give rise to a straight line, and this is seen in Figure 15-1(b) to be the case.

The arbitrary nature of the integration constant C may bother the reader, and actually there is a way to be more precise. It is possible to eliminate C by considering the integration to be carried out within certain limits, to be determined by practical aspects of a growth experiment. Let us choose to consider only a portion of a growth curve (Figure 15-2) in a region of the curve where measurement is convenient. Let N_0 be the amount of cell material when $t = 0$ (t_0) and consider the amount to have increased to N_1 at time t_1. Then the integration shown above may be carried out within these limits to give the actual increase of N during the interval:

$$\int_{N_0}^{N_1} \frac{dN}{N} = k \int_{t_0}^{t_1} dt$$

This integration yields

$$\log_e N_1 - \log_e N_0 = k(t_1 - t_0)$$

347

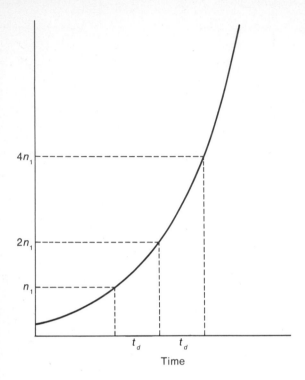

or, since $t = 0$ at t_0,

$$\log_e N_1 - \log_e N_0 = kt_1$$

This may be written

$$\log_e \frac{N_1}{N_0} = kt_1 \qquad (15\text{-}3)$$

or, raising both sides of the equation to the base e,

$$\frac{N_1}{N_0} = e^{kt_1} \qquad (15\text{-}4)$$

Thus, we have obtained an exponential equation for the growth of cell material (Equation 15-4), and it may be written in the completely equivalent logarithmic form of Equation 15-3. These equations represent nothing more than the mathematical consequence of the previously mentioned fact that cell (and all) growth has an autocatalytic character.

It should be recognized that these equations are of considerable generality in the natural world, since a number of inanimate processes also exhibit kinetic properties that are exponential. For example, these relations hold rigorously for the process of radioactive decay provided there be a sign change in the expression, that is,

$$\frac{dN}{N} = -k\,dt$$

Radioactive decay exhibits kinetic properties of a sign different from growth owing to what might be termed "inverse autocatalysis," where, as decay occurs, the amount of material left to decay diminishes.

The doubling time

Since an atom of a radioactive isotope decays with a specific probability, it is possible to describe the *half-life* of that isotope as a characteristic feature of it. The half-life is the time required for one half of a given initial sample to undergo radioactive decay, and a little thought will convince the reader that the half-life is independent of the amount of isotope present. A similar quantity may be defined in the case of growth, which, since the polarity of growth and decay are opposite, is termed the *doubling time*. It is written t_d and is the time required for a population of cells to double in weight or number. While the actual growth rate is dependent on the number of cells present, the doubling time is not, and is a characteristic of a given type of cell under given conditions of environment. The reader should satisfy himself as to the truth of this last remark by careful examination of the growth equations, perhaps working out a specific example. It is possible to derive an expression for t_d from Equation 15-3. Since

$$\log_e N_1 = \log_e N_0 + kt_1$$

and since, at the end of the doubling time,

$$N_1 = 2N_0$$

and

$$t = t_d$$

we may write

$$\log_e (2N_0) = \log_e N_0 + kt_d$$

from which

$$\log_e N_0 + \log_e 2 = \log_e N_0 + kt_d$$

or

$$\log_e 2 = kt_d$$

Finally, solving for t_d,

$$t_d = \frac{\log_e 2}{k} \tag{15-5}$$

This is really very useful, since $\log_e 2$ is known to many places (0.693...). Thus it is possible to obtain k, usually known as the *growth-rate constant*, from information about the doubling time. It is clear that the growth-rate constant, like t_d, is a characteristic of the growth of a given organism under given conditions.

The rate of exponential growth

Whereas the half-life is characteristic and immutable for a given isotope, the doubling time (or growth constant, k) may alter over a wide range in a given organism. Cells grow more rapidly under favorable conditions than they do under adverse ones. A cell population will exhibit optima of pH, temperature, ionic strength, and nutrient concentration, and frequently these optimal conditions are instructive with regard to the ecology of a given cell type. For example, a bacterium with a strongly acid pH optimum for growth would be likely to have originated in such an environment. Bacteria that are human pathogens and have been isolated from a human "culture medium" are likely to have complex nutritional requirements reflecting the wealth of compounds available in their natural medium. It is significant that such a pathogenic bacterium, *Corynebacterium diphtheriae*, which has been isolated from the human throat, grows at an optimum rate at about human body temperature (37°C) and in a medium rather like that required for a human cell in culture, containing numerous amino acids, vitamins, and other growth factors.

It is often possible to grow cells under conditions where the growth rate may be shown to depend upon the concentration of only one nutrient, the others being present in excess. When this is the case, it is possible to examine the relation between growth rate and nutrient concentration. It is interesting that, in such instances, growth obeys a kinetic relationship quite like that found with simple enzyme systems. For example, if k is the specific growth constant under conditions of measurement, k_{max} is the maximum growth constant attainable, s is the concentration of the limiting nutrient, and K_s is a constant analogous to the Michaelis constant of Equation 3-7 in Chapter 3, then

$$k = \frac{k_{max}s}{K_s + s}$$

The similarity between the relation of growth rate and nutrient concentration and that between enzyme catalysis rate and substrate concentration undoubtedly reflects the decomposition of the net process of growth into individual enzyme-catalyzed reactions.

The limitation of growth rate

The knowledge that growth is, by its nature, exponential leads to some conclusions about the range of validity of the treatment above. Clearly, if a pair of cells divide, and their progeny divide 1 doubling time later, and so on, then their numbers describe an exponential progression of the form $2, 4, 8, 16, 32, 64, \ldots$. Since the exponential nature of growth brings one to very large numbers very quickly after a finite number of generations, one should expect the progeny of a given cell literally to take over the world. It has been calculated that the mass of a population of *E. coli* originating from one cell dividing every 30 minutes would in a few days come to equal that of the solar system. That is, however, at least one catastrophe that does not seem to occur. In other words, something intervenes to limit the infinite expansion of populations, and the exponential character of growth is maintained only as long as environmental conditions permit. It is interesting to follow the behavior of a population of cells growing in a finite volume and with a finite supply of nutrients and to consider the ways in which population growth is finally limited.

Figure 15-3 shows the idealized growth of a population of cells growing in a constant volume. We imagine that, at $t = 0$, cells are placed in a new medium and allowed to grow. One observes first a period when growth does not commence: this is termed the *lag period*. The lag period may be described simply as the time required for the cells to adjust their internal chemistry to the tasks of utilizing a new medium, which is probably in part a question of the synthesis of new induced enzymes appropriate to the nutrients available. As one might predict, when cells that have been growing actively are placed in a medium of a composition similar to that in which they have been residing, the lag period is diminished or undetectable.

Following the lag period, the cells begin to grow at an increasing rate until growth becomes exponential. During the exponential phase of growth, population changes can be described in terms of the equations presented earlier. It is, however, of finite duration, and the population passes through a period of declining growth rate until the stationary phase is reached, where net growth has ceased. The decline in growth is due to the exhaustion of nutrients in the medium or to the formation of toxic products of cell metabolism or both. Thus, the addition of nutrients that are limiting will give rise to additional growth, as will the removal of waste products from the medium. In many instances, it appears that the stationary phase represents a sort of equilibrium where there is some

Figure 15-3   Growth of a cell population in a finite volume.

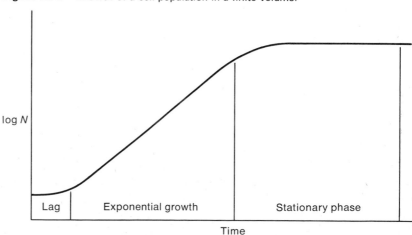

growth, balanced with and dependent on the dissolution of some of the cells of the population.

It is important to realize that the term *growth medium* should be used in a sense wide enough to include not only the medium of cultured cells (*in vitro*) but also the medium formed by the totality of a multicellular organism. It is clear that cells growing in such a complex medium experience many of the same limitations on growth, related to nutrition and to the removal of wastes, as do isolated cells. Such ecological considerations are of the greatest importance in relating the results of cell physiology to the function of multicellular organisms.

Systems for continuous growth

The preceding comments about the growth of a real population of cells and the limitations to infinite expansion of a population are valid as long as the cells exist in a constant volume of medium with no replenishment of nutrients or removal of toxic products. It is possible to devise a situation where these restrictions are avoided and where growth may be maintained in the exponential phase for extended periods of time.

Such a device for the continuous cultivation of cells is known as a *chemostat* and depends upon the continuous addition of new medium to a population of cells growing in a container. The addition of medium at once provides new nutrients and leads to the dilution of waste products and cells, so there is no reason why a declining phase of growth should be observed. The addition of medium is at a rate such that the formation of new cells by cell division is exactly balanced by the removal of cells, so that although the portion of the population remaining in the growth tube grows exponentially, the mass of cells per unit volume remains constant. Apparatus of this sort has proved of great value in studying the growth of cells from a variety of organisms, although bacterial growth has been most extensively examined. It might be argued that the continuous-flow situation of the chemostat is somewhat artificial and that results obtained in this way are not likely to be of great relevance to the study of cells under conditions normal for them. In fact, the steady-state condition achieved in a flow system is in some ways not unlike that of the organism as a whole, and the possibility of looking at cells under conditions not changing with time represents a unique opportunity in the investigation of growth and metabolic events related to it. Moreover, the steady-state condition accurately reflects some natural habitats in which cells find themselves, excellent examples being the bacteria in a human intestine or microorganisms in a river.

One of the advantages in studying growth under conditions of continuous flow is the precise measurement of growth rate, dN/dt, the growth-rate constant, k, and the doubling time, t_d, that becomes possible. To see how this occurs, one must consider the design of such a system, outlined in Figure 15-4. Sterile medium drops, or is pumped, at a controlled rate into the growth tube. As medium is added, an equal volume of medium

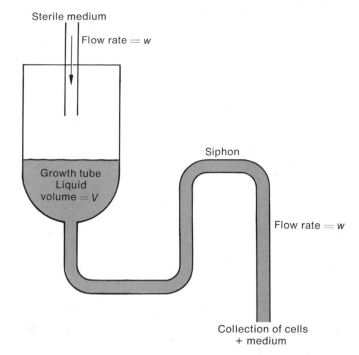

Sterile medium

Flow rate $= w$

Growth tube
Liquid
volume $= V$

Siphon

Flow rate $= w$

Collection of cells
+ medium

Figure 15-4 Chemostat.
Additional details, not shown,
include devices for maintaining a
constant flow rate, temperature,
and degree of aeration.

and cells is siphoned off and may be collected. Thus, owing to the action
of the siphon, the volume in the growth tube remains constant, and a
steady state is possible, where the cell density remains constant with time.
The quantitative usefulness of the system stems from the constant volume
and the equality between the rate of dilution of cells and that of new cell
formation. Let V be the volume of the growth tube in milliliters and w be
the flow rate through the whole system in milliliters per hour. Obviously
both numbers may be obtained with great accuracy. We know that the
bacteria are growing inside the tube at a rate determined by the equation

$$\frac{dN}{dt} = kN \tag{15-6}$$

In addition, the bacteria are being diluted out at a rate equal to

$$\frac{w}{V}(N)$$

so that the net growth rate of the population in the tube is given by

$$\frac{dN}{dt} = kN - \frac{w}{V}(N) \tag{15-7}$$

or

$$\frac{dN}{dt} = \left(k - \frac{w}{V}\right)N$$

From this it is clear that if the flow rate, w, is too great for a given value
of V, then $k \ll w/V$, and the cells will be eventually diluted to a cell

density of zero. In other words, it is quite possible to slosh medium through the growth tubes so rapidly that cells will be washed away. A more interesting situation occurs when the flow rate is somewhat reduced. There is (under suitable conditions) a wide range of flow where the cells will grow just as fast as new medium is added, in other words, where some component of the medium can limit growth of the population. At this steady state the population in the growth tube reaches a constant value determined only by the rate of flow. Under this condition

$$\frac{dN}{dt} = \left(k - \frac{w}{V}\right)N = 0 \qquad\qquad (15\text{-}8)$$

and it is clear that

$$k = \frac{w}{V} \qquad\qquad (15\text{-}9)$$

From k it is possible to determine the doubling time t_d, as described above.

Thus, we find a third means of determining the growth-rate constant k, the first two being via Equations 15-5 and 15-6. We said that k was a property of a population under a given set of conditions, and this is demonstrated clearly by our ability to change k, within a certain range, by simply altering the rate of flow.

In considering the usefulness of continuous-flow devices it is necessary to concentrate one's attention on the means whereby growth rate is actually limited. There are, in principle, two distinct classes of continuous-growth experiments and, although the preceding discussion is applicable to both, there are significant differences. In the first place, it is possible to perform such experiments where growth is actually limited by making the concentration of one nutrient in the medium low with respect to the requirements of the organism. Growth is thus limited by the concentration of one identifiable component of the system and by the flow rate. As we have seen, a steady state can be attained over a range of flow rates, and the mass of cells (N) in the growth tube at equilibrium is independent of flow but is a function of the concentration of limiting nutrient.

The second class of continuous-flow devices is called a *turbidostat* and differs from the limiting nutrient method in placing an external constraint on growth. The system is a classical chemostat (Figure 15-4) but with provision for monitoring the cell population with a photoelectric cell (see Figure 15-5). A feedback loop is included whereby it is possible to set the device for a specific turbidity, which is maintained by regulation of the flow rate. Thus, if the density of cells falls below the pre-set figure, the flow of nutrient medium is stopped until the cells catch up. In a turbidostat, a growth medium is employed that is not limiting to growth at the flow rates used, so that the feedback loop imposes the only control. In continuous-flow experiments, the nature of the results obtained is frequently a function of the manner in which growth is limited, with different situa-

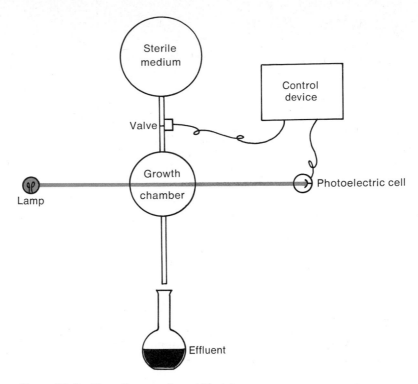

Figure 15-5   Flow diagram of a turbidostat.

tions obtaining, depending on whether growth is limited externally by means of a turbidostat or by means of limiting the concentration of any one of a number of nutrients.

Continuous-flow systems have been used for a wide variety of physiological and genetic experiments, including studies of nutrition and mutagenesis. The ability to control growth rate in a continuous fashion has led to the study of the relationship between growth rate and the composition and enzymic content of cells, and a number of interesting findings have been obtained. For instance, the composition of bacterial cells is often quite similar over a range of growth rate, with the important exception that a direct correlation exists between the growth rate and the content of ribonucleic acid (RNA), especially that RNA associated with ribosomes. Finally, the availability of continuous-flow systems enables biologists to obtain practically unlimited amounts of cells, grown at constant and defined conditions, for use in subsequent experiments. It is only necessary to collect the cells from the overflow tube in such a way that they are preserved in the required condition. This is often done by allowing them to flow at once into a flask maintained at the temperature of liquid nitrogen. In this way, continuous-flow systems have played a role in the study of properties of a wide assortment of cells, including bacteria (where they were first employed), algae, fungi, and, indeed, animal cells grown in culture.

Synchronous growth

Increase in cell number is a quantum-like event, in that it takes place in a stepwise manner. There is no intermediate state between one and two cells. Thus one might expect a growth curve to look like a flight of stairs, with the population doubling each time the doubling time rolls around. This, of course, does not usually occur, because cells divide at random times within the population, so that cell division is spread over the whole doubling period. Even if we start out with one cell, by the time a few cell divisions have occurred the divisions are well spread out and the growth curve is smooth. This is a pity, because it would be very useful in the study of growth to be able to examine a population of cells, all about to divide. It is difficult to adduce metabolic features of cell division if the population is random with respect to it.

Fortunately, there are ways to synchronize cell division in a large population of cells so that things happen at the same time, at least for a while. These approaches, which have been applied principally to bacterial and protozoan studies, depend, with one exception, on interfering with the normal function of cells to such an extent that division is held in abeyance until conditions return to normal, at which time division takes place all at once in the population. For example, the withholding of an important cell constituent (such as thymine, required for DNA synthesis) prevents division of cells that are unable to synthesize it, and, when cells thus treated are returned to a thymine-containing medium, cell division occurs in synchrony. The first few divisions remain in step, but soon the time of division becomes increasingly random and the growth curve smooths out (see Figure 15-6). A similar experiment may be performed, not by starvation, but by cooling and then rewarming cells about 10°C in a short time. It is interesting that synchrony may also be produced

Figure 15-6 Growth of a population whose cells initially divide in synchrony.

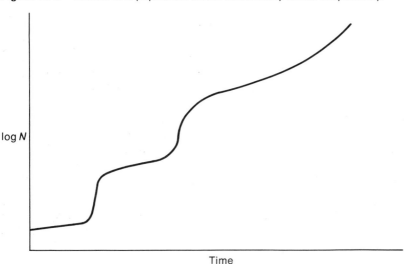

by placing bacterial spores into a suitable medium, after which germination and division take place initially in a coordinated fashion. This method is of limited applicability, since most cells do not form spores. Clearly, all methods described thus far for establishing synchrony of cell division depend on tormenting cells, or at least, altering their physiological state in a drastic way, hardly the way to examine "normal" cell division. The one technique that avoids this difficulty to a good measure depends on the rapid filtration of bacteria using calibrated filters to separate cells about to divide (the largest cells in the population) from those that have just divided (the smallest ones). Thus, one is able to obtain a measure of synchrony without fiddling around with the physiology of the cells, and one may feel that the cells are in a somewhat more natural state.

It is important that the ability to synchronize the time of cell division in a population need not influence the overall growth rate, illustrating a fundamental independence of growth rate and cell division. We have earlier defined growth as an increase in cell mass. Cell division obviously reflects an increase in cell number. In most populations, over a sufficient interval of time, the two are in close correspondence, since the average mass/cell ratio remains within narrow limits. However, it is possible, by manipulating the growth conditions of cells (or allowing conditions to become highly unfavorable) to observe growth for a considerable period in the absence of cell division. In such cases, the mass/cell ratio must increase and, in that of bacteria, the cells are seen to become extremely long. Similarly, it is possible to grow cells under such nutritional conditions that they are unable to make some important component such as DNA or RNA at a rate consistent with the growth of the cell as a whole. Thus, as the cell grows in total mass, the protein content of the population may increase while the DNA/protein ratio may actually decline. Clearly, this situation cannot continue for very long and must lead to a decline in viability. However, while it lasts, it is known as *unbalanced growth* and provides some opportunities to study the role of various cell components on the growth process.

Metabolic aspects of growth

For a number of reasons, both practical and theoretical, the study of biochemical events related to growth has been vigorous, especially as might relate to the problem of the control of growth rate. When considering the results of investigations, it is important to make a clear distinction between aspects of growth per se and those of cell division. For example, there is a great deal of information, largely derived from synchronous growth studies, related to changes in the chemical composition of cells in different stages in their cycle of cell division and without reference to the rate of growth itself. One may observe a cyclic change in the percentage composition of cells with respect to protein, RNA, or DNA, with a period related to the time of division. At the present time, such changes are somewhat difficult to interpret, as there is some question as to how much of the change is correlated with the cycle of division and how much with the mistreatment of cells. Results differ, depending on the method used to induce synchrony.

Equally interesting are the results of studies on the relationship between cell composition and growth rate. By means of the chemostat it is possible to vary the growth rate continuously over a considerable range by adjusting the flow rate. Thus, there is an infinity of possible rates of growth below the maximum rate attainable, and, correspondingly, of possible cellular metabolic states. Thus, it is of great interest to change the growth rate and to examine the cell composition on the grounds that any component uniquely associated with growth should bear a special relation to its rate.

When such experiments are performed using bacteria, a complication in the form of varying amounts of nuclear material per cell under different conditions interposes difficulties in the interpretation of results. In other words, it is possible to obtain conditions where unbalanced growth occurs to the extent that the bacterial nuclear material is replicated at a rate differing from that of cell division, and it is probable that the amount of a component should be expressed in terms of DNA, rather than on a per cell basis. If this is done, one observes that the protein/DNA ratio shows a slight correlation with growth rate, whereas the cell mass/DNA ratio is more significantly increased by an increase in growth. Most of the increase is accounted for by the increase in ribosomes as the growth rate goes up. Ribosomes (and ribosomal RNA) appear especially sensitive to the growth state of cells, and there appears to be an exponential relationship between them, on one hand, and the DNA, on the other. Since ribosomes are known to be the site of most cell protein synthesis, and since active growth must include active protein synthesis, it is not surprising that the rate of protein formation is also correlated with growth rate. It seems clear that ribosomes exist in the cell under conditions of constant size and composition, so that the increase in total ribosomal material (or RNA) must be due to an increase in the number of ribosomes.

Thus, the number of ribosomes in a cell appears dependent upon the growth rate of a population. It would be extremely interesting to know if the growth rate is itself dependent upon the ribosome content (to consider if the causation might work both ways). Most of the evidence about this point comes from kinetic studies in which cells are shifted from one growth rate to another by alterations in temperature or medium and the changes in cell components noted. Suffice it to say that in such experiments RNA is always the first component to change, so that, when cells are jumped to a higher growth rate and the rate of RNA formation increases as rapidly as it is possible to measure, the other components (DNA and protein), as well as total cell mass, begin to increase at the new rate only after the RNA level has approached a new steady state. Thus, although it is still impossible to say what component is reponsible for the regulation of growth, or indeed if any one component can have such a role, it is clear that RNA is a likely candidate for at least a major influence.

The regulation of growth in eukaryotic cells

In general, growth of cells from higher organisms has much in common with that of the much more extensively studied bacteria, with, however, a considerable increase in complexity. Bacteria (and other microorganisms) have evolved under conditions that select for maximum growth under a wide range of environmental situations. The premium on rapid growth is greatly lessened in multicellular organisms, where cells must remain in proper proportion to the mass and structure of the organism as a whole. The situation in which a population of cells within a multicellular organism begins to proliferate more rapidly than the organism itself is highly destructive to normal functioning, an excellent example being the growth of a tumor. Cells of a tumor may be regarded as having escaped the regulation required for multicellular organization, although the exact nature of that regulation is, unfortunately, not well understood.

An added level of complexity in growth of eukaryotic cells comes from the complicated character of their structure. For example, we saw in Chapter 1 that mitochondria and chloroplasts are, to a degree, independent of nuclear control and may be regarded as undergoing replication within the cell much as the cell itself replicates within a population. It appears likely that one aspect of the regulation of cell growth resides in maintaining the correct number of such energy-producing organelles per cell, so that there must be an interaction between cell and organelle division. In this connection, the antibiotic chloramphenicol has been shown to inhibit growth of animal cells apparently by selectively interfering with the replication of mitochondria.

Cell division in eukaryotic cells

Growing cells divide, thus maintaining their constant mass and composition. Division occurs as often as every 20 minutes in the case of certain bacteria under optimal conditions or every 12 hours in that of very rapidly dividing plant or animal cells. In a multicellular organism, different cells exhibit widely different division times. For example, in an adult human being, nervous tissue exhibits little or no cell division, while tissue such as bone marrow, actively engaged in formation of circulating blood cells, contains many rapidly dividing cells.

In the case of prokaryotic cells, cell division occurs by pinching off a cell in its middle to produce two daughter cells. The constriction is accompanied by the synthesis of new cell wall and membrane and is preceded by division of the nuclear region. Indeed, division of the nuclear region occurs about midway between cell divisions, while replication of DNA appears to occur continuously during the entire cycle. There is no visible apparatus for the partition of nuclear material into the two daughter

 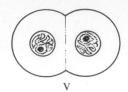

I   II   III   IV   V

Figure 15-7   Mitosis in an animal cell: a diagram. I. *Interphase*: The nuclear envelope is intact, enclosing a resting nucleus, which contains chromatin, and a discrete nucleolus. Centrioles are seen in the cytoplasm. II. *Prophase*: Nuclear envelope disappears and chromatin begins to condense into chromosomes, each composed of a pair of identical chromatids. Centrioles are increasingly surrounded by spindle fibers and move toward opposite regions (poles) of the cell. III. *Metaphase*: Chromosomes align themselves on the cell equator. IV. *Anaphase*: Chromatids move toward the poles of the cell. Each chromosome contributes one of its chromatids to each pole. V. *Telophase*: Finally, chromatids become transformed into chromatin; nucleolus and nuclear envelope reappears. Content of the cell becomes partitioned into two daughter cells by furrowing and pinching-off of the plasma membrane. In general, a plant-cell mitosis differs from the above in the absence of centrioles and with respect to telophase, where furrowing is replaced by deposition of a new cell membrane–cell wall complex on the equatorial plane.

plasma membrane in such a manner that growth of the membrane could pull the daughter strands of DNA apart.

Eukaryotic cells, on the other hand, exhibit a characteristic pattern of cell division that appears largely required for the partition of nuclear material into the two daughter cells. Since eukaryotic cells contain about 1,000 times as much DNA as bacteria and since failure to partition the DNA equally between daughter cells would result in frequent inviability and genetic instability, a mechanism for the exact partition of DNA is of the greatest importance. The universal eukaryotic mechanism is called *mitosis* and is diagrammed in Figure 15-7.

A great deal of attention has been paid to the initiation of mitosis, since it would represent an important regulatory step in cell growth. Little is known, however, about the triggering event, except that there is some evidence that the nucleolus may be centrally involved. Mitosis requires energy, probably in the form of ATP, and is inhibited by respiratory inhibitors, such as carbon monoxide or cyanide. Mitosis is likewise inhibited by inhibitors of DNA synthesis, such as *mitomycin C*, and inhibitors of RNA synthesis, such as *actinomycin D*. Finally, mitosis is inhibited in a specific fashion by *colchicine*, which appears to prevent formation of the mitotic spindle (see Figure 15-7). It is interesting that colchicine does not prevent replication of chromosomes, so that as it inhibits cell division, the number of chromosomes per cell increases in steps of doubling.

Understanding of mitosis has been aided by the recent isolation of intact mitotic apparatus from dividing cells by mild disruption of cellular structure. The spindle is made up of protein, with a small amount of RNA bound to some of it. The spindle is composed of microtubules, and spindle fibers are oriented with respect to the centroles.

The regulation of cell division

It is worth examining in a little more detail some features of the regulatory events underlying the division of eukaryotic cells. At first examination, there are a number of conditions that must obtain if the growth of cells and the partition of their genetic material (through mitosis) be held in proper balance. For instance, cells maintain a ratio between the mass of the cytoplasm and that of the nucleus that changes in a cyclic fashion during a cycle of cell division, but which, averaged out over many divisions, is necessarily constant. The cellular contents of the several nucleic acids of the nucleus are also maintained in coordination with each other. Moreover, the location of DNA in chromosomes must undergo a complete cycle before cell division can occur. In other words, the DNA must be brought to a state like that just before the previous division. Finally, each protein in the nucleus and the cytoplasm must be maintained at a constant level in the cell (provided conditions remain constant), so that a doubling must occur for them to be replenished. The complexity of these various requirements renders the matter of regulation difficult to study, as there are so many relationships that could, in principle, trigger cell division. It may be added that the idea of regulation by the ratio of cell mass to something else (such as nuclear or DNA mass) is not only a possibility, in principle, but enjoys some experimental support. Thus, when cytoplasm is removed from amoebas by means of microsurgery, cell division is prevented, and this inhibition can be maintained for a long time by repeated removal. A role for the cytoplasm in initiating division is also supported by the observation that several nuclei in the same cytoplasm can be seen to divide in synchrony, so that at least a part of the regulatory mechanism must be cytoplasmic.

For example, it appears self-evident that the DNA content of cells must be constant over a number of divisions, and it is tempting to argue that this constancy supports a regulation of division based on the ratio of DNA to something else. Indeed, the constancy is an experimental fact, as resting nuclei contain a quite characteristic amount of DNA, as seen in Table 15-1. The figures given in the table refer to the *diploid state*, and haploid nuclei, found in germ cells, contain exactly one half as much. The reader will note the lack of correlation between DNA content and apparent complexity of the animal. It would be difficult to imagine a frog to be over twice as complex (in terms of information) as the reader of this page. On the other hand, from a genetic-informational point of view, it is reasonable that a magic kiss could transform a toad into a maiden, while the kissing of a duck could not be recommended for the production of the same effect. The underlying reason for the variation of DNA content in different, but similar, organisms (for example, frog and toad) is not understood.

The replication of DNA prior to cell division requires a substantial fraction of the time between divisions. There is evidence that this replication proceeds in an asynchronous fashion, occurring at different times in different parts of the chromatin material. Immediately following division,

Table 15-1   *The amount of DNA*
            per diploid nucleus
            in several animals

There is thus no clear relation between the observed
complexity of the animal and the DNA content.

Animal	DNA ($\times 10^{-12}$ g)
Man	6.0
Pig	5.1
Dog	5.3
Mouse	5.0
Duck	2.2
Toad	7.3
Frog	15.7

there is a period where no DNA is replicated; during this period, the
introduction of radioactive precursors of DNA (such as tritiated thymidine) into cells does not lead to uptake into the chromosomes. This
period is followed by a long one, during which DNA synthesis does occur,
as we said, in an asynchronous fashion. Finally, there is a second quiet
period when synthesis has ceased and only after it does, does cell division
actually commence. Thus, it seems somewhat unlikely that DNA content
per se could be the trigger for cell division, since the latter occurs rather
long after DNA replication is done. On the other hand, it is quite clear
that DNA formation must precede division, since any treatment of the
cell that interferes with replication also inhibits mitosis. Thus, the use
of radiation in the treatment of tumors appears to be active at the level of
DNA integrity and division. Purine and pyrimidine base analogs that
interfere with DNA synthesis are likewise effective in preventing mitosis.

One should also consider the matter of the regulation of DNA synthesis
prior to cell division. There is some evidence coming from studies with
plants that the initial event in the process is the rapid and transitory
synthesis of a single enzyme, *thymidine phosphokinase*, which catalyzes the
reaction

Thymidine + ATP $\rightleftharpoons$ Thymidine phosphate + ADP

where the phosphate goes to the deoxyribose part of the molecule.
(Thymidine is simply the pyrimidine thymine + deoxyribose and is,
hence, an immediate precursor of DNA.) It appears that prevention of
the synthesis of this enzyme, by the addition of inhibitors of RNA or
protein synthesis, leads to the inhibition of cell division as well.

It should be clear that the regulation of DNA replication is of great
importance not only in coordinating cell division and cell growth but in
completely preventing cell division in many cells from multicellular
organisms. Mitosis is a relatively rare event in most tissues, but the
potential must remain, since cell division occurs very rapidly when

wound healing or regeneration occurs, or when the tissue harbors neoplastic cells.

DNA synthesis is only one of the events that must precede cell division, and the period between the end of DNA replication and actual mitosis is occupied with other processes that must also precede mitosis and which provide additional opportunities for its regulation. For example, the assembly of the mitotic apparatus must necessarily be a complex matter, as must be the condensation of chromatin into chromosomes. Regulatory aspects of these processes are presently being studied in an intensive manner in the hope that one will, one day, be able to manipulate mitosis in cells to such ends as the induced regeneration of tissues and the control of neoplastic growth.

The role of chromosomes

The genetic material of eukaryotic cells is partitioned into daughter cells upon mitosis in the characteristic vehicle of the chromosome. These cells appear as solid bodies in the light microscope and possess a characteristic structure, so that the individual chromosomes of a human being may be identified and classified. Damage of certain, specific chromosomes is associated with certain genetic diseases, such as Down's syndrome (a form of mental retardation).

Chromosomes viewed with the electron microscope appear considerably more diffuse and are seen to be composed of fibers of chromatin. These fibers are of such a diameter that they could quite possibly be strands of individual DNA double helix or DNA in association with basic protein. A chromosome is of the order of 10 microns in length and about 1 micron in width. If the chromatin strands could be teased out into a single strand, it could be as much as several centimeters in length. Obviously, the vast amount of genetic material in eukaryotic cells requires that DNA be packaged in such a compact fashion. This situation should be considered in contrast to the single molecule that comprises the chromosome of bacteria.

The manner in which the strands of DNA are organized to form the morphology of the chromosome is not well understood. It has been suggested that the chromosomes are formed by loops or open strands of DNA and protein extending from a central core, which has been regarded as either protein or DNA, depending upon the particular investigator proposing the model. In fact, there is little evidence suggesting such a configuration, and the best present indications are that the chromosome may be viewed as a highly folded fiber that contains both DNA and protein without either portion being concentrated in any particular region. It would be imagined that the specific structure of a particular chromosome would be the consequence of the specific interactions between different regions of the strand, the chromosome folding into a shape that is the most thermodynamically favorable one. Some chromosomes show characteristic banding, the bands at one time having been thought to represent the genes. It now appears that the actual situation is somewhat more complicated. There is a correlation between gene and band, but it is not necessarily a one-to-one relation. Some bands appear to contain more than a single genetic locus and some may contain none. These bands can be explained on the basis of the folded-strand approach as alternating regions of tight and loose coiling, the tightly coiled portions containing the genetic loci.

Often chromosomes exhibit a region called the *kinetochore* (or *centromere*), which is either more strongly condensed or is a constriction. This region serves as the locus of attachment for the spindle fibers during the *anaphase* of mitosis (see Figure 15-8).

Oocytes from various vertebrates contain chromosomes with a particular structure consisting of loops attached to a central strand. These *lampbrush* chromosomes have been studied intensively for a long time and have led to interesting views of chromosome function. For one thing, the loops appear to consist of fibers whose structural intactness depends on DNA. Addition of the enzyme DNAase to the chromosomes leads to breakage of the loops, while that of proteolytic enzymes does not. Cells containing these chromosomes carry out particularly active RNA synthesis, and it appears that the loops are the sites of synthesis of both RNA and protein. Thus, radioautographic experiments indicate the loops as the regions where radioactive amino acids and radioactive RNA precursors (such as ^{3}H-uridine) are incorporated. RNA synthesis there is inhibited by actinomycin D, suggesting that it is a DNA-dependent synthesis. The loops may be interpreted as regions of the DNA–protein complex where the strands are exposed to the surrounding nucleoplasm for greater ease in copying. This interpretation is also consistent with the identification of the *puff* regions of many chromosomes as regions of active DNA-dependent RNA synthesis. It would thus appear that the expression of genes through the synthesis of mRNA often requires that the tight packing of the chromatin be relieved and the individual strands presented to the exterior. One is tempted to say that chromosomes (and parts of chromosomes) can exist in one of two states, tight coiling and extension, and the former state is associated with the requirements of cell division, while the latter is associated with those of mRNA synthesis. During most of interphase, the requirement for RNA synthesis throughout the genetic map leads to the entire chromosome's assuming the more open configuration, so that the individual chromosomes are no longer readily identifiable in the nucleus. Lampbrush chromosomes are those in which different regions assume different configurations. Chromosomal puffs are transitory and represent transitions to the open configuration in regions related to the synthesis of particular mRNA molecules.

Finally, it has been observed that the chromosomal fibers of the embryonic honeybee are permanently attached to the nuclear envelope. This situation is reminiscent of the close association of the bacterial chromosome and the cell's plasma membrane. The significance of the association of chromosome and envelope in the case of the honeybee is not understood; it may be that the phenomenon is more general, but with the binding being reversible in most cases and, hence, not obvious

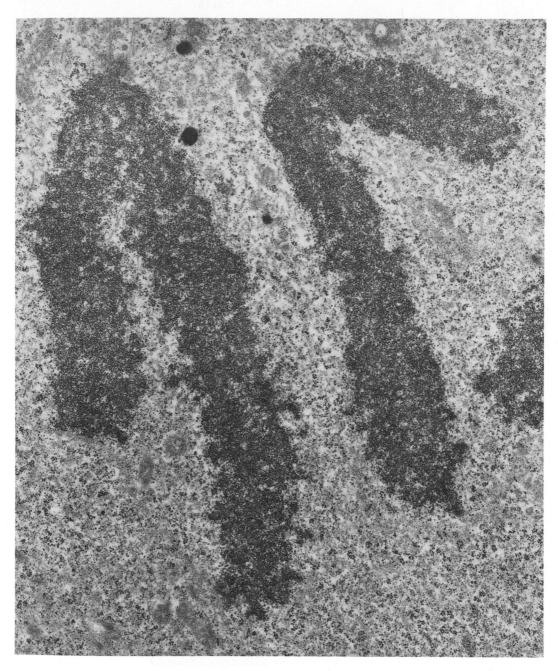

Figure 15-8   Anaphase chromosomes from the rat kangaroo (*Potorus tridactylus*). Fibroblast cells from this organism were grown in culture and fixed with glutaraldehyde and osmium tetroxide. Sections were stained with uranyl acetate and lead citrate. Magnification is ×19,800. Note the spindle fibers leading away from the crook in each chromosome (the kinetochore). (Courtesy of Dr. R. B. Brinkley.)

to the electron microscopist. The fibers, because they are attached to the nuclear envelope in the honeybee cells, are particularly easy to observe, as they do not become as tangled as usual. It develops that the fibers have dimensions that could be explained by their being a central core of DNA surrounded by protein. When they are treated with the proteolytic enzyme trypsin, the dimension changes to that of naked DNA. Finally, when they are digested with DNAase, the fibers disappear altogether.

Meiosis

We have stressed the maintenance of a constant amount of genetic material per cell through cycles of mitosis. There is one occasion when cells must, in fact, reduce their complement of DNA, and that is the formation of haploid cells, cells that will later fuse with another haploid cell to produce the original diploid condition. Thus, when an egg and sperm cell fuse their nuclear material to form a zygote, there is a necessary doubling of genetic material, and it is clear that the germ cells must in the course of their development be rendered haploid. This occurs through the process of *meiosis* or reduction divisions. The essence of the process is the occurrence of two divisions of the nuclei, while the chromosomes divide but once. Whereas in the diploid cell there were two copies of each chromosome, after meiosis there is only one. During the prophase of meiosis, the corresponding chromosomes come to lie in close contact, so that there are two pairs of chromatids in each instance. At the end of the two nuclear divisions each nucleus contains, therefore, a single copy of each chromosome. Genetic mixing can occur at this level associated with the phenomenon of crossing over, which is the exchange of part of each of a pair of corresponding chromatids. The greater complexity of meiosis, when compared to mitosis, is reflected in the longer prophase period that precedes it. The controlling features that inform nuclei that they are to undergo meiosis instead of ordinary mitosis are not well understood.

Suggested Reading

Books

Bresnick, E., and A. Schwartz, *Functional Dynamics of the Cell*, Academic Press, New York, 1968, Ch. VIII.

DuPraw, E. J., *Cell and Molecular Biology*, Academic Press, New York, 1968, Chs. 18–19.

Gunsalus, I. C., and R. Y. Stanier (eds.), *The Bacteria*, vol. 4, Academic Press, New York, 1962.

Harris, R. J. C., *Cell Growth and Cell Division*, Academic Press, New York, 1963.

Levine, L., *The Cell in Mitosis*, Academic Press, New York, 1963.

Malek, I., and Z. Fencl (eds.), *Theoretical and Methodological Basis of Continuous Culture of Microorganisms*, Academic Press, New York, 1966.

Monod, J., *Recherches sur la croissance des cultures bacteriennes*, Hermann, Paris, 1942.

Watson, J. D., *Molecular Biology of the Gene*, 2nd ed., W. A. Benjamin, New York, 1968, Ch. 10.

Articles

Eagle, H., "The sustained growth of human and animal cells in a protein-free environment," *Proc. Nat. Acad. Sci. U.S.*, **46**, 427 (1960).

Maaløe, O., "The nucleic acids and the control of bacterial growth," in *Microbial Genetics*, W. Hayes and R. C. Clowes (eds.), Cambridge University Press, Cambridge, 1960.

Monod, J., "The growth of bacterial cultures," *Ann. Rev. Microbiol.*, **3**, 371 (1949).

Novick, A., "Growth of bacteria," *Ann. Rev. Microbiol.*, **9**, 99 (1955).

Payne, W. J., "Energy yields and growth of heterotrophs," *Ann. Rev. Microbiol.*, **24**, 17 (1970).

Sakai, H., "Contractile properties of protein threads from sea urchin eggs in relation to cell division," *Int. Rev. Cytol.*, **23**, 89 (1968).

van Uden, N., "Kinetics of nutrient-limited growth," *Ann. Rev. Microbiol.*, **23**, 473 (1969).

16 DNA and protein synthesis

In earlier chapters our examination of a number of aspects of the life of the cell led us to discuss selected examples of regulatory mechanisms. Of course, these examples represent a small fraction of those known. The proportion of known to unknown regulatory mechanisms could be compared to an iceberg, where the bulk is still undiscovered. Our concern in this chapter will be with the genetic forms of control that occur in cells, and we shall again be forced to adopt the tactic of considering only a small fraction of what is actually known. We shall begin by considering what is meant by genetic control (and therefore by the genetic code) and shall examine the operation and control of protein synthesis as an example of "reading out" that code.

Genetics is the study of the continuity between generations of cells and the mechanisms behind it. The element central to all modern genetic theory is that there is a master plan (genotype) that somehow determines the observed properties (phenotype) of organisms. The genotype is believed to reside largely in a single molecular species, deoxyribonucleic acid (DNA). Whether the *totality* of the genetic code resides in DNA or whether it represents one among several forms of information storage is an important question at the present time and will be discussed in the last section.

Why DNA is known to be genetic material

At one time, in the 1930's, when the essential chemistry of proteins and their role as enzymes were becoming understood, nucleic acids represented something of an enigma in cell chemistry. They had been known as cell components for about a century, but their function was completely obscure. Then, over a period of about a decade, a series of observations was made which, together, indicated that the nucleic acids play a central part in the genetic regulation of cell activities. For one thing, it appeared from histochemical studies that DNA was located in the nucleus and specifically concentrated in the chromatin material, already known to involve mechanisms of genetic control. Furthermore, it developed that there was, in a given organism, a constant amount of DNA per nucleus and that diploid and haploid cells of that organism differed by a factor of 2 in this regard. Since diploid cells contain twice the number of genes found in the haploid state, this provided an additional suggestion that DNA had something to do with heredity. However, all evidence remained somewhat circumstantial until a single experiment was performed that at once indicated clearly that DNA was the chemical entity in which genetic information was somehow encoded.

Seldom is a single experimental approach a turning point in science, but the discovery of bacterial transformation in 1944 by Avery, McLeod, and McCarty was such an instance. Their experiment made use of two strains of *Diplococcus pneumoniae*, which differed in the ability to make a functional capsule external to the cell wall. They showed that DNA, isolated from capsule-producing cells, when added to a culture of non-capsule cells, gave that population the ability to produce capsules. In other words, the information required to manufacture capsules was transferred from one population to another in the form of chemically pure DNA. In addition, it was observed that the enzyme deoxyribonuclease (DNAase), which specifically hydrolyzes DNA, completely destroyed the transforming ability. Additional transformation experiments were performed using a variety of other bacteria and resulting in the transfer of a number of genetic traits.

The general structure of nucleic acids

The demonstration of an informational role of nucleic acids at once leads to questions about the connection between their structure and the nature of the code. We shall uncover this connection, first, by considering some general features of nucleic acids and, second, by considering the structure and informational function of particular nucleic acids.

Nucleic acids are polymers that, on hydrolysis, yield 5-carbon sugars, purine or pyrimidine bases, and phosphate. Examples of these components are illustrated in Figure 16-1. Furthermore, there is a common pattern in which these components are assembled, and this basic nucleic acid structure is shown in Figure 16-2. The polymer thus is composed of a homogeneous repeating backbone of alternating sugar and phosphate, to which is attached a series of nucleotide bases, which may differ from each other. The appropriateness of nucleic acids to the task of bearing information about heredity resides in the nucleotide bases projecting from the backbone, which may be thought of as letters of the code that determines cell properties.

There are important differences between DNA and RNA, as well as between the various types of RNA, which play roles in the cell to be discussed later. All DNA contains deoxyribose as its major sugar component and all of it contains thymine, whereas RNA contains uracil.

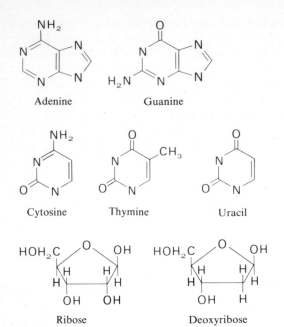

Figure 16-1 Common components of nucleic acids.

Adenine

Guanine

Cytosine

Thymine

Uracil

Ribose

Deoxyribose

Figure 16-2 (a) Generalized structure of a nucleic acid, composed of an alternating phosphate–sugar backbone with projecting nucleotide bases. (b) A dinucleotide, showing locations of linkages among sugars, phosphate, and bases. This geometry is applicable to nucleic acids of any size. Note that in this example the sugar is ribose.

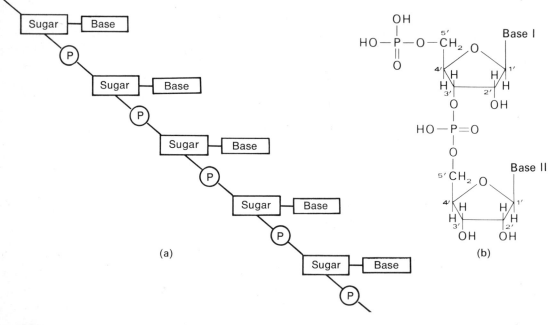

(a)

(b)

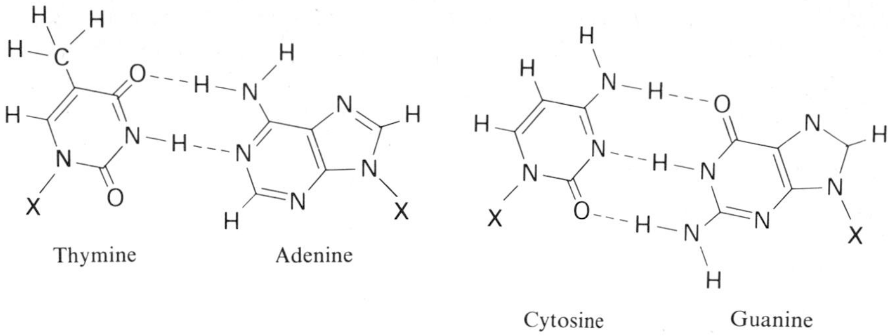

Figure 16-3 Base pairing in DNA. *X* stands for the point of attachment of the nucleotide base to the deoxyribose of the backbone. Hydrogen bonds are denoted by dashed lines.

As we shall see, an important feature of all sorts of nucleic acids is their ability to form intramolecular bonds in such a manner as to take on a specific tertiary structure (that is, configuration in space). These bonds are mostly of the same hydrogen-bond class that was seen earlier to play an important role in determining protein structure. In the case of nucleic acids, the hydrogen bonds not only govern structure of the individual molecule but enable distinct molecules to bind together. Also, hydrogen bonding in nucleic acids enjoys a degree of specificity that is of great importance in the informational role of these molecules.

We saw that nucleic acids contained a limited number of purine or pyrimidine bases that protrude from the phosphate–sugar backbone. The capacity of these bases to form hydrogen bonds with each other determines the conformation of the nucleic acid polymer and allows for completely specific binding between different molecules. The geometry of the bases is such that thymine and adenine fit well together and hence form relatively strong bonds, as do cytosine and guanine. In RNA, where there is no thymine, adenine binds specifically to uracil. This high degree of binding selectivity is at the basis of all functions of nucleic acids and is known as *specific base pairing*. The geometry of two such base pairs is illustrated in Figure 16-3. The energy of a single hydrogen bond is much lower than that of a covalent bond, but in the case of base pairing, geometry favors formation of two or three hydrogen bonds, whose energies are additive as far as binding between the two molecules is concerned.

The structure of DNA

It is well known that DNA is a very large molecule (with a molecular weight of more than 10^8 daltons), which is usually formed from the combination of two chains wound together. The two chains are wound together in the now famous *double helix* (which is illustrated in Figure 13-4), and the force that binds them together is that of hydrogen bonds between base pairs. Since there is specificity between bases with regard to pair formation, and since all the bases in a DNA molecule are paired, there are equal amounts of adenine and thymine in a sample of DNA. Likewise, the amounts of guanine and cytosine are equal. Indeed, these regularities of base composition were of importance in leading to the original formulation of the helical picture for DNA. Additional information that led to the helical formulation was largely from crystallographic (X-ray diffraction) studies of DNA. The reader is directed to *The Double Helix*, which is a popular, but highly illuminating, account of the development of the modern view of nucleic acids, written by James Watson, who should (and does) know about it. He is also the author of a less popular account of nucleic acids and related matters called *The Molecular Biology of the Gene*; both books are listed in Suggested Reading and are highly recommended.

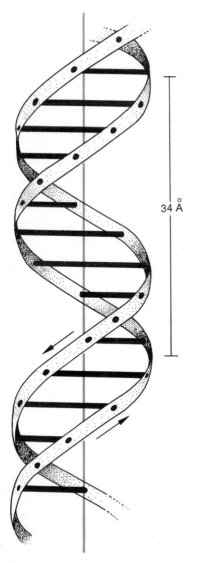

Figure 16-4 Configuration of the DNA double helix. Locations of some hydrogen bonds are indicated by dashed lines. Arrows denote the sense of the two chains—i.e., the direction of the 3'–5' linkages. Note that they run in opposite directions.

34 Å

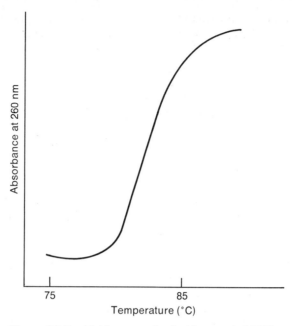

Figure 16-5   Melting curve for double-stranded DNA.

An important aspect of DNA structure is the vectorial character of each chain, which can be said to have a direction or polarity. This is because the attachment of phosphate to sugar is not symmetrical. Thus, each ribose (or deoxyribose) has a phosphate attached to the position-3′ and the position-5′ carbons,[1] so that termination of the chain must always leave a free 3′ position at one end and a free 5′ at the other. Even if the chain is a closed loop, which is often the case, the 3′–5′ asymmetry gives it a unique direction (see Figure 16-2). In the case of the double helix, the two individual chains are opposed to each other in direction (see Figure 16-4).

A great deal of information about the relationship between the two strands of the helix has come to us from "melting" experiments, where hydrogen-bond breakage, and thus chain separation, is effected by heat. When a solution of DNA is heated slowly to a critical temperature (the so-called melting point), which is about 80°C, the two strands separate, and any property that reflects the double-stranded character, such as viscosity or ultraviolet absorbancy, is seen to change. Such a melting curve is illustrated in Figure 16-5. Similar changes can be induced by lowering the pH or altering the concentrations of cations. If the treatment to achieve strand separation is sufficiently gentle, the two strands can be allowed to recombine. Thus, if they have been separated by gentle heat treatment, gradual cooling will allow them to come together again and the absorbancy (or viscosity) will return to that of the initial double-stranded figure. This property is called *annealing* and signifies the

---

[1] The prime on the position number differentiates the position in the sugar from that in the nucleotide base.

spontaneous nature of chain pairing. Recently, it has been possible to do melting experiments where a mixture of DNA from two related sources was used and to anneal chains from the two sources together. It is interesting that the degree of association that can be obtained between strands from different sources reflects the closeness of their phylogenetic relationship as determined by other criteria. This is due to the requirement for base pairing between the two strands; if organisms (or viruses) are closely similar, they may be expected to have similar genetic systems, and this, as we shall see, implies similar DNA base sequences. If the bases form a similar sequence, there will be extensive regions where base pairing and hence binding can occur between corresponding chains.

Finally, there is a class of DNA, which has been isolated from certain bacterial viruses, that exhibits neither the kind of melting behavior described nor regularities in base composition. This nucleic acid is DNA in that it contains thymine and deoxyribose instead of uracil and ribose, but in molecular structure it is single-stranded. When this DNA is replicated within a sensitive host bacterium, it exists in a temporary double-stranded form, but the subsequent generation of viruses contain the single-stranded DNA again.

The structure of RNA

Ribonucleic acid, RNA, contains ribose in place of DNA's deoxyribose and generally uracil in place of thymine. RNA is single-stranded, lacking the helical structure of DNA, and does not exhibit the one-to-one ratio between homologous bases (for example, between adenine and thymine). There are, however, regions of the single strand where base pairing can occur within the molecule, and sequence determinations of various RNA's have led to the proposal of structures with considerable intramolecular pairing. As we shall see (Figure 16-6), a frequent configuration of such RNA molecules is a cloverleaf pattern dictated by the base sequences and by the usual base-pairing relationships.[2]

With the exception of single-stranded DNA, DNA from a wide spectrum of sources exhibits a similar molecular structure and hence similar properties. RNA, on the other hand, is quite varied, and there are several quite different sorts of RNA with differing properties and different (and characteristic) roles in the cell. For example, *transfer RNA* (tRNA) represents a class of small RNA molecules with molecular weights of approximately 25,000 daltons. The small size of tRNA prevents it from being easily centrifuged, and it remains in the "soluble" fraction of cells when ribosomes and other organelles are sedimented. This accounts for its old name *soluble RNA*, which is now little used. The sequences of a number of tRNA's are known and an example is given in Figure 16-6.

---

[2] In RNA, which contains no thymine, the base pairs are adenine–uracil and cytosine–guanine.

All tRNA molecules studied have extensive regions of base pairings, so that the homologous base ratios approach 1. All have a similar sequence at the 3' end consisting of *cytosine–cytosine–adenine*, and we shall see below that, when tRNA functions in protein synthesis, the 3' end is the site of attachment of amino acids.

Messenger RNA (mRNA) is a class of RNA molecules that are quite unstable (and therefore difficult to study) and quite heterogeneous. The molecular weight of mRNA ranges from 50,000 to 5,000,000 daltons, and it is large enough to be seen in suitable preparations as a thin strand in the electron microscope. It does not possess many regions where intramolecular base pairing can occur and so is not very much folded back on itself, but it does form base pairs with certain regions of DNA from the same organism, and we shall see shortly that this *homology* is far from trivial.

There are several different RNA species associated with ribosomes and these are called *ribosomal RNA*. The structures of some of these are known and it is clear that a number of regions of base pairing exist. There are types of RNA associated with mitochondria and chloroplasts, and present evidence suggests that they are similar to categories of RNA described already, as these organelles possess ribosomes (and therefore ribosomal RNA) and also probably mRNA and tRNA. Finally, many viruses contain (single-stranded) RNA as their genetic material, and this material is quite variable. For example, RNA from tobacco mosaic virus forms few intramolecular base pairs, whereas an animal virus, adenovirus, leads to production of RNA with extensive base pairing and a structure similar to the cloverleaf pattern of tRNA (Figure 16-7).

The replication of DNA

Since DNA serves as the genetic material of cells, it must be able to replicate, since, by cell division, a single cell can give rise to an infinite number of progeny, all of whom must contain the same (or similar) genetic information. Thus, when cells divide, DNA must divide as well, and in such a fashion as to conserve the genetic information that it contains with a minimum number of mistakes (mutations). The double-stranded structure of DNA is ideally suited to the necessity for replication, for, in principle, all that is required is for the two strands to separate, allow a homologous copy to be made on each by base pairing, and then to go their separate ways. In fact, this is exactly what happens, and the precise nature of the base-pairing relationships enables the replication to proceed in an error-free manner. A summary of DNA replication is given in Figure 16-8, and the reader is directed to Suggested Reading for a more complete description.

An interesting problem in DNA replication is that of topology, that is, the manner in which the two strands are taken apart so that the new copy may be made. Since the strands are twisted together in a true helix, it is necessary to untwist them to carry on replication and the process of untwisting includes the separation of two chains that have a strong total

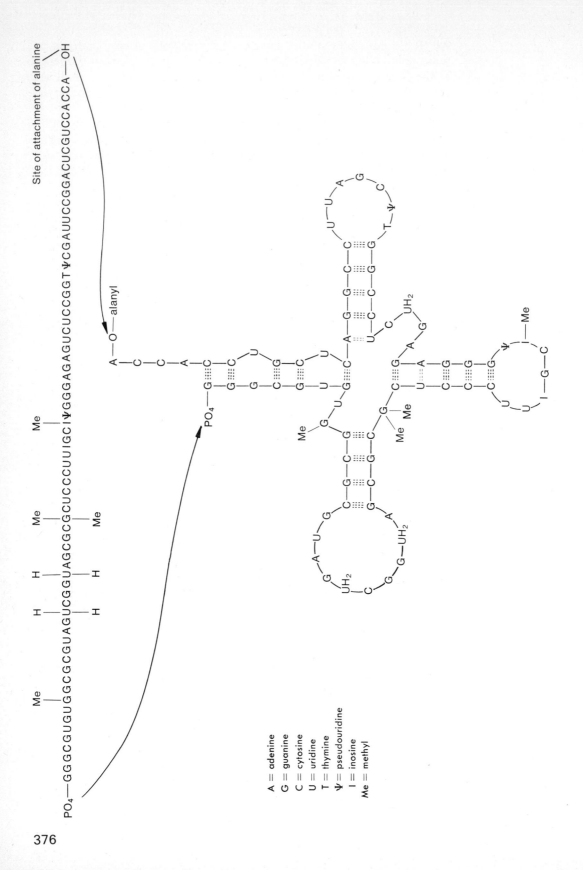

376

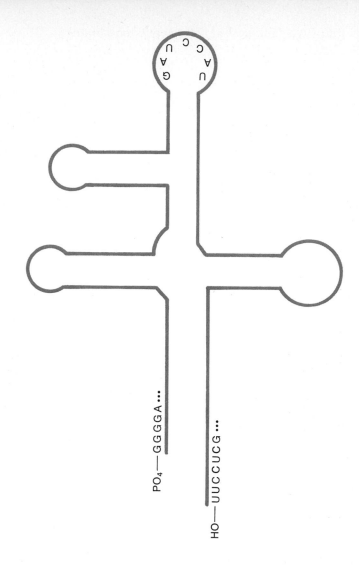

PO₄—GGGGA...

HO—UUCCUCG...

Figure 16-6 Base sequence of yeast alanyl-tRNA. At the top the sequence is shown in linear form, as determined experimentally. Below that it is deformed into the cloverleaf pattern believed to represent the actual conformation. This transformation is based on the maximizing of the number of intramolecular base pairs, as indicated by double or triple lines between bases in the cloverleaf. The combined hydrogen-bonding energies of these base–base interactions together render the cloverleaf-type structure thermodynamically favorable. This structure was proposed by T. H. Jukes in *Biochem. Biophys. Res. Commun.*, **24**, 744 (1966). Note that it includes both thymine and uracil in contrast to DNA which contains only thymine and other forms of RNA which contain only uracil.

Figure 16-7 Cloverleaf representation of an RNA associated with infection by adenovirus. This type of RNA is isolated from cultured animal cells infected with the virus. The complete structure is known, but only a portion of the sequence is shown here. There are regions with strong similarity to regions in several ordinary transfer RNA molecules. The complete sequence may be found in the original paper: K. Ohe and S. M. Weissman, *Science* **167**, 879 (1970).

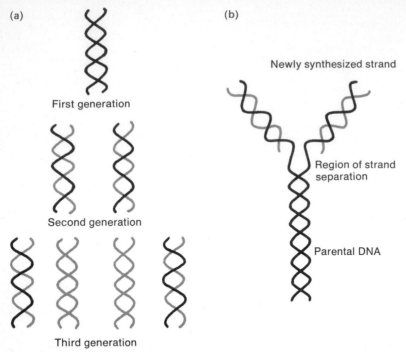

Figure 16-8 Outline of DNA synthesis. DNA replication is termed semi-conservative, meaning that, in each double helix, there is one strand from the previous generation and one synthesized *de novo*. In (a) the *de novo* strand is indicated by a grey line. Note that in two generations, one half of DNA will be completely new and one half will contain an original strand. (b) Strands are separated prior to formation of a new strand on each one. The parental strand serves as template for the synthesis of the second (new) one.

energy of binding (by hydrogen bonds and electrostatic forces) to each other. It appears that, at least in bacteria, the energy required to separate the two strands comes from cell respiration, since replication of double-stranded DNA is inhibited by the respiratory inhibitor cyanide, while replication of single-stranded DNA from a virus is not. An added problem in DNA replication is the circular nature of many DNA molecules, which must be taken into account in explaining strand separation. The probable mechanism for this is given in Figure 16-9.

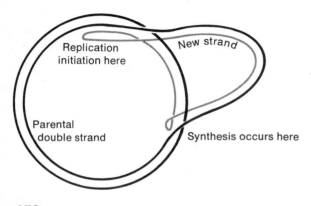

Figure 16-9 Outline of synthesis of circular DNA. This model, where the point of synthesis moves away from the initiation point as the two loops separate, is due to the experiments and their interpretation by J. Cairnes (see Suggested Reading).

The genetic code

To understand how DNA can serve as genetic material, consider the information required to synthesize a particular protein. Obviously, the problem of synthesizing a whole cell—which is the real genetic question— is much more complex and cannot be rationally approached at the present time. It has been often suggested, and is at least a partial truth, that the synthesis of a cell is, in fact, the summation of the regulated synthesis of all the cell enzymes. In other words, if the cell can manufacture the correct amounts of the correct enzymes at the correct time, all else will fall into place. This is undoubtedly an extreme view, but the fact is that protein synthesis is a major duty of a genetic system. It is also clear that protein synthesis is a good system for studying the output of genetic information, since the exact nature of the output is understood: the necessity of producing the exact sequence of amino acids to give the protein its necessary properties, such as enzymic activity.

We shall consider in the following section the mechanism (and information pathway) of protein synthesis. Here we need only assert that there is a special coding relationship between the information content of DNA and the sequence of amino acids in a protein. In other words, the code located in the DNA molecule can, according to specific rules, be mapped onto the sequence of amino acids of a protein. It is therefore necessary to know the rules and to understand the specific nature of the *message* contained in the DNA. These relationships are completely known, at least in the systems that have been most extensively studied, and they may be summarized as follows:

1. In the case of DNA, the symbols of the code are the individual nucleotide bases, and the "words" that they form consist of groups of three such bases. As in our (human) language, the order of letters within a word is required to specify meaning.

2. The location of a particular word in the DNA chain specifies the location of a particular amino acid in a protein. The correspondence between the DNA code and a particular protein is illustrated in Figure 16-10. It is clear that there must be a region of the DNA molecule that carries information for the synthesis of a particular protein, and that it is divided up into three-base regions that denote the location of each individual amino acid. Since the triplets are densely packed on the DNA strand—that is, there are no bases in between—and since they do not overlap, it is important that the reading of the code begin at the correct base. Thus, if the code were read in units of three bases but began one base too far along, the whole reading would be out of phase and no triplets would be read correctly. As we shall see, the genetic code includes preventative measures against this possible defect, as there are triplets that denote *begin* and *terminate* and so define limits of a particular message.

It is interesting that there are three bases in a word and four different letters (bases) in either DNA or RNA. This means that there are 4^3 or 64 different words available to encode the 20 (or so) amino acids found in

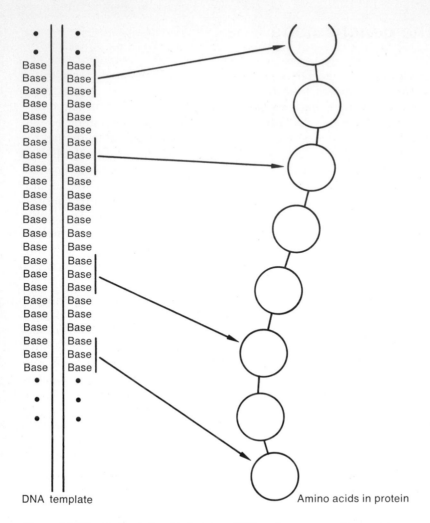

Figure 16-10 Code relationship between DNA and a corresponding protein (intervening steps involving RNA and several enzymes are omitted). Note that three bases encode the location of one amino acid. In this diagram, neither bases nor amino acids are identified; Normally four bases would be involved and as many as 20 amino acids included in the protein.

proteins. Note that a two-letter word would be inadequate, leading to 4^2 or 16 different words. It will also be seen below that the code is degenerate, with more than one word denoting a given amino acid.

DNA resides in the nucleus of eukaryotic cells, and most protein synthesis occurs in the cytoplasm, at the ribosomes. Thus, the information transfer from DNA to finished protein must include a spatial transfer of information through the nuclear membrane, and it appears that this occurs via messenger RNA. Messenger RNA (mRNA) is synthesized in the nucleus and passes to the cytoplasm, where, as we shall see, it participates in protein synthesis. The synthesis of mRNA is believed to occur on one

380

of the strands of DNA in such a manner that the bases for the RNA are lined up with the bases of the DNA by the usual base-pairing arrangement. DNA thus serves as a pattern, or *template*, for mRNA synthesis, and the mRNA must bear a close similarity to the strand of DNA *not* being copied, since both are related to the copied strand by the base-pairing relationship. The information content of mRNA must be identical to that of the DNA on which it was made, since each triplet word is transformed to a corresponding *antiword* by the base-pairing rule. The synthesis of mRNA on its DNA template is known as *transcription*.

Just as one strand of the DNA molecule serves as a template for the synthesis of mRNA, so does a mRNA molecule serve as a template for the synthesis of a particular protein. While a DNA molecule is divided up into many regions corresponding to individual proteins, the mRNA appears to encode the sequence of just a single protein or, in some instances, several proteins that are under a common regulatory mechanism. The information in a collection of DNA triplets, each specifying the location of an amino acid in a protein, has been transformed by base pairing to information in a collection of mRNA triplets, forming the direct template for manufacturing the protein. The code upon which this template is based is known with increasing certainty and is shown in Figure 16-11. This code was elucidated in experiments with bacteria but, as far as evidence exists, most features apply to all genetic systems.

Figure 16-11 Genetic code. Note that the third base in each codon is less specific than the first two. The codons UAA, UAG, and UGA are all termination codons (denoted, Stop), causing the completed polypeptide chain to be released from the polyribosome. The codon AUG represents either methionine or formylmethionine, in the latter instance serving as a chain-initiation signal.

UUC	Phe	UCU	Ser	UAU	Tyr	UGU	Cys
UUC	Phe	UCC	Ser	UAC	Tyr	UGC	Cys
UUA	Leu	UCA	Ser	UAA	Stop	UGA	Stop
UUG	Leu	UCG	Ser	UAG	Stop	UGG	Trp
CUU	Leu	CCU	Pro	CAU	His	CGU	Arg
CUC	Leu	CCC	Pro	CAC	His	CGC	Arg
CUA	Leu	CCA	Pro	CAA	Gln	CGA	Arg
CUG	Leu	CCG	Pro	CAG	Gln	CGG	Arg
AUU	Ileu	ACU	Thr	AAU	Asn	AGU	Ser
AUC	Ileu	ACC	Thr	AAC	Asn	AGC	Ser
AUA	Ileu	ACA	Thr	AAA	Lys	AGA	Arg
AUG	Met	ACG	Thr	AAG	Lys	AGG	Arg
GUU	Val	GCU	Ala	GAU	Asp	GGU	Gly
GUC	Val	GCC	Ala	GAC	Asp	GGC	Gly
GUA	Val	GCA	Ala	GAA	Asp	GGA	Gly
GUG	Val	GCG	Ala	GAG	Glu	GGG	Gly

"Reading out" the genetic code: protein synthesis

Just as a computer without a means of output is utterly useless, the biological code is meaningless unless it can transform the blueprint of a protein into the actual molecule with speed and accuracy. An error, such as a mutational error in the code itself or a mistake in transcription, will lead to an enzyme with one or more wrong amino acids, probably producing abnormal activity. The general pathway of *readout* of the genetic code was given above as passing from DNA to mRNA to protein. Further details are instructive.

The pathway of information transfer in protein synthesis is seen to occur in two steps, the first being the transcription of the DNA code into the base sequence of an RNA molecule. Then, it remains to transform that information into the correct amino acid sequence of a protein. The first step might be described as the DNA-linked synthesis of mRNA and appears to take place in the nucleus of higher cells, where most of the DNA resides. As stated before, the information transfer takes place by base pairing, such that the DNA serves as template for the synthesis of the RNA. The actual synthesis involves the formation of covalent bonds between the nucleotides, and an enzyme manages this noninformational aspect. As we have seen, the sequence of bases in the mRNA molecule was determined as a result of base pairing with those of the parent DNA. Thus, there should be a special relationship between the base contents of the DNA and the resultant RNA. In the simplest case, where the DNA template is single-stranded viral DNA, there is a very close correspondence between composition of the DNA and the complementary RNA such that, for example, the guanine content of the DNA equals the cytosine content of the RNA.

There is an additional form of specificity in the synthesis of mRNA. The cellular DNA is supposed to contain information for the synthesis of all the proteins a cell can make, but the cell does not necessarily form all those proteins at once. Many bacterial cells form certain enzymes only in the presence of their substrate, and, in cells of higher organisms, embryonic differentiation appears to include a great amount of selectivity as to what enzymes are synthesized and when. Since a mRNA molecule directs the synthesis of a single protein (or a few closely related proteins), it is reasonable that the regulation of synthesis be at the level of mRNA synthesis— *transcription*. As we shall see, available evidence points to such transcription as the locus of regulation in protein synthesis.

Thus, a specific mRNA is formed on the DNA template and serves, itself, as the template for a specific protein. Since the chief locus for protein synthesis is the ribosome, often located on the endoplasmic reticulum, it is clear that the mRNA must migrate from the nucleus into the cytoplasm. That done, the only remaining step is the translation of the code, still in the form of nucleotide base sequences, into an actual protein molecule. Since amino acids have been shown to have no special affinity for the template themselves, an additional step is necessary, which might be termed modification of the amino acid to enable it to find its place. The modification takes place in two steps, the first of which is often called *amino acid activation*, where an enzyme specific to each amino acid catalyzes the addition of an adenosine monophosphate unit to the amino acid:

Amino acid + ATP $\rightarrow$ Amino acid—AMP + pyrophosphate

Then, the same "activating enzyme" catalyzes the transfer of the amino acid to a tRNA molecule:

Amino acid—AMP + tRNA $\rightarrow$ Amino acid—tRNA + AMP

There is a specific tRNA for each of the 60 or so codons found on mRNA. It will be recalled that all tRNA molecules studied exhibit a 3′ terminal sequence of cytosine–cytosine–adenine to which the amino acid is attached. Also, the tRNA molecules studied all contain a greater or lesser proportion of uncommon nucleotide bases (that is, other than adenine, uracil, cytosine, or guanine).

The role of tRNA is to allow the amino acid to become aligned at the correct position on the mRNA molecule. The recognition occurs because a region of the tRNA molecule, consisting of three bases, bears a base sequence complementary to the three bases of the correct word of the

Figure 16-12 Transfer RNA molecules bearing amino acids to the correct location on the mRNA template. Each tRNA molecule is specific to one particular amino acid and is bound at the correct location for that amino acid by hydrogen bonding between base pairs.

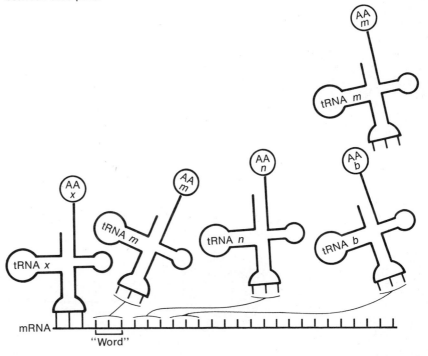

mRNA code. This triplet, which binds to the *codon* of mRNA, is known as the *anticodon*. This means that the problem of recognizing the amino acids is shifted to the activating enzyme (which brings the amino acid together with the appropriate tRNA), which, because of ordinary enzyme specificity, is able to catalyze the reaction of the amino acid with the proper tRNA. The tRNA then lines up on the template (see Figure 16-12) and places the amino acid in the correct relation to the others, likewise lined up by means of their tRNA's. Thus, the correct order of the amino acids is established, and the remaining problem is to synthesize the peptide bonds connecting them, certainly the simplest part of the whole procedure.

Although the activation of amino acids takes place in the soluble fraction of the cytoplasm (the part not easily sedimented), the actual synthesis of protein occurs largely at the ribosomes. This presents certain problems of a topological nature, as the template was said to be the linear chain of the mRNA molecule. An outline of how the ribosome is involved in scanning the template is given in Figure 16-13, where it is seen that the function of the ribosome is to allow the proper orientation of mRNA and

Figure 16-13 Relation of ribosomes to mRNA and tRNA. (a) Three ribosomes are seen scanning a single mRNA strand with polypeptide chains of increasing length issuing from them. (b) A tRNA molecule bearing amino acid 1 is shown bound to the ribosome and to the mRNA strand by base pairing. Amino acid 1 is already attached to the growing peptide chain by enzymic formation of a peptide bond. A more detailed view of the process is given in Figure 16-14.

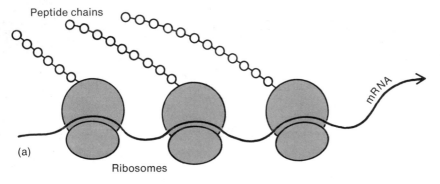

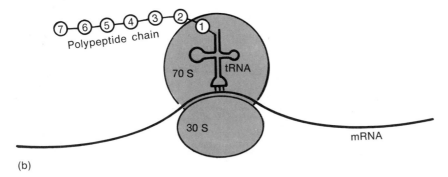

tRNA-amino acids, so that peptide-bond synthesis may occur. A number of ribosomes are seen to be scanning the same mRNA chain, a possibility deduced from the fact that ribosomes that are active in protein synthesis appear to be in the form of linear aggregates (polyribosomes), looking in electron micrographs rather like beads on a string. This simple view of protein synthesis obscures our many areas of ignorance, which include the details of the enzyme system responsible for peptide-bond closure and the role of the large amount of ribosomal RNA, which appears to be quite without information content in the sense of specifying amino acid sequence. Note that, since no specificity for a given protein resides in the ribosomes, the same population of ribosomes may work toward the formation of different proteins, depending on what mRNA molecules may be present in the cytoplasm.

Additional features of ribosome structure and action are of interest. For example, the structure of a ribosome depends on the presence of Mg^{2+}, with the complete ribosome dissociating into two subunits at low concentrations of the ion. Table 16-1 illustrates this event in the cases of ribosomes from prokaryotic and eukaryotic cells. In addition, recent evidence suggests that there are two attachment sites for tRNA on a given ribosome, one that initially bears the tRNA attached to the growing peptide chain and the other bearing the tRNA that carries the next amino acid to be added to the chain. This arrangement is diagrammed in Figure 16-14. Other features of protein synthesis are not yet completely worked out. For example, not only is ATP required for amino acid activation, but the trinucleotide guanosine triphosphate (GTP) is required for chain elongation at the ribosome, acting in a capacity not yet understood. Moreover, several soluble (nonribosomal) proteins are required for chain elongation and termination, and their exact roles are likewise presently unclear.

Finally, a property of mRNA that appears significant in the context of regulation is its frequently short half-life in microorganisms. When the

Table 16-1   Dissociation of ribosomes on removal of magnesium

This effect can be produced by addition of the chelating agent, ethylenediamine-tetraacetate (EDTA). Size of ribosomes and of fragments thereof is measured with the ultracentrifuge.

Class of ribosome	Size (Svedberg units)	Size of subunits (Svedberg units)
Eukaryotic	80 S	60 S + 40 S
Prokaryotic	70 S	50 S + 30 S

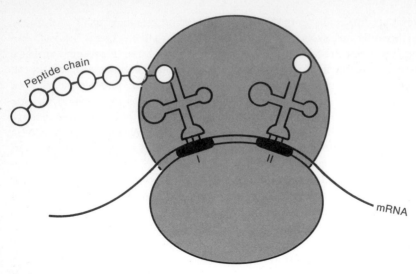

Figure 16-14 Two ribosomal sites for tRNA binding. Site I is the peptidyl-tRNA site, II is the aminoacyl-tRNA site. A cycle of peptide chain elongation involves the peptide chain being attached to tRNA (or the last amino acid to have been added), as shown. The next amino acid bound to its tRNA binds to site II, a process that requires GTP and a protein factor. Then the peptide chain is transferred to the amino acid at site II. Site I is thus seen to contain only a tRNA, which is freed to react with another amino acid molecule by means of the activation enzyme. The final state of elongation is the translocation of the now enlarged tRNA-peptide chain to site I so that the whole sequence of events can be repeated.

synthesis of mRNA specific to a given protein ceases, the free mRNA in the bacterial cytoplasm soon disappears, so that no more of that particular protein will be manufactured until new mRNA is formed at the DNA template. This transient character of mRNA is obviously suited to the rapid control of enzyme synthesis characteristic of microorganisms. In contrast, mRNA appears much more stable in cells of higher organisms, suggesting that, in such cases, different sorts of regulatory mechanisms must be in operation.

The regulation of enzyme synthesis in microorganisms

A great deal of information has been available for many years about the ability of microorganisms to regulate the formation of their enzymes, and it is worthwhile considering aspects of this sort of control. We shall first examine some of the regulatory phenomena that are known and then place them in the context of mRNA synthesis.

For many years, cases have been known where the presence of a certain enzymic activity depends upon the prior exposure of the organism to the enzyme's substrate. For example, early in this century a fungus was found that could hydrolyze the disaccharide sucrose only when the organism had been previously grown in its presence. Enzymes that are synthesized only in response to the presence of their substrate have come to be called *induced enzymes*, and those that are not thus inducible (those synthesized equally well in the presence or absence of substrate) are called *constitutive*. At the present time, there is great research interest in induced enzymes not only because of their obvious role in the regulation of cell metabolism but also because they represent a nongenetic response to the environment, with obvious overtones in the area of cell differentiation. In addition, induced enzymes have turned out to be very interesting from the point of view of studying protein synthesis, as they represent an opportunity to turn the synthesis of a specific protein on or off as necessity dictates.

Much of our present knowledge about induced enzymes comes from the elegant studies carried out by workers at l'Institut Pasteur in Paris over the past two decades. This work has largely centered on one organism, *E. coli*, and one enzyme, β-galactosidase. Although this is only one of many known induced enzymes, knowledge of it is more complete than in other cases, and we shall use it as our example for the class of induced enzymes as a whole. The action of β-galactosidase is the hydrolysis of a β-galactoside, of which the best known example is lactose. Under the influence of the enzyme, lactose is hydrolyzed to yield one molecule each of glucose and galactose:

The presence of the enzyme enables *E. coli* to use lactose as a carbon and energy source and so is required for growth when other carbon compounds are lacking. When cells in which β-galactosidase is inducible are grown in the absence of lactose, no enzyme is formed. Then, when the cells are placed in a new medium in which lactose is the sole carbon and energy source, growth is seen to resume only after a considerable lag period. During this time, the cells are synthesizing β-galactosidase and must make a finite amount of it before lactose can be used and growth begin.

When such experiments were first performed, there were few prior instances where cell metabolism was known to be affected by external factors in a nongenetic (temporary) way, and explanations were sought to explain the results in terms of selection. Thus, it was said that the action of the "inducer" was really to select for those few β-galactosidase-containing cells in the population, enabling them to grow to the exclusion of the others. Such selection would appear to the external observer as if lactose were stimulating the production of the enzyme, whereas actually it was only selecting for those cells that had the enzyme already. This ambiguity was resolved by the discovery of compounds that could induce activity of the enzyme while not serving as either its substrate or as a carbon source for the cell. It was found that thiomethylgalactoside (TMG),

$$
\begin{array}{c}
\text{CH}_2\text{OH} \\
\text{HO} \quad \text{O} \quad \text{S}-\text{CH}_3 \\
\text{OH} \\
\text{OH}
\end{array}
$$

was able to stimulate the formation of β-galactosidase in a cell population without being metabolized by the cell, thus providing no selective advantage for cells already containing the enzyme. In this fashion it became clear that the inducer acted by directly stimulating the synthesis of the enzyme and not by any form of selection. Such enzymic induction, where the inducing compound is unused by the cell and another carbon (and energy) source is available for growth, is called *gratuitous induction* and has proved a valuable means for studying the inductive process. In addition to the inducibility of a cellular enzyme, such as β-galactosidase, regulation of substrate utilization can occur at the level of permeation of the substance into the cell. Thus, there are in the plasma membrane specific, inducible transport proteins called *permeases* that permit transport of many substances into bacteria, and the regulation of permease synthesis occurs in a manner quite similar to that of an inducible enzyme involved in ordinary metabolic reactions.

The repression of enzyme synthesis

We have seen that the growth of cells in the presence of a compound can lead to the production of enzymes required for its utilization, a device enabling cells to make only the enzymes needed under existing environmental conditions. This sort of control is exerted at the *beginning* of a metabolic chain of reactions, enabling the cell to utilize an external source of material for the synthesis of required cell components (Figure 16-15).

In addition, a second species of control exists in cell metabolism, which is a type of mirror image of enzyme induction and which prevents a cell from synthesizing components rendered unnecessary by their presence in the extracellular medium. For example, a cell may be able to synthesize enzymes leading to the production of an amino acid or, say, a purine,

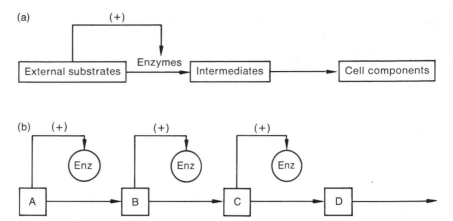

Figure 16-15   Regulation of metabolism by induced enzymes. (a) Enzyme induction, where an externally available substrate induces synthesis of the enzyme required for its utilization. In some cases both the enzyme and an associated permease are induced. (b) Sequential induction, a special case of the phenomenon where induction of the first enzyme of a metabolic sequence leads to the formation of a product (b), which, in turn, induces the second, and so on.

which are required by the cell for growth and viability. The enzymes are required only so long as there is no other source of the amino acid or purine. However, when the compound exists in the medium, it is no longer necessary for the cell to make it, and, in fact, the continued synthesis of the compound becomes redundant and wasteful. Clearly, an additional means of regulation is advantageous.

Control of the synthesis of enzymes of a metabolic sequence by the final product is widespread and is known as *enzyme repression*. It is diagrammed in Figure 16-16. It is very important to remember that enzyme repression is the repression of enzyme *synthesis*, and not activity, as in the case of product inhibition. Control of enzyme synthesis tends to yield a much more total control, and, as we said, regulation of enzyme activity is a kind of fine adjustment. On the other hand, the response of a pathway to the inhibition of activity by a product is likely to be more rapid than a repression mechanism, which often depends on dilution of the existing enzyme by means of cell growth without new enzyme synthesis. In this sense, the repression of synthesis takes care of the long-term metabolic needs of the cell, while inhibition may smooth out fluctuations and keep material flowing evenly through the pathway over shorter time intervals.

Figure 16-16   Outline of enzyme repression.

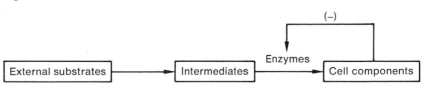

The mechanism of induction and repression

It has been considered likely for a long time that induction and repression were, somehow, aspects of the same molecular process. Thus, induction has been thought of as reflecting the removal or inactivation of an intracellular repressor. We are now in possession of a detailed model for these regulatory processes, proposed by Jacob and Monod (see Suggested Reading), a model that regards regulation as occurring at the level of transcription—the synthesis of mRNA. This model, the *operon model*, is illustrated in Figure 16-17 and has the following essential features:

Genes that are subject to induction–repression control are located on the genetic map in clusters called *operons*, which consist of structural genes and one each of *operator gene*, *promoter gene*, and *regulator gene*. The structural genes are those that direct the synthesis of cellular proteins, as described above. There will be as many structural genes as there are proteins under a common regulatory constraint. In the case of the much-studied β-galactosidase operon, there are three structural genes, one each for the enzyme itself, the galactoside permease, and a galactoside transacetylase. The *operator gene* is the site of regulation for the structural genes; if the *repressor* is bound to the operator, the structural genes will not become transcribed to form mRNA; if the operator is free of the repressor, synthesis can occur. Repressor is a protein and is the product of the *repressor gene*. The addition of inducer to a cell leads to the inducer's becoming bound to its specific repressor molecule, and the inducer–

Figure 16-17 Operon model. This cluster of genetic loci is concerned with the synthesis of three cellular proteins; this synthesis is under a single regulatory system. For repression to occur, the repressor must be bound to the operator gene. In a case where the proteins are inducible, the inducer binds to repressor protein to render it inactive. Where end-product repression occurs (see Figure 16-16), the end product of the pathway leading to repression serves as a corepressor, probably leading to a conformational change in the repressor, in turn enabling it to bind to the operator site. See text for additional details.

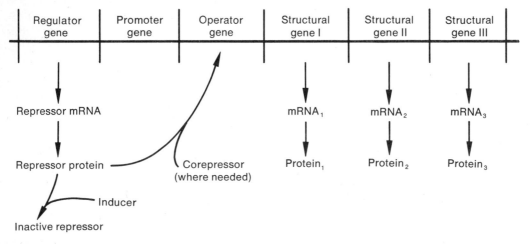

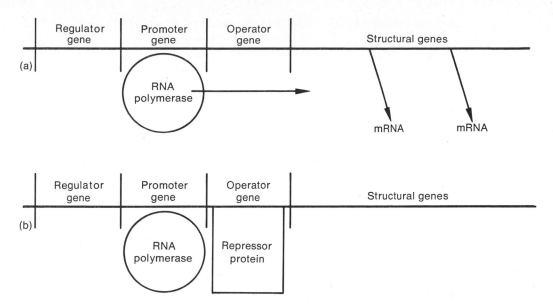

Figure 16-18 Model for the role of the promoter locus. (a) RNA polymerase binds to the promoter prior to moving along the structural genes, where it catalyzes synthesis of mRNA. (b) Motion of the polymerase is blocked by repressor, bound to the operator locus.

repressor complex is inactive on the operator gene, so that the structural genes can be expressed through the synthesis of mRNA.

When an enzyme is subject to repression by an external metabolite, the metabolite can be thought of as a co-repressor, which activates the internal protein repressor, possibly by an allosteric mechanism. In this case, it might be imagined that repression at the operator gene requires the binding of a complex of metabolite and repressor, rather than repressor alone. Finally, the concept of *promoter gene*, lying between repressor locus and operator locus, is a relatively new one and based only on studies with one operon in *E. coli*. However, the idea fits very well with what we know about the operator gene and is widely believed to be of more general occurrence. The promoter gene is considered to be the site of attachment for the enzyme that catalyzes transcription, *RNA polymerase*. This enzyme appears to attach at the promoter locus and then move along the DNA chain, as it traverses the structural genes, forming bonds between adjacent nucleotides in the growing mRNA. Since the operator lies between the promoter and structural genes, it is possible that the binding of repressor protein to the operator simply blocks passage of the polymerase molecule on its way to the structural genes.

The operon approach has the virtue of explaining induction and repression by a single mechanism and by accounting for many features of the regulatory process. It has recently received strong added support with the isolation of the internal repressor associated with the β-galactosidase regulatory system and the direct demonstration of its ability to repress. The repressor turned out to be a protein, as predicted by the theory, and to have a molecular weight of about 200,000 daltons.

Finally, another dimension has been added to the study of bacterial enzyme regulation with the recent (1969) demonstration of a role for adenosine-3',5'-monophosphate, *cyclic AMP*, whose structure is

Glucose is a corepressor in the β-galactosidase system and, when added to cells of *E. coli*, prevents the synthesis of the enzyme. For this reason, when cells are grown in the presence of glucose and lactose, the glucose is used preferentially until it is exhausted, at which time the lactose enzymes become depressed and the cells, after a short lag, begin to utilize lactose. This observation may be explained in terms of the operon hypothesis, with no apparent difficulty. However, glucose has also been shown to lower the intracellular concentration of cyclic AMP (cAMP) in cells of *E. coli*. More recently, it has been demonstrated that the addition of cyclic AMP itself to cells can overcome repression by glucose, so that repression probably reflects changes in the cAMP level within the cell. The site of action of cAMP in the protein synthesis apparatus appears to be the promoter gene, since a point mutation at that locus leads to insensitivity to intracellular cAMP.

It appears that regulation by cAMP is superimposed upon the control exerted by a protein repressor (described before). It may be that cAMP is required for binding RNA polymerase at the promoter locus. In any case, it is evident that a general regulatory mechanism is being uncovered, as cAMP repression has been observed in a number of different bacteria. In addition, the manner in which glucose alters cAMP levels within the cell is unclear. Its level is both a function of production by the *adenyl cyclase* reaction,

$$ATP \rightleftharpoons cAMP + pyrophosphate$$

and its removal by the phosphodiesterase reaction,

$$cAMP \rightarrow AMP$$

and glucose could act to influence either one.

The discovery of a regulatory role for cAMP in bacterial protein synthesis is especially interesting in view of the increasing evidence that it is an important control substance in eukaryotic cells that are subject to regulation by hormones. There is a growing collection of different hormones that influence cAMP levels in animal cells, including insulin, epinephrine, glucagon, and vasopressin. Likewise, there is a growing list of enzyme systems that are influenced by cAMP in specific cells, including muscle phosphorylase and various enzymes of sugar metabolism (see Chapter 8). The enzymes influenced by cAMP in a specific tissue are a function of tissue type, so that a single hormone might exert a similar effect on the level of cAMP, but quite opposite effects on enzyme activity in different tissues. The present view is that cAMP constitutes an *intracellular messenger* that responds to signals from hormones (*intercellular messengers*) or, in the case of repression in bacteria, to the extracellular concentration of metabolites such as glucose.

The control of protein synthesis in eukaryotic cells

The operon approach to regulation has come entirely from studies with bacteria and cannot be applied in a blanket fashion to eukaryotic organisms. There are instances where enzyme induction is known in such cells, but they are relatively few and relatively ill-characterized, owing to the technical difficulties in studying cells in multicellular organisms. Cells grown in culture present the best chances for such studies, and some examples of induction or repression have been uncovered. For example, the enzyme glutamine synthetase, which catalyzes the reaction

$$\text{Glutamate} + NH_3 + \text{ATP} \rightleftharpoons \text{Glutamine} + \text{ADP} + \text{phosphate}$$

has been shown to be repressed in a mammalian cell culture by a product of the reaction, glutamine. It does not follow, however, that the mechanism for repression in prokaryotic and eukaryotic cells is identical.

An important distinction between different types of control resides in the nature of mRNA from different cells. The mRNA associated with inducible (or repressible) enzymes in bacteria is distinguished by having a very great instability and, hence, a short half-life. Since free mRNA is rapidly broken down, the control of protein formation can reside in the regulation of mRNA synthesis. Cessation of mRNA formation will thus rapidly lead to complete elimination of protein formation.

In higher organisms, on the other hand, several lines of evidence indicate that mRNA is quite stable, so that sensitive regulation of protein synthesis cannot reside in mRNA production alone. Clearly, such regulation would involve excessive time lags. There is now a growing body of evidence that regulation in higher organisms is subsequent to the manufacture of mRNA—at the level of translation of the RNA code into the final protein molecule. For example, the antibiotic actinomycin D inhibits mRNA formation but, under certain conditions, fails to prevent protein synthesis. In these circumstances protein synthesis can be imagined to occur by means of extant, and relatively stable, mRNA. Such protein synthesis in the absence of new mRNA synthesis appears subject to regulation of an inductive nature, including stimulation by hormones in suitable tissue.

This regulation can only occur at a site subsequent to mRNA production—the "reading out" of the code represented by the mRNA molecule. There is evidence that the mechanism for such regulation involves the masking of the mRNA code by a protein coat and that stimulation of activity results from the removal of the mask, but sufficient information about such matters is not yet available.

In addition, there is evidence suggesting that a class of basic proteins called *histones* acts to prevent protein synthesis in eukaryotic cells by binding to DNA and thus preventing DNA-dependent mRNA synthesis. It is, however, not known precisely how specific this interaction is for a single region of the DNA chain. If inhibition is general, it is unlikely that it reflects a control mechanism *in vivo*.

Whatever the mechanism of the regulation of protein synthesis may turn out to be in the case of higher organisms, it is clear that it will be very actively studied during the next few years and equally clear that results will be of greatest interest to a wide variety of biologists. Such regulation is central to many questions in the control of growth, which is of great practical interest. Likewise, regulation of enzyme synthesis appears to have great significance in the study of differentiation. It is widely believed that all cells of an organism have the same DNA informational content, all having descended from a common ancestor (the zygote). In contrast, the different cells exhibit widely differing properties, including enzymic activities, so that regulation points exist between DNA and protein formation, and it is reasonable to suppose that these regulation points are influential in controlling the direction of differentiation.

Information transfer between nucleus and cytoplasm

The mechanism of protein synthesis that we have been describing includes the transfer of information from the nucleus, where most of the cell's DNA occurs, to the cytoplasm. The chemical form of this transfer is the migration of the appropriately named messenger RNA. Many features of living cells become explicable in light of this transfer of mRNA. For example, the mechanical removal of a nucleus from *Amoeba* leads to a decline in protein synthesis, which appears to reflect the breakdown in previously synthesized mRNA. The alga *Acetabularia* suffers a similar decline in ability to synthesize proteins on removal of its nucleus, but the decline is much slower. This difference may indicate a longer average half-life of its mRNA and may well reflect the contribution of its chloroplasts, which contain their own DNA and mRNA.

The dependency of cytoplasm on the nucleus is mirrored by a requirement for cytoplasm to maintain the structure of the nuclei. When isolated nuclei are devoid of cytoplasm, they soon begin to deteriorate, and attempts to culture nuclei independently have been universally unsuccessful. In a way, nuclei exhibit a stronger dependence upon cytoplasm than the reverse. The eggs of sea urchins have had their nuclei removed and then have been induced to divide by manipulation of the salt content of the medium. These can undergo several divisions before they begin to degenerate. Thus, it appears that cellular division requires neither mitotic chromosome partition nor any other nuclear function. This experiment also suggests that the unfertilized egg contains sufficient mRNA to maintain protein synthesis for several rounds of cell division. This mRNA must be quite stable to survive for so long (up to 72 hours).

The adequacy of the genetic code

The view of information transfer in protein synthesis outlined above appears to explain much of what is known about enzyme formation in cells. Although many details remain to be worked out, it seems increasingly unlikely that the general outline will prove invalid. There appears ample justification for the statement that DNA bears information required for the ordering of amino acids in proteins and that information transfer is by means of the special relationships due to the geometry of nucleotide bases. It is thus tempting to argue that DNA is *the* genetic material—that it carries all information required for the synthesis of cells—and many biologists appear prepared to make this assertion. It is one thing, however, to say that information transfer and storage take place as we have described them; it is quite another to say that there are no other paths for such transfer. The world of the cell is exceedingly rich in phenomena, all of which might be said to require that sort of information transfer for their continuance. The question is whether or not the spartan mechanism of information transfer through DNA and mRNA to protein is sufficient to provide for the richness and complexity of cells.

For example, consider the matter of the location of apparatus for making proteins. The preceding description places protein synthesis at the ribosomes and views information as passing from the nuclear DNA to the cytoplasmic ribosome. In fact, this pattern appears untrue for a portion of total cellular protein synthesis—that associated with closed organelles, such as chloroplasts or mitochondria. For example, protein synthesis *within* mitochondria is known to occur, and it is likely that the information for it is encoded in a separate mitochondrial DNA, which has been demonstrated both chemically and by means of electron microscopy. Similar observations have been made using chloroplasts.

The observation of genetic material in the form of DNA closely associated with cell organelles should not be surprising in view of the indications that mitochondria and chloroplasts (as well as other organelles) possess a degree of genetic autonomy. For example, *Euglena* chloroplasts appear to be self-replicating, and it is possible to "cure" their cells, after which no more may be made. There is also considerable evidence that yeast mitochondria are at least partially responsible for their own formation, a view supported by the subsequent observation of mitochondrial DNA. Such autonomy should also not be totally unexpected, in view of the obvious problems in imagining that ribosomal protein synthesis could

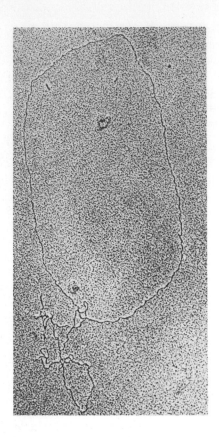

Figure 16-19 Circular DNA from mitochondria. Preparations of such DNA have molecular weights of about 10×10^6 daltons (as compared to about 1×10^8 daltons for the T2 bacteriophage of *E. coli*). (Electron micrograph courtesy of Dr. A. Kroon.)

be the means for the formation of enzyme molecules *within* a closed mitochondrial membrane, or those comprising that membrane. Thus, there may be more than one pathway for information transfer within cells, just as there appear to be sites of information storage in the form of extra-nuclear DNA. In fact, the situation is further complicated by recent evidence that suggests that some proteins of mitochondria are synthesized from mRNA from the nucleus and others from mRNA originating at the mitochondrial DNA. For example, the enzymes of the outer mito-chondrial membrane, as well as most of the nonmembrane-bound intra-mitochondrial enzymes, are synthesized in the cytoplasm, under nuclear control. After synthesis, they are transfered to their mitochondrial locus. On the other hand, the insoluble proteins of the inner membrane, such as succinate dehydrogenase, appear to be synthesized by the mitochondrial system on the basis of information contained in the mitochondrial DNA.

Finally, one should probably maintain an open mind as to the possibility of cellular-information storage and transfer not involving DNA. At the present time, it appears likely that protein synthesis proceeds from information encoded in DNA, and the basic assumption of molecular biology that information transfer is one-way—that protein cannot serve as a template for nucleic acid synthesis—is valid. A second assumption should be considered with some care, and that is the widely held view that the only genetic information necessary for the production of a new cell is that required for the production of the right amount of the right proteins (enzymes). In other words (putting it rather crudely), if one could make all the right enzymes, the cell would flop together, owing to the intermolecular forces between the various components. This approach, an extreme statement of *the principle of self-assembly*, is probably correct in most (or all) cases, but the evidence is too limited to enable one to embrace it in an uncritical fashion. Until it turns out to be possible to reconstitute from isolated compounds cell organelles in their normal morphology and function, the case for total self-assembly of cells must remain somewhat speculative.

Suggested Reading

Books

Cantoni, G. L., and D. R. Davies, *Procedures in Nucleic Acid Research*, Harper & Row, New York, 1966.

Chantrenne, H., *The Biosynthesis of Proteins*, Pergamon Press, Oxford, 1961.

Cold Spring Harbor Symposia on Quantitative Biology, vol. 33, *Replication of DNA in Microorganisms*, 1968.

Cold Spring Harbor Symposia on Quantitative Biology, vol. 35, *The Mechanism of Protein Synthesis*, 1969.

Davidson, J. N. (ed.), *Biochemistry of Nucleic Acids*, John Wiley & Sons, New York, 1969.

Florkin, M., and E. H. Stotz (eds.), *Nucleic Acids*, vol. 8 of *Comprehensive Biochemistry*, American Elsevier Publishing Co., New York, 1963.

Ingram, V. M., *The Biosynthesis of Macromolecules*, W. A. Benjamin, New York, 1965. A primer.

Maaløe, O., and N. O. Kjeldgaard, *Control of Macromolecular Biosynthesis*, W. A. Benjamin, New York, 1966.

Roodyn, D. B., and D. Wilkie, *The Biogenesis of Mitochondria*, Methuen & Co., London, 1968.

Stahl, F. W., *The Mechanics of Inheritance*, Prentice-Hall, Englewood Cliffs, N.J., 1964.

Watson, J. D., *The Double Helix*, Atheneum Publishers, New York, 1968.

Watson, J. D., *The Molecular Biology of the Gene*, 2nd ed., W. A. Benjamin, New York, 1968.

Woese, C. R., *The Genetic Code*, Harper & Row, New York, 1967.

Zubay, G. (ed.), *Papers in Biochemical Genetics*, Holt, Rinehart and Winston, New York, 1968.

Articles

Burton, K., "Sequence determination in nucleic acids," in *Essays in Biochemistry*, P. N. Campbell and G. D. Greville (eds.), vol. 1, Academic Press, New York, 1965, p. 57.

Cairns, J., "The bacterial chromosome," *Sci. Am.*, **214**, 36 (Jan., 1966).

Cohen, N. R., "The control of protein biosynthesis," *Biol. Rev.*, **41**, 503 (1966).

Crick, F. H. C., "The genetic code: III," *Sci. Am.*, **215**, 55 (Oct., 1966).

Crick, F. H. C., "The origin of the genetic code," *J. Mol. Biol.*, **38**, 367–79 (1968).

Gilbert, W., and B. Müller-Hill, "The lac operator is DNA," *Proc. Nat. Acad. Sci. U.S.*, **58**, 2415 (1967).

Holley, R. W., "The nucleotide sequence of a nucleic acid," *Sci. Am.*, **214**, 30 (Feb., 1966).

Jacob, F., and J. Monod, "Genetic regulatory mechanisms in the synthesis of proteins," *J. Mol. Biol.*, **3**, 318 (1961).

Nomura, M., "Ribosomes," *Sci. Am.*, **221**, 28 (Oct., 1969).

Novelli, G. D., "Amino acid activation for protein synthesis," *Ann. Rev. Biochem.*, **36**, 449–84 (1967).

Ochoa, S., "Synthetic polynucleotides and the genetic code," in *Informational Macromolecules*, H. J. Vogel, V. Bryson, and J. O. Lampen (eds.), Academic Press, New York, 1963.

Smillie, R. M., and N. S. Scott, "Organelle biosynthesis," *Progress in Molecular and Submolecular Biology*, **1**, 136 (1969).

Sonneborn, T. M., "Does preformed cell structure play a role in cell heredity?" in *The Nature of Biological Diversity*, J. M. Allen (ed.), McGraw-Hill Book Co., New York, 1963, pp. 165–221.

Taylor, D., "Chloroplasts as symbiotic organelles" in *International Review of Cytology*, G. H. Bourne and J. F. Daniell (eds.), vol. 27, Academic Press, New York, 1970.

17 The physiological action of viruses

Viruses are small particles of nucleic acid surrounded by a protein coat, and they both originate in living cells and produce dramatic effects upon the cells that they invade. Virology considers viruses in the context of their being genetic systems and with relation to the pathogenic properties of many of them. We shall largely ignore these important areas on the grounds that they are well treated in many specialized books (see Suggested Reading) and shall center on the effects that viruses produce in the cells that they invade. It will become evident that viruses are by no means always lethal to cells and that, in a sense, they should be regarded as a part of a cell–virus *supracellular system*. It will also become clear that, because of the changes that they produce in cells, viruses are notably effective probes for cell function, especially with regard to the physiological mechanisms underlying genetic continuity and exchange.

The structure of viruses

The most obvious structural feature of viruses is their extremely small size and their relative lack of complexity when compared to cells themselves, or with most cell organelles. We noted that viruses are composed of a *core* of nucleic acid, which may be either DNA or RNA (but not both). The core is surrounded by a *coat*, which is mainly protein and which serves to protect the nucleic acid. The coat also possesses other functions related to the invasion of a cellular membrane and is the locus of any antigenic properties that the virus might possess. Very small plant and animal viruses often exhibit exceedingly simple morphology where the core is encased in a regular geometric array of identical protein subunits, called *capsomers*. In contrast, many bacterial viruses (bacteriophages) exist in a more complex configuration with a tadpole-like head and tail arrangement. Typical spherical and phage types of cirus are shown in Figure 17-1.

Virus particles are observed in a rather wide range of size, but all smaller than even the simplest prokaryotic cell. Indeed, the size of viruses

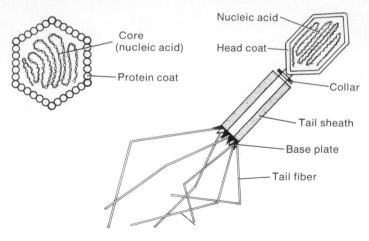

Figure 17-1 Construction of viruses. Left, a relatively simple spherical virus, with a protein coat made up of identical subunits and a nucleic acid core. Many plant, animal, and bacterial viruses show this general configuration. Right, a tadpole-shaped bacterial virus of considerably greater complexity. In the bacterial virus of this sort, the tail complex represents a modification for attachment to the host, as well as a device for injection of the DNA into the host cytoplasm.

used to be the sole criterion for their identity, as they were effectively defined as infectious particles too small to be seen with the light microscope and small enough to pass through fine filters that would stop cells. Availability of electron microscopes led to the ability to measure the dimensions of viruses and the consequent discovery that there is a large range in size, over more than one order of magnitude. Thus, a very small bacterial virus might be about 200 Å in diameter, with human polio virus being about the same size. On the other hand, another human virus, the agent of smallpox, is about 2,500 Å across and the rod-like tobacco mosaic virus can be as much as 3,000 Å in the long dimension.

There is very little relationship between size and complexity in viruses, as a very large plant virus, such as *tobacco mosaic virus*, may contain only RNA and a single species of protein, repeated some 2,000 times. In this instance, each protein subunit has a molecular weight of about 17,000 daltons, and the subunits are arranged in a helical manner around the RNA thread in the core, which contains about 6,000 nucleotides. In the case of a small RNA bacterial virus, such as F2, there are only about 180 subunits arranged in a cubical array, each subunit having a molecular weight of about 14,000.

At first sight, it might be thought that the genetic information for a given virus would consist simply in the amount of nucleic acid sufficient to specify the structure of the subunit proteins. In the case of the tobacco mosaic virus, there would be a single gene just as there is a single coat protein. In the case of a fairly complex bacteriophage there might be a dozen or more coat (and other) proteins, such as those of the head, tail, tail fibers, and so on. It turns out that, in either case, a major part of the genetic information carried by the virus relates to the manufacture not of coat proteins but of proteins that are not a part of the virus particle at all; rather, they are manufactured only in the host cell pursuant to the synthesis of new virus.

The virus-cell interaction: lytic bacteriophage

We shall introduce the discussion of the ways in which cells and viruses interact by describing what is perhaps the simplest form of interaction, that where the production of new virus leads to the destruction (lysis) of the cell involved. It should not be construed, however, as indicating that this is the most common form of interaction; it is merely the one best known, for the simple reason that it is the least complicated. This form of interaction has the virtue of being very clear-cut in effect, since the effect from the point of view of the cell is total destruction. This produces some highly practical benefits to the experimenter. For instance, the most important technique for identifying and counting many types of viruses is by means of a *plaque count*, in which a suspension of viruses at suitable dilution is added to a population of cells grown on the surface of some nutrient solution (such as nutrient agar, when the cells are bacterial). Each virus particle enters a cell, induces the cell to carry out a cycle of virus production, after which the cell breaks down to yield a "burst" of new viruses, which repeat the process with other cells. The net effect is that, where each virus originally attacked a cell, a large number of cells are destroyed, leaving a clear area on the surface of the cell population (see Figure 17-2). These may be counted and exist on an approximately

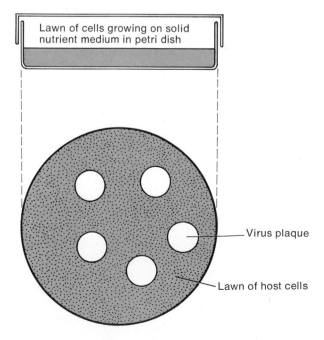

Figure 17-2 Plaque-count technique. See text for details.

one-to-one relation to the number of infective viruses in the suspension. This technique depends on the killing of the cell by the virus and when this does not occur, the technique is useless. It might be added that the technique of plaque counting is of value not only in the case of phage but also with viruses that attack higher organisms.

The "life cycle" of a virus may be divided into two phases, depending on whether the virus nucleic acid is inside or outside the cell. When the virus exists as a free particle, it is called a *virion* and consists of the nucleic acid core and the coat, as described above. When the virus is within the cell, it is, on the whole, engaged in replication and is, for a considerable period, invisible to the electron microscope, existing only as replicating strands of DNA or RNA. In the case of many bacteriophages, the penetration of the host cell is only made by the nucleic acid and the protein coat remains outside. In the case of most viruses, however, the entire virion enters the cell, but only the nucleic acid may be said to replicate.

Let us turn to a cycle of reproduction of a bacterial virus such as the T2 bacteriophage. In this instance, the virion is of the tadpole sort of morphology and contains DNA as its genetic material. A cycle of infection, reproduction, and release is illustrated in Figure 17-3. The first event that we shall consider is the adsorption of the virion to a sensitive bacterial cell of the species *E. coli*. It should be emphasized that only certain strains of a species will be likely to be sensitive to a given virus and that it is possible to obtain mutations of the cell that are quite unaffected by the virus. Indeed, early work on bacterial viruses was stimulated by the hope that the viruses would turn out to be clinically useful antibacterial agents. This hope was blasted by the discovery that bacteria could quickly mutate to insensitive forms, or, more exactly, that the effect of the viruses was to select for any insensitive mutants present in the population.

Figure 17-3   Cycle of virus infection and replication in the case of the T2 phage of *E. coli*. Details are given in the text.

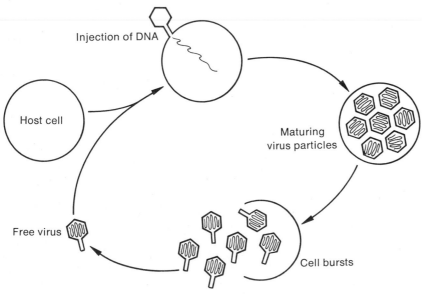

The interaction between a bacteriophage and a sensitive cell begins with the adsorption of the virus to the cell by means of the tail. The phage tail has a rather complex structure consisting of a core, contractile sheath, base, and several tail fibers. It appears that the virus attaches to the bacterium by means of the tail fibers, which bring the base of the tail into contact with the cell wall. There are enzymes located on the base that lead to the rapid erosion of a small hole in the cell wall and, finally, plasma membrane. Then, the tail-sheath proteins contract, and the phage DNA is literally injected into the cytoplasm of the bacterium.

As soon as the bacterial cell contains the phage DNA, a series of events occurs that can be summarized by saying that the viral DNA takes over the direction of the host cell's metabolic apparatus and redirects it toward the synthesis of virus. First, it can be shown that a number of phage-specific mRNA molecules are synthesized at the phage DNA. These include mRNA encoding enzymes required for the synthesis of phage DNA (see below) as well as enzymes that break down any bacterial mRNA molecules extant in the cytoplasm. Indeed, the viral DNA includes, in some instances, coding for an enzyme that destroys the host DNA, so that any subsequent mRNA and, therefore, protein synthesis that occurs will be that required for virus production. The complete nature of the virus-induced alteration in the cell's metabolism is seen in the case of T4 phage, where cytosine in the DNA is completely replaced by 5-hydroxy-methylcytosine (which, however, possesses the same base-pairing specificity as cytosine). Since there is no hydroxymethylcytosine in the normal cell of the host *E. coli*, an early metabolic function that must be coded in the phage DNA is the synthesis of this compound, a synthesis that must occur before any DNA replication can take place.

The synthesis of phage-specific mRNA molecules leads rapidly to the synthesis of enzymes related to phage production, the latter protein synthesis occurring by means of the ribosomes already existing in the bacterium. Since rapidly growing cells contain many more ribosomes than resting cells, it is not surprising that phage production is more rapid and effective in the former case. When the necessary enzymes have been made, the process of phage DNA replication can begin, so that the host cell is rapidly filled with phage DNA.

Later phases of phage replication consist in the synthesis of coat and tail protein, again utilizing phage DNA as the template for mRNA. The phage proteins, once synthesized, assemble spontaneously into the characteristic shape of the completed virus with the head forming around the DNA first. When the virion is completed, the final step in the replicative cycle is the lysis of the bacterial cell from within, through the synthesis of an enzyme (lysozyme) that breaks down the cell membrane and cell wall. The cell bursts and releases a population of new virions that can number as many as several hundred in the case of some smaller bacteriophages. The basic experiment that illustrates a cycle of replication by a lytic virus is called a single-step growth experiment, and is illustrated in Figure 17-4. If the cells are disrupted artificially (as by ultrasonic vibration) before the lysozyme breaks them from within, infective particles are released, unless breakage occurs too early in the cycle.

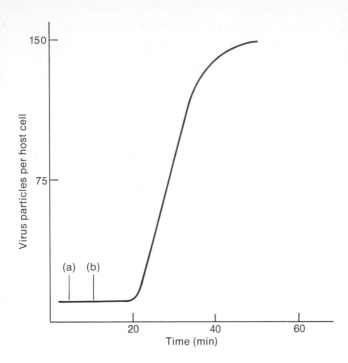

Figure 17-4 Single-step growth curve for a bacteriophage. (a) denotes the time of addition of phage to an exponentially growing bacterial culture. (b) indicates the point when active virus particles are first produced upon artificial disruption of the cells by ultrasonic oscillation (see text).

The replication of the viral nucleic acid

In the case where the virion contains double-stranded DNA, its replication is not significantly unlike that of the DNA of its host cell. Thus, strand separation appears to precede replication, each individual strand serving as a template for the synthesis of the corresponding partner. There are however, bacterial viruses that contain single-stranded DNA and still others (and indeed many viruses of higher organisms) where the genetic material is RNA. In these cases, different mechanisms must be involved in replication.

For example, in the case of the bacterial virus ΦX174, the DNA is single-stranded and the first step in replication is the synthesis of a double-stranded replication form. This undergoes the usual cycle of replication in a fashion quite like that of ordinary DNA replication. When a sufficient number of copies are formed, the complementary strands of DNA are discarded and the virus coat is formed around the copy of the initial single DNA strand.

A similar mechanism underlies the replication of viruses that contain single-stranded RNA. Some bacterial and animal viruses and all plant viruses that have been studied contain RNA as their genetic material In these instances, replication of the single RNA strand occurs, leading to a double-stranded *replication form of RNA*, a process taking place at the bacterial ribosome. This form leads to the formation of numerous copies, in a fashion analogous to that employed by a single-stranded DNA virus. It is interesting that, in this case, the RNA acts to replicate itself and serves as mRNA as well, in the sense that virus-specific proteins are formed that use it directly as template.

It should be added that much of the present interest in viruses that contain single-stranded DNA or RNA comes from the small size of their genetic strand. Thus, in the case of ΦX174, there are about 6,000 nucleotides, so that there is "room" to encode about 2,000 places in the amino acid sequence of its proteins. This amounts to about eight proteins of moderate size and, in fact, eight genes have been identified. Of these, four encode proteins of the virion coat and three have been identified with replicatory functions. Thus, the relations between the genetic endowment and functioning of this virus are very nearly understood, a level of knowledge not obtained with larger genetic systems.

The replication of viruses that interact with eukaryotic cells

In general, the study of bacterial viruses has led to results that are applicable to other viruses as well, although with a number of differences. The main features of invasion and replication are the same, at least in cases where the action of the virus is lytic. On the other hand, the details are often quite different. For example, there are few examples of plant or animal viruses that share the tadpole-like structure of the T series of bacteriophages mentioned above. Thus, invasion does not occur by means of the injection of viral genetic material into the host cell but rather by means of passive entry through imperfections in the cell surface or by eroding the cell surface so that the entire virus particle can enter.

The structure of most plant and animal viruses is simpler than that of the tadpole-like phages mentioned above. Either they are rod-shaped, with a cylinder composed of protein subunits surrounding a central core of nucleic acid (as in the case of tobacco mosaic virus or certain insect viruses), or they approach being spherical as do many plant and animal viruses. When a virus appears spherical, it is generally, in fact, a regular polyhedron such as an icosahedron (which has 20 triangular "sides"). Some animal viruses possess added complexity of structure in being surrounded by a membrane derived from the host cell membrane. Thus, the virus (which carries an ordinary protein coat) leaves the host by a process in which it carries some of the plasma membrane with it, and it may be that the invasion of a subsequent cell occurs by means of fusion of the membrane around the virus with the plasma membrane of the cell being attacked.

When eukaryotic cells are attacked by viruses, it is often the case that a new generation of viruses is produced but without necessary destruction of the host. The host cell thus can liberate virion particles over a considerable period of time, and the number of particles (*burst size*) can become quite large, exceeding 1,000. When cells of higher organisms are invaded by viruses, the location where virus replication takes place can become quite varied. Experiments with radioactive nucleic acid precursors indicate that the pox viruses replicate in the cytoplasm of the sensitive animal cells, while most other DNA viruses replicate in the nucleus (where DNA precursors would be expected to be much more plentiful). Most RNA viruses turn out to replicate in the cytoplasm of the host, a finding that is in keeping with the role of ribosomes in their replication. It is likely that the synthesis of the protein coat of all categories of virus occurs in the cytoplasm and that the final assembly of virion occurs there as well.

Temperate viruses

The category of lytic viruses that we have examined represents one end of a continuum with regard to the harmful effect that they produce. We have seen that a number of viruses produce a new generation of virions without, however, destroying the cell. At the extreme end of this scale of harm to the host, we find the temperate viruses that kill their host cells only very infrequently. These are exemplified by the temperature bacteriophages as well as, perhaps, certain tumor viruses.

We will again employ bacterial viruses as examples of the more general category *temperate virus*. The temperate phage that has been studied most extensively is the λ phage of *E. coli*, and its "life cycle" will now be described. In the first place, the fundamental action of the temperate virus is the insertion of its genetic material (which has always proved to be DNA) into the genetic map of the host cell. In other words, the virus exists in two main conditions, that of the virion with the DNA core and the protein coat, and the intracellular form, the *provirus*, which is effectively a locus on the host chromosome. In the case of a bacterial virus, the provirus is also called a *prophage*, and a cell carrying a prophage in its genetic makeup is called *lysogenic* for reasons that will shortly become plain. When the provirus is incorporated in the host genetic material, it divides as the host DNA is replicated. In fact, it behaves exactly like an ordinary gene (or locus) on the host genetic map. We shall see shortly that the presence of a provirus in an otherwise normal cell can produce significant alterations in the physiology of the cell.

If a cell carrying a provirus is subjected to any one of a number of stimuli, such as ultraviolet light, ionizing radiation, or addition of a mutagen, the provirus is *induced* to dissociate itself from the host genetic map and to begin replication in a manner quite like the replication of an ordinary lytic virus. When this occurs, the effect on the host cell is decidedly harmful, as the cell lyses, yielding a burst of virus particles. These particles can again enter cells of a suitable strain and lead to the lysogenic state. An important property of lysogenic cells is their immunity to subsequent attack by the same virus. Thus, if one wishes to perform an experiment in which a provirus is inserted into a cell, the cell must be without that provirus to begin with. This suggests that the presence of the provirus changes some of the properties of the cell in a manner that makes subsequent

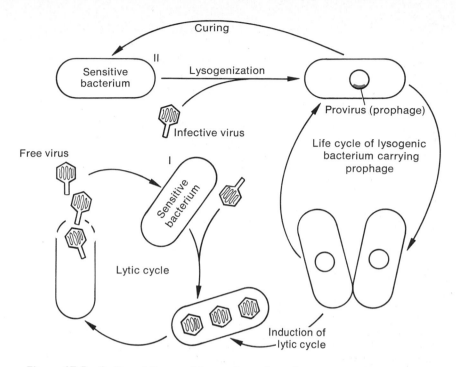

Figure 17-5 Outline of the possible transformations of a temperate virus. Two fates of a sensitive bacterium infected by temperate virus are denoted I and II. In the case of I, the virus simply undergoes replication leading to a burst of new viruses and lysis of the cell. In the other case (II), the genetic material of the virus becomes integrated into that of the host (lysogenization) and undergoes division as the cell divides. In this integrated form the genetic material of the virus is called a provirus. Irradiation of a cell carrying a provirus with ultraviolet light leads to induction of a lytic cycle and eventual (i.e. after a few minutes) production of infective viruses. These may then attack another sensitive cell with either lysis (I) or lysogenization (II). On occasion, a lysogenic cell loses its prophage without entering a cycle of phage development and lysis. This is called curing.

attack unlikely. This alteration probably influences the cell surface, among other things, since entry of subsequent viruses is prevented. Figure 17-5 shows the several states of temperate viruses and their relationships to the action of a lytic virus.

Although in this book we do not intend to discuss viral genetics in any great detail, it is worth mentioning that an interesting genetic (and topological) problem exists with regard to the place of the provirus in the host genome. Mapping studies as well as electron-microscopic evidence indicate that the DNA of both a temperate bacterial virus (such as λ) and a bacterium (such as *E. coli*) is circular, that is, without a free end. With this sort of geometry, it is not immediately obvious how the circular provirus can be incorporated into the much larger circular bacterial chromosome. A considerable amount of evidence now suggests that the inclusion is the result of a sort of crossing over, where recombination between the two rings leads to formation of a single, larger ring, as shown in Figure 17-6. We shall see shortly that this recombination is of

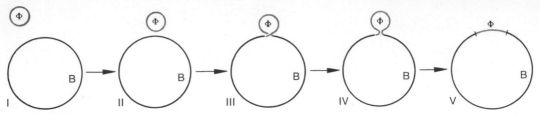

Figure 17-6 How a circular phage genetic system (denoted Φ) can become integrated into a larger circular bacterial genetic system (denoted B). Step III represents a kind of recombination, or crossing-over, between bacterial and phage genome.

considerable importance with regard to the ability of the provirus to influence the physiology of the host cell.

Defective temperate viruses

Lysogenic cells sometimes mutate to a condition where they retain their immunity to subsequent phage infection but are unable to give rise to active virus particles when induced by physical or chemical treatment. These mutants are located in the prophage genetic material and render the prophage unable to encode some of the normal viral functions. For instance, when such defective prophage-carrying cells are induced, they frequently lyse but release virus parts (such as phage tails and headpieces) rather than intact infective viruses. In these cases, it appears that the defect precludes the synthesis of one of the coat or other indispensable proteins, while allowing formation of the rest.

Occasionally, a few normal virus particles are produced by the induction of presumably defective prophage. This appears to be the result of a rare back mutation to the active state. In some cases, it is possible to obtain active viruses from a defectively lysogenic cell by superinfecting with a second phage that is not so closely related to the first as to be excluded by immunity. This second, *helper virus* is then able to provide the missing function and leads to a mixed population of viruses, both the defective virus and the helper virus, in the lysate. Obviously, the defective viruses are still genetically lacking in whatever function was damaged and when they are used to infect additional cells, the helper virus technique must be used again, if active viruses are to be produced. Also, the use of the helper virus only is effective when the function being "helped" is one in common between the two viruses. A tail-less phage could not help a defective prophage produce intact tails.

Transduction

It has been known for a long time that temperate phage can transfer bacterial genetic material between different populations. For example, a population of cells that are both resistant to a particular antibiotic and

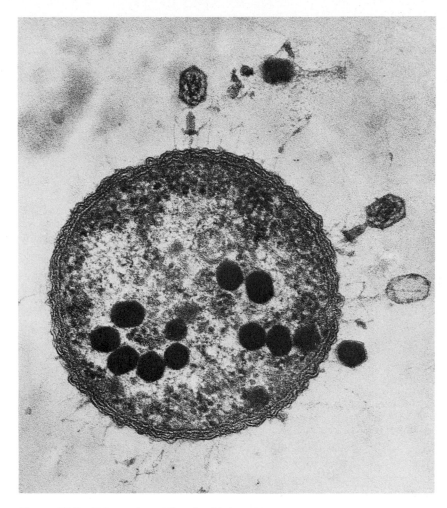

Figure 17-7 Thin section of *E. coli* cell infected by
T2 phage. Intact phages are shown external to the
cell adsorbed to the cell wall. Note attachment by
the tail region of the phages. Note also that of the
three phages, one has an empty head; the other two
still contain DNA. At least one phage has succeeded
in injecting DNA into the cell, as a number of
developing phage heads can be seen within.
Magnification is about × 110,000.
(Courtesy of Dr. L. D. Simon.)

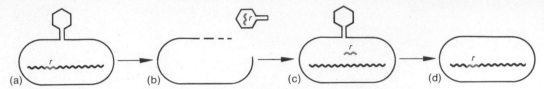

Figure 17-8 Outline of transduction. (a) A phage attacks a host cell. A portion of the cell's genetic map is shown on which a drug-resistance locus is denoted *r*. (b) The cell bursts. Of the phage produced, a few carry the *r* locus. (c) One of these attacks a cell that does not carry the *r* locus. The cell becomes lysogenized, and the *r* locus becomes incorporated into the cell's genetic map. A suggested mechanism underlying this event is given in Figure 17-9.

lysogenic with a particular prophage can be induced to lyse, and then the cell-free virus suspension can be used to infect a second population of cells. If these cells are sensitive to the antibiotic in question, there is a finite possibility that these cells that have gained a prophage will also turn out to have gained antibiotic resistance (see Figure 17-8). The likelihood will be greater in proportion to the proximity of the prophage and the antibiotic resistance locus on the bacterial genetic map. In some cases, more than one bacterial function can be transduced; in such cases the

Figure 17-9 Probable basis for high-frequency transduction. This diagram shows the temperate phage genetic material (gray line) being incorporated in the host genome (black line). When the prophage leaves the host genome, the dissociation occurs in a slightly asymmetrical fashion, producing some interchange between host and virus. (a) Virus genetic material becomes aligned on that of the host. There is probably a specific site on the host genome where it can bind by base pairing. (b) Recombination occurs, leading to the prophage state (c). (d) Dissociation occurs with a second recombination. The crossing-over point is slightly displaced from the initial one, so that the viral genome now contains some host DNA and vice versa (e). Thus, the virus has lost some of its genome and is defective. It has also gained some bacterial genetic material, which it transduces with high efficiency.

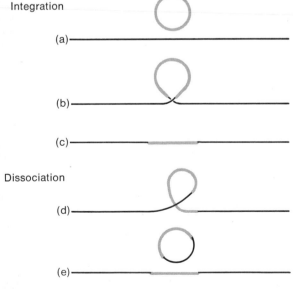

Integration

(a)

(b)

(c)

Dissociation

(d)

(e)

greater distance between the two bacterial markers, the less the frequency of their *cotransduction*. It is also possible for one phage to transduce a second prophage, an occurrence that is not odd in view of the integration of prophages into the bacterial genetic material and the consequent identity of bacterial and viral loci.

In general, transduction occurs at a very low frequency, say of the order of one transduced marker in 10^6 virions. However, certain phage types are able to transduce particular loci at much higher frequencies, a phenomenon called *high-frequency transduction*. Examination of this situation has revealed the phage in question to be defective, and the relation between their defect and the power to transduce is instructive. In the first place, the defectiveness of high-frequency transducing virus requires that there be also a helper virus present in the bacterium so that any active phage will be available to effect the transduction. In the second place, it appears that the defect in the phage and the transduction of the particular locus (or loci) carried results from some sort of genetic interchange between the bacterial and viral genetic material. Thus, because the virus and host genetic systems have undergone some sort of recombination, the virus has lost part of its map to the host and the host has, likewise, lost part of its map (the part being transduced) to the virus. These relationships are illustrated in Figure 17-9. Such recombination between host and virus suggests that their genetic material is very similar and that, in a sense, the virus and host are made for each other.

Physiological effects of virus replication

We saw earlier that the net effect of the attack on a host cell by a lytic virus is the capture of the host metabolism by viral functions. The viral genetic system subverts the ordinary activities of the cell for the nefarious (from the point of view of the cell) activity of viral production. However, in the case of many plant and animal viruses, the nature of the take-over is much more subtle and less a case of outright piracy. For one thing, animal and plant cells are much larger than the bacterial counterpart and the virus particles are, in many instances, much smaller. Thus, the needs for their production are much more compatible with the continued activity of the host. Infection of a bacterium with a lytic phage leads to death of the cell, while many animal viruses lead to less drastic results, an observation of possible comfort to sufferers with the common cold, who at least, do not face lysis.

Microscopic examination of plant and animal cells that are infected with a variety of viruses reveals a wide range of effects. If the virus is a particularly baneful one, the effects can be those generally observed with a dying cell, such as a rounding-up of the cell outline, destruction of some of the membrane systems, increase in vacuoles, and displacement of the nucleus from its normal location in the cell. In some cases, viruses replicate to the extent in the host cell that they can be seen to fill much of its interior with a highly regular array of particles that are clearly viruses in

411

one stage of assembly or another. For example, polio virus forms a compact array that is so regular that it is often termed a "crystal." Similar formations have been observed on a wide variety of plant and animal virus-infected cells.

Animal DNA viruses that replicate in the nucleus sometimes produce characteristic changes there. For instance, cellular inclusions that may be virus particles in the process of formation are noted in the nuclei of infected cells. Moreover, some viruses, including *herpes* and *polyoma* types, produce changes in the host chromosome, including the formation of intrachromosomal bridges, as well as chromosomal breakage. There is some evidence, from fluorescent antibody staining, that suggests that the coat protein from *reovirus* binds selectively to the mitotic spindle of the host cell. The meaning of this observation is presently unclear, especially in light of the identity of reovirus as an RNA virus, which would be expected to have no special connection with the process of mitosis.

We have already seen that a major feature of the physiological change produced by lytic viruses is the capture of the raw materials for nucleic acid synthesis as well as of some of the enzymic machinery, including the ribosomes, which are all diverted to the production of virus genetic material and, later, protein. In cases where the nucleic acid of the virus is different from that of the host, containing, for example, hydroxymethyl-cytosine, in place of cytosine, enzymes related to the synthesis of the viral component must be made where they did not exist previously in the host. In the case of animal viruses, similar subversion of nucleic acid metabolism occurs. In the case of polio virus, changes in RNA and protein synthesis can be measured, even under conditions where the virus is itself unable to replicate. It is also clear that the introduction of an RNA virus into a cell interferes with the synthesis of DNA even though there is no requirement for viral DNA synthesis. The decline in DNA synthesis is, in such cases, reflected in the inhibition of subsequent mitosis, although its elimination usually requires a considerable period to take effect.

Poliovirus and meningovirus, both of which contain RNA as their genetic material, lead to abrupt inhibition of RNA synthesis in an infected cell. Inhibition of protein synthesis prevents the virus-induced inhibition of RNA synthesis, suggesting that the process requires synthesis of an "inhibitory" protein. The inhibition of RNA synthesis is largely (or entirely) that of ribosomal RNA formation, and the production of ribosomes ceases as well. The ribosomes that remain in the cell are found to be no longer associated with host mRNA but rather with the viral RNA, presumably in order to synthesize viral-specific proteins.

Interferon

An effect of many different animal viruses is to induce their host cells to produce a material called *interferon*, which blocks the reproduction of many different viruses in many different cell types. This material is a small protein with a molecular weight of about 20,000 daltons and is, for a protein, uncommonly heat-stable. Interferon, although produced in response to a wide selection of viruses, is species-specific, in that interferon from one organism fails to protect cells of another from viral attack. Needless to say, the action of interferon is being actively investigated. Unfortunately, it has not yet become the basis of practical antiviral therapy nor is the mechanism of its action understood. It is clear that interferon does not prevent entry of viruses into treated cells but interferes in some manner with replication. It is effective against both DNA and RNA viruses, so that it must act on some process common to the replication of both. The best indications at the present time are that interferon does not prevent the synthesis of nucleic acid components nor the formation (transcription) of mRNA from viral DNA. The binding of viral RNA to the host-cell ribosomes appears to be severely curtailed whether the RNA be the genetic material of an RNA virus or mRNA transcribed from viral DNA, in the case of a DNA-containing virus. The production of viral-originated proteins is, as expected, severely inhibited in the presence of interferon, an inhibition that leads to the inhibition of net virion production.

It should be added, finally, that interferon acts in extremely small amounts; 4 nanograms provides detectable inhibition. These low concentrations suggest that interferon acts in a catalytic manner, in that a few molecules can promote the formation of greater amounts of material that actually produces the inhibition.

Interferon, being a protein, appears to have its structure encoded in the host cell's DNA and to be synthesized at the ribosomes on a mRNA template. Indeed, inhibitors of protein synthesis or mRNA synthesis prevent formation of interferon in a manner that suggests that the normal protein-synthesizing apparatus of the cell is employed. The mechanism whereby the synthesis of interferon is triggered by the presence of virus is unknown, but it may involve binding between a viral component and a cellular repressor substance that ordinarily prevents mRNA synthesis related to the formation of the interferon molecule.

Interferon thus provides a possible approach to the treatment of viral diseases, provided problems of species specificity, production, and administration can be dealt with. Interferon may also be of interest as an inhibitor relating to ribosome function. Finally, its production and action should be regarded as a mechanism that cells have evolved in protecting themselves against potentially harmful viruses. Another example of protein synthesis in response to viral infection will be encountered below in the case of the synthesis of diphtherial toxin.

Physiological effects of a provirus on cells

We have seen that the introduction of a prophage into a cell is able to influence the nutritional or physiological attributes of the cell by means of the process of transduction. Transduction is, essentially, the transfer of genetic information between one host cell and another by a virus able to lysogenize the cells. There is an additional sense in which the presence of a provirus is able to affect the physiology of the host cell (other than taking over the host's metabolism in the event that replication occurs following induction). Thus, the very presence of certain proviruses within a cell alters the properties of that cell in ways that do not appear to have anything to do with the replication of the virus or with the host cells in which the virus had been previously grown (as would be expected in the case of transduction).

Thus, when some bacteria are lysogenized with certain temperate viruses, their pigment content or colony form are altered. In these cases, one speaks of *lysogenic conversion* of the cells to the altered state. An additional example of this phenomenon is the alteration in the antigenic properties of the cell wall lipopolysaccharides of the bacterium *Salmonella*, on introduction of a specific (episelon) prophage. Only in the presence of that prophage do the cells form a certain *polysaccharide synthetase* enzyme that forms a linkage between the sugars mannose and galactose. When the prophage is present, several other activities are repressed. It is not likely that the prophage-induced changes are connected with phage production or immunity, since mutants of the prophage that have lost the ability to convert nonetheless reproduce in a normal fashion.

A final, and very interesting, example of lysogenic conversion is the case of the diphtheria organism *Corynebacterium diphtheriae*, a bacterium that is harmful to human cells through the production of a specific toxic protein. This protein, with a molecular weight of about 70,000 daltons, is produced only by cells of the bacterium that contain a prophage called *beta*. Cells that do not contain the prophage do not manufacture the toxin but can be converted to the state of toxigenicity upon lysogenization with beta prophage. Again, it appears unlikely that the toxic protein is related to the normal production of properties of the phage system itself. There exist mutants of the phage that do not confer toxigenicity, and these are able to replicate in a normal fashion. Finally, it is interesting that there are some mutants of a related prophage that are defective in failing to produce normal phages upon induction with ultraviolet light. It appears that the presence of such a defective prophage is connected with the ability of the host cell to manufacture much higher concentrations of toxin than is usually the case. In the cases that have been studied, the presence of defective prophage is correlated also with alterations in the bacterium's growth rate and oxidative enzymes. The cells containing a defective prophage grow at about one third the rate of normal cells and contain only cytochrome *b* instead of a complete respiratory chain. The relationship between these alterations is not presently fully understood, but we shall see below that changes in metabolism and growth rate in cells that harbor viruses may also account for some features of tumor initiation.

Episomes

The apparent ability of prophages to confer new capabilities on host cells causes one to reflect on the relationship between provirus and cell. It thus appears that not only does the cell provide a locus for virus replication when it occurs and a site on the bacterial chromosome when it does not, but the provirus provides the cell with some genetic information that influences cell function. As a unit for the transfer of cellular genetic information, the virus has a number of advantages, including its infectivity, which enables it to transfer the information rapidly within a population. It is also obvious that the information borne by the provirus must be something that is, under most circumstances, nonessential for survival of the cell, since cells frequently do not contain a particular provirus. Finally, the provirus has the interesting property of existing in two states in the cell, integrated within the bacterial chromosome (provirus) and free, when it is in the process of replicating, following induction. It turns out that viruses are not unique in possessing these properties. Several other genetic units have been described in bacteria (and suggested elsewhere) that are infectious in a population and exist either integrated into the host chromosome or free in the cell.

These have been defined as *episomes* and include the mating factor of *E. coli* and *Pseudomonas aeruginosa* as well as some drug-resistance genes in the same species. The mating factor (also called F particle) determines the ability of a particular cell to behave as a donor or acceptor of genetic material in a bacterial genetic interchange that has this sort of polarity. When it is absent, the cell is an acceptor; when present, a donor (or "male"). When it is present, free, in the cytoplasm of the cell, the donor transfers genetic material with a low frequency, while its incorporation in the chromosome, in a manner completely like the incorporation of the prophages, leads to a much higher frequency of transfer. This factor is further similar to a temperate virus in its ability to carry some of the host genetic material from one cell to another. This is, of course, quite analogous to transduction and is often termed *F-duction*. There is evidence that other episomes, such as the drug resistance factors, may be regarded as F factors that have acquired additional information relating to other matters. If we are to regard these genetic elements as comprising a part of the normal genetic endowment of cells (and this appears to be the case), then we are not on very safe grounds in regarding viruses (or at least temperate viruses) as anything very different. Likewise, it is somewhat suspect to call temperate viruses episomes and cellular genetic elements, while the lytic mutants of the same virus are regarded as invaders of the cell and quite unrelated to normal cellular function. Rather, present evidence leads to the position that temperate viruses should be regarded as genetic elements not distinct from normal cell function, and the lytic viruses may best be considered to be aberrant forms that have, somehow, escaped cellular control.

Effects of viruses on cellular membranes

Viruses that invade cells must necessarily pass through the plasma membrane to enter the interior where they are replicated. We have seen that certain bacterial viruses have a complex tail–tail fiber apparatus for

415

effecting their entry into the cell. Other viruses appear to be able to erode cell membranes and to enter in that manner. Still others appear to be passive, awaiting entry through a break in the cell membrane or perhaps entry in a pinocytotic vesicle. It is generally the case that viruses produce characteristic changes in cellular membranes and that, at least in one instance, these changes provide an interesting tool for cell study.

Addition of a high titer of virus to cells often results in destruction of the cells without any production of new virus. This is called *lysis from without* and reflects damage to the cell membrane and, in the case of bacteria, to the cell wall. It would appear that the large number of virus particles eroding the cell membrane produce damage so great as to be irreparable. A similar destruction of cellular membranes has been noted in the case of infection of mammalian cells by poliovirus. This RNA virus replicates in the cytoplasm, and one indication of its presence there is the destruction of the membranes of mitochondria (presumably resulting in a decline in cellular-energy production).

A final action of viruses on cellular membranes is the very interesting property of certain animal viruses to promote the fusion of cells as a result of coalescence of their plasma membranes. For example, viruses of the measles and herpes group lead to the fusion of cells to the extent that many cells end by containing several nuclei. This property of some viruses has been exploited in the study of cells by effecting the fusion of cells with different properties. For example, cells grown in culture have been fused together by the addition of a virus (called Sendai virus), and the fusion process occurs even between cells originating in quite different organisms and tissues. There is an additional sense in which viruses influence the properties of cell plasma membranes (in a nondestructive way), where the surface properties are altered on infection by a virus. Cultured cells alter their ability to adhere to a glass surface or to each other when certain viruses are introduced. Some of these changes appear to be special properties of *oncogenic* (tumor-producing) viruses, and will be mentioned below in that context.

Physiological effects of tumor viruses

Neoplastic diseases are conditions under which a population of cells within an organism fails to function in conformity with the organism as a whole and grows rapidly at the expense of the remaining cells. Thus, a hard tumor grows more rapidly than the cells around it and, in the case of a malignant tumor, also invades the surrounding tissue. When the tumor erodes blood vessels, tumorous cells may enter the blood stream and migrate to remote parts of the organism where the cells serve as foci for secondary (*metastatic*) tumors. A malignant tumor is defined quite reasonably as one that, if allowed to develop unchecked, will finally kill the organism in which it grows.

Important categories of neoplastic conditions include the *sarcomas*, which are tumors derived from mesodermal cells, and *carcinomas*, which originate from endodermal or ectodermal cells. In addition, *leukemia* is a neoplastic disease of the blood-forming organs in which an abnormal and excessive population of white blood cells (leukocytes) is produced. A vast literature has grown up in the field of physiology and biochemistry of tumors (rather like a malignant tumor in that it has invaded many other areas), but the firm conclusions about the problem as a whole have been few. One problem that confounds the study of hard tumors is the degree to which they outgrow their blood supply, so that studies often reveal only that the cells of the tumor are in rather bad condition, owing to deficient respiration, nutrition, or waste removal.

One feature of tumors that has gained wide acceptance and been the basis for many studies of biochemical regulation is the observation, first pointed out by Otto Warburg many years ago,[1] that the oxidative activity of tumor cells is depressed and that glycolysis, as measured by lactic acid production, is increased. It appears that many such cells shift in their energy production from aerobic (mitochondrial) ATP synthesis to anaerobic glycolysis. This observation has led to many studies of the regulation of cell metabolism (see Chapters 5 and 6) but has not been of much practical benefit. It should be said that further analysis of the possibility of defective respiration has led to a welter of conflicting results. For example, the level of NAD and other respiratory coenzymes have been said to be lowered in mitochondria from tumors, while studies of oxidative ATP synthesis suggest that both oxidation and phosphorylation are intact in isolated tumor mitochondria. If anything, both oxidation and ATP synthesis operate at a more rapid rate than in normal mitochondria, so that the meaning of the observation by Warburg (and others) is unclear. In any case, there is little evidence to suggest that impairment of oxidative activity of cells is the primary cause of the neoplastic transformation of cells. It is safe to say that, in any case, neoplastic diseases, such as tumors and leukemias, are not a single disease but may often be remote in their causation and underlying physiology. Such diseases are, after all, known to be produced by radiation of a number of types, chemical *carcinogens*, such as the tars found in petroleum and in tobacco smoke, and by a number of viruses to be introduced below. It should not be surprising that a wide range of causative factors might produce an equally wide range of diseases but all with the unifying feature of rapid cell growth.

There are a number of practical and ethical reasons why tumors and related diseases are not best studied in the intact organism, and tissue culture has been employed with increasing frequency. Thus, the basic experiment in tumor virology has been the demonstration that a given virus can transform cultured cells to a malignant state. Clearly, the nature of that malignant state must be described differently in the cultured cell and the intact organism. The main criteria that may be used to call a cell line "malignant" are the following:

1. There is an increase in the growth rate of the cells. This is, in part, through the loss of what is called *contact inhibition*, where cultured cells

---

[1] In the 1920's. For a more recent discussion, see, O. Warburg, "Origin of Cancer Cells," *Science*, **123**, 309 (1956).

stop growing when their plasma membranes come into contact with other cells. Thus, normal (untransformed) cells form a monolayer on the surface where they grow and do not pile up on each other in additional layers. Malignant cells do not experience this sort of inhibition and so form large populations with many layers.

2. Normal cells frequently can only divide for a set number of divisions in culture after which they stop growing and die. Malignant cells do not suffer this termination and continue dividing as long as nutrient is provided and suitable conditions maintained.

3. A number of metabolic changes occur, some of which are apparently reflected in the synthesis of new enzymes. These are detectable as new antigens (which may not necessarily be identical with enzymes only). The existence of these antigens, which are presumably tumor-specific, has led to considerable interest in using antibodies against these antigens as antitumor agents. This tactic has not yet been of great usefulness.

4. A final and important criterion for the malignant transformation of a cultured cell is that such cells have frequently been shown to give rise to malignant tumors when transplanted into a suitable experimental animal.

A growing number of viruses have been isolated and studied which cause the growth of neoplastic cells in experimental animals. These include both DNA- and RNA-containing viruses, and the demonstration of their tumor-producing (*oncogenic*) ability has been carried out with a wide variety of experimental animals. Table 17-1 shows the major categories of oncogenic viruses and their hosts. While there is no doubt whatsoever about the ability of certain viruses to induce neoplastic disease both in the laboratory and in nature, it is not certain that all diseases involve viruses. Indeed, there is at present no case of unambiguous identity of a virus as the causative agent in a human tumor, although there is quite a lot of indirect indication that this can be so. The closest example available appears to be the ability of human *adenoviruses*, which are associated with minor respiratory disease in man, to cause malignant tumors when injected into newborn rodents. It is not certain that this virus can produce similar effects in human beings.

Table 17-1   *Major categories of oncogenic viruses*

Class of virus	Genetic material	Host organism
Papilloma virus	DNA	Rabbit, other mammals
Polyoma virus	DNA	Mouse, primates
Adenovirus	DNA	Primates, including man*; cattle, birds
Herpes virus	DNA	Mammals, birds, amphibians
Avian leukemia	RNA	Birds
Murine leukemia	RNA	Mice
Various leukemia and sarcoma viruses	RNA	Cat, rodents

* Occurs, but has not been shown to induce tumors in man. See text.

The effective history of tumor virology could be said to have originated in 1911 with the discovery by Peyton Rous of a virus-caused sarcoma in chickens. This *Rous sarcoma virus* (RSV) has been studied ever since and has provided much of the basis for our knowledge in the field. When cells are infected with RSV, they continue to grow and release new virus particles slowly but at a constant rate (about one virion per cell per 5 hours). The viruses appear to be assembled in the region of the plasma membrane, and antibody against the virus protein reacts with maturing virions. In many instances, the tumors induced by this virus appear themselves to be virus-free, a finding that somewhat muddied the waters concerning the causation involved. This result has become more understandable with the finding that RSV is a defective virus in the manner of a defective prophage mentioned above. The replication of complete virus requires a second, helper virus called the Rous-associated virus (RAV), and the presence of both is required for the complete production of RSV virions. It should be added that the RAV is, by itself, an oncogenic virus, causing a sort of leukemia in chickens.

It was mentioned that RSV matures at the plasma membrane of its host cell. In fact, this virus (and related ones) ends by becoming surrounded by a vesicle derived from the cell surface. Thus, the virus must produce important changes in the plasma membrane, and it might be wondered if those alterations have anything to do with the ability of the virus to promote neoplastic changes. It will be recalled that one of the significant effects of tumor viruses is the alteration of the cell surface that leads to a loss of contact inhibition. It is not presently known if there is any connection between these facts of the viral invasion. It might be added that the action of a virus such as RSV leads to some recognizable biochemical changes in the cell surface, including increased synthesis of acid mucopolysaccharides. This appears to be the result of an increase in the requisite enzymes, such as *hyaluronic acid synthetase*. Again, this finding suggests the importance of the cell plasma membrane in the events leading to a malignant transformation.

Figure 17-10 Production of Rous sarcoma virus with an outer coat formed from the host plasma membrane. A vesicle is formed at the cell surface in which the virus is incorporated.

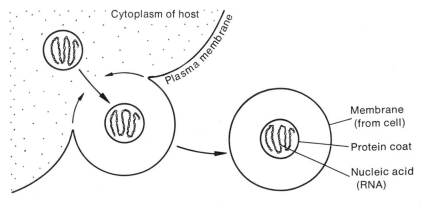

The Rous sarcoma virus contains RNA as its genetic component. There is evidence that the intracellular form of the virus (analogous to the prophage of bacterial lysogeny) is, however, DNA. For one thing, the replication of the Rous virus is inhibited by inhibitors of DNA synthesis. Recently, enzymes have been described that occur in the virus particle and which catalyze the copying of RNA into DNA so that the mechanism for creation of the DNA provirus is assured.

An example of an oncogenic virus that contains DNA and about which much is known regarding effects on cellular function is the group of *polyoma viruses*. These are approximately spherical in shape with 72 protein subunits and a molecular weight of about 25,000,000 daltons. The DNA is in the form of a closed circle similar to that of many bacteriophages and bacteria. The polyoma viruses have been extensively studied, in part because of their small size and relative uncomplexity. With a diameter of about 0.5 nanometer, these are among the smallest DNA viruses, and they contain enough DNA to encode only a total of about 1,500 amino acid positions. It is likely that the virus genetic map includes less than 10 genes. This virus produces both lytic and neoplastic responses in sensitive cells, the latter being much more infrequent.

Since polyoma virus can be replicated in cells that are not actively making DNA themselves (such as cells under the control of contact inhibition), they appear able to induce the formation of the enzymes required. It is significant that the infection with virus, in this case, leads to a stimulation of viral and host DNA. This is in striking contrast to the action of many animal viruses that lead to a rapid inhibition of host DNA synthesis as part of their take-over of the host cell's metabolism. The stimulation of DNA synthesis includes the increase of a number of enzymes, which are listed in Table 17-2. These enzymes must be encoded by the host cell, as there is insufficient genetic material in the polyoma DNA to account for their synthesis. Thus, the most reasonable explanation for the increases in enzyme activity is the removal of some sort of regulatory mechanism, so that the cellular genes for these enzymes can be expressed.

Additional changes in the cell produced by infection by the polyoma class of viruses center about the plasma membrane. There are differences in concentrations (or locations) of mucopolysaccharides on the cell surface, as well as those of a number of other membrane components. There has been some interest in the idea that transformed cells should exhibit

Table 17-2   *Some enzymes that are increased when cultured cells are infected with polyoma viruses*

Enzyme	Maximum relative increase over uninfected cell
DNA polymerase	7-fold
Thymidine kinase	55-fold
Cytosine diphosphate reductase	4-fold
Dihydrofolate reductase	3-fold

some different permeability properties, but, in this case, there is no evidence—the ability of polyoma-transformed cells to transport amino acids being unaltered.

Suggested Reading

Books

Cairns, J., G. S. Stent, and J. D. Watson (eds.), *Phage and the Origins of Molecular Biology*, Cold Spring Harbor Laboratory of Quantitative Biology, 1966.

Cohen, S. S., *Virus Induced Enzymes*, Columbia University Press, New York, 1968.

Cold Spring Harbor Symposia on Quantitative Biology, vol. 35, *Transcription of Genetic Material* (1970).

Luria, S., and J. E. Darnell, *General Virology*, 2nd ed., John Wiley & Sons, New York, 1968.

Stent, G., *Molecular Biology of the Bacterial Viruses*, W. H. Freeman and Co., San Francisco, 1963.

Watson, J. D., *The Molecular Biology of the Gene*, 2nd ed., W. A. Benjamin, New York, 1970.

Articles

Calendar, R., "The regulation of phage development," *Ann. Rev. Microbiol.*, **24**, 241 (1970).

Dulbecco, R., "Cell transformation by viruses," *Science*, **166**, 962 (1969).

Dulbecco, R., "The induction of cancer by viruses," *Sci. Am.*, **216** (April, 1967).

Green, M., "Oncogenic viruses," *Ann. Rev. Biochem.*, **39**, 701 (1970).

Huebner, R. J., and G. J. Todaro, "Oncogenes of RFA tumor viruses as determinants of cancer," *Proc. Nat. Acad. Sci. U.S.*, **64**, 1087 (1969).

Rapp, F., "Defective DNA animal viruses," *Ann. Rev. Microbiol.*, **23**, 293 (1969).

Wood, W. B., and R. S. Edgar, "Building a bacterial virus," *Sci. Am.*, **217** (July, 1967).

18 The physiology of cellular differentiation

For practical reasons we often discuss cellular function as if time did not exist and cells carried out their many activities at a constant level for all eternity. Indeed, the necessary plan of a book such as this one is to consider each cellular topic as somehow independent from most of the other cellular topics and, especially, independent from the life history of the cell. The reason for this tactic is the apparent inability of the human mind to perceive more than approximately one aspect of a complicated matter at one time. Since the reader has discovered thus far that the cell *is* a complicated matter, a textbook must issue a disclaimer to the effect that cells, in fact, possess life histories and, indeed, in many instances change dramatically as time passes. The purpose of this chapter is the issuance of exactly that disclaimer: Cells do change and often that alteration is directly related to the specific functions that a cell must carry out. For us, the most interesting example of cellular changes will be *differentiation*, and the following sections will describe something of what is known about its underlying physiological and biochemical mechanisms.

The study of differentiation

There is some variation in the meaning attached to the term *differentiation* by scientists of differing viewpoints. We shall use the word in a fairly wide context to indicate changes that occur to members of a collection of cells such that the cells, having been alike, become different in some way that can be detected. An example of differentiation can be seen in the case of slime molds belonging to the family *Acrasieae*. These organisms exist during one phase of their life cycle in the form of a population of amoeboid cells, which are undifferentiated in the sense that they cannot be told apart. These *vegetative cells* under suitable environmental conditions can form an aggregated mass that gradually differentiates into a stalk-and-fruiting body that is composed finally of spores, each able to give rise to a new colony. The cells that form the stalk (see Figure 18-1) not only do not

Figure 18-1 Differentiation of a slime mold belonging to the family Acrasieae. Individual cells aggregate (a) before forming a fruiting complex (b) composed of base, stalk, and fruiting body.

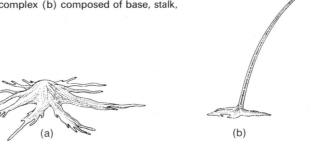

(a)   (b)

engage in spore formation, but they exhibit the special function of secreting a cellulose sheath that gives the stalk sufficient strength to support the fruiting head. Thus, the original population of cells differentiates to form two subpopulations with importantly different characteristics. The reader will immediately appreciate the simplicity of the slime-mold "system" for the study of differentiation, and, indeed, great interest in slime molds has been entertained by developmental biologists, with some important results that will be touched on later.

Other examples of differentiation can be much more complicated but, nonetheless, of central importance in biology. For instance, differentiation as applied to the development of multicellular organisms is of importance in considering the fundamental nature of such organisms and includes many areas of special practical significance to man, including the causes of birth defects, regeneration of damaged organs and, as we shall see, neoplastic diseases, which include cancers. The study of differentiation as it relates to, say, mammalian development differs from that of the slime molds in the much greater complexity of the whole business and in the close coupling of the separate processes of growth and differentiation. Thus, a multicellular organism develops from a single cell, the fertilized ovum. As subsequent proliferation of cells takes place by means of cell division, the process of differentiation is superimposed upon it so that the progeny of the single ovum are quite unlike each other. The original ovum must, therefore, carry the information necessary to construct its own structure, including all of its enzymes, but, in addition, it must carry all the information to specify the total structure and function of its many classes of progeny. As an example, a fertilized human ovum probably does not contain either the protein hemoglobin or the protein myosin, but its descendents that form red blood cells must contain the information required for the former, while those that become muscle must manage the latter. Thus, in a sense, differentiation must be thought of in the context of cellular information (see Chapter 16), in that all genetic information must be extant in the parental cells prior to differentiation, which, then, is the selective expression of that information. The central issue of the whole area is the mechanism whereby genetic information is selectively expressed in such an orderly manner as to produce a functioning (and recognizable) multicellular organism.

Much of what we know of differentiation is a function of the choice of a system in which to investigate it, a choice that is sometimes rather accidental. Thus, the biology of the chick embryo is probably much better understood than that of the human embryo for reasons of greater availability as well as convenient location of the chick embryo—external to the mother hen. Likewise, our information about the development of the chick and the frog is available in much greater detail than in the case of most other organisms. Owing to certain features of early development in the *Echinodermata*, it has been convenient and rewarding to study these organisms in great detail, and much information about the biochemistry of early development has come from studies using starfish and sea urchins.

The differentiation that occurs in the embryo of a multicellular organism is necessarily complex and difficult to manipulate, and recent years have seen increasing emphasis on studies employing cells in culture. We have seen earlier that the technique of growing cells in pure culture can provide benefits in terms of the ease of controlling the environment. In principle, one of the best ways to study differentiation would be to examine cultures of identical cells and see what changes in their environment would lead to relevant changes. In fact, this approach has recently begun to yield important results using cells from a variety of sources, including animals, higher plants, and even bacteria. In the latter case, a number of investigators have been interested in the process whereby certain bacteria form spores, and they have endeavored to examine the process as a "model" for more complex sorts of differentiation.

Self-assembly

Whether one is interested in the construction of a single cell or in the formation of complex differentiated organisms from a collection of cells, one must be concerned with the formation of complex structures from simpler subunits. We have seen (in Chapter 16) that the synthesis of protein occurs on the basis of information encoded on a strand of DNA and that there are mechanisms for turning protein synthesis on or off. We have said relatively little about the next stage of cell manufacture, which is the process whereby the proteins (and other molecules) are brought together to form membrane systems and, finally, the cell itself. The prevailing opinion about this matter among biologists is summarized as the *principle of self-assembly*, and, in its most extreme form, this theory has it that the cell need only synthesize the basic molecules, which will then fall together in the proper manner. To put it more exactly, the reason that the cellular basic molecules flop together to form the intricate structure of the intact cell is that the cell, in fact, represents the most thermodynamically favorable configuration in which they can exist. The extreme view that this is all that there is to the assembly of cells is certainly unproved, but there is no question but that self assembly is of significance and, indeed, that it operates on several levels.

We saw in an earlier chapter that, although proteins were linear polymers of the various amino acids, they could be represented in three dimensions of space as possessing regions of helical structure (secondary structure) and that the whole business could be wound on itself in a highly specific way to give an overall globular structure. Such winding could bring different parts of the polypeptide chain together so that distant amino acids could participate in the same active, catalytic site. The driving force for the winding of the protein into its secondary and tertiary structure is, as we saw, the high probability of that structure, a probability that stems from the positions of the many intramolecular forces that occur between different parts of the polypeptide chain. In other words, the winding of the protein into its configuration occurs in a spontaneous way—that is, the event has associated with it, a negative free energy.

Not only does a protein assemble itself into the most probable conformation, but many proteins have been shown to possess an additional level of structure: quaternary structure. The active form of such proteins is an aggregate of several individual polypeptide chains, the *protomers*. Under suitable conditions, the protomers aggregate to form the active *oligomers*, which are held together by forces entirely like those maintaining tertiary structure. Indeed, extremes of pH or ionic strength that would lead to unfolding of tertiary conformation also cause disaggregation into protomers. For example, muscle aldolase (Chapter 5) contains four protomers that dissociate under conditions of acid pH. If the pH is not too extreme, the effect is reversible; that is, the oligomeric form is regained when the pH is returned to neutrality. Since the protomers unwind under the same conditions that are required for dissociation of the oligomers, at acid pH, the following equilibrium can obtain:

$$\text{Oligomer} \rightleftharpoons \text{Protomer}_{native} \rightleftharpoons \text{Protomer}_{unwound}$$

As in the case of tertiary structure, under cellular conditions, the most probable state is the oligomeric one; in both cases, the configuration is maintained by the presence of many noncovalent bonds, which, added together, render the native configuration stable.

A still higher level of protein structure that illustrates the role of self-assembly even better is the formation of functioning multienzyme complexes of the sort described in Chapter 4. For example, the fatty acid synthetase system of yeast was seen to consist of a complex of seven different enzymes that assemble themselves spontaneously under proper conditions. Each of these individual enzymes consists of three polypeptide protomers, and it is interesting that not only the protomers are inactive but the individual enzymes are inactive, unless they are aggregated together to form the whole complex.

It is clearly a long way from the self-assembly of a multienzyme complex to that of an entire cell, or even of a morphologically recognizable part of a cell. Indeed, the application of the principle of self-assembly to gross cellular structure is very appealing (in view of a scarcity of obvious alternatives) but largely unproved. There are, however, a few cases of apparent spontaneous assembly at levels of structure intermediate between relatively simple multienzyme complexes and entire cells, of which the best example is probably the reconstitution of membrane structure from resolved components.

Thus, enzymes from mitochondrial membranes, such as the cytochromes (Chapter 6), may be highly purified and then recombined, together with phospholipids, to form complexes that are similar to the original membrane when viewed in the electron microscope. Moreover, such reconstituted membranes are enzymically active in carrying out the oxidation of certain metabolic intermediates, although, thus far, efforts

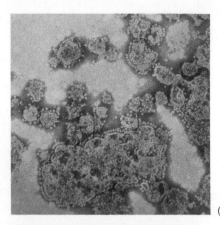

(a)

Figure 18-2 Reconstitution of active mitochondrial fragments from coupling factor F_1 and smooth vesicles. (a) Fragments (submitochondrial particles) that are able to synthesize ATP exhibit knobs that are identical with F_1. (b) Vesicles from which F_1 has been removed. (c) When purified F_1 is added back to the vesicles in (b), a native configuration is regained and the vesicles can, once again, synthesize ATP. (Negatively stained preparations, courtesy of Dr. E. Racker.)

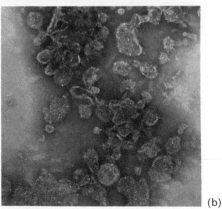

(b)

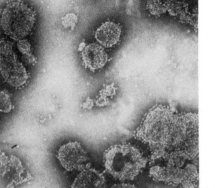

(c)

Table 18-1   Reconstitution of an enzyme system able to oxidize succinate*

Oxidation was measured using an oxygen electrode. The complete system included succinate dehydrogenase; cytochromes b, c, c_1, a, and a_3; and ubiquinone and phospholipid.

System	Rate of oxidation of succinate (% of complete system)
Complete	100
Cytochrome c_1 omitted	15
Phospholipids omitted	29
Succinate dehydrogenase omitted	12

* The components were combined following the technique described by S. Yamashita and E. Racker, in *Journal of Biological Chemistry*, **244**, 1220 (1969).

to couple such oxidations with ATP synthesis have been unsuccessful, even in the presence of added coupling factors (Chapter 6). An example of such a reconstitution experiment is shown in Table 18-1. It is interesting that, while assembly of polypeptide chains into their active configuration is rapid, often requiring only a few seconds, the assembly of a reconstituted respiratory chain may require several hours at an elevated temperature (37°C). This slowness probably reflects the greater order of magnitude of possible configurations that exist and the large number of noncovalent bonds required for an active complex. It is important to add that the rate of oxidation carried out by such a reconstituted respiratory chain system can be of the same magnitude as that carried out by intact mitochondria when measured on a protein basis (specific activity).

Finally, the principle of self-assembly can be seen to be operative in the case of mitochondrial or chloroplast membrane fragments prepared by sonic oscillation. If such fragments are prepared so as to be able to carry out ATP synthesis, examination with the electron microscope reveals small knobs on the membrane surfaces, the knobs being the coupling factor F_1 (ATPase—see Chapter 6). The coupling factor may then be removed and purified, leaving vesicular membranes devoid of knobs. When purified coupling factor is added to such "naked" membrane, the knobs assemble themselves in a manner similar to their original configuration on the membrane surface. It is significant that the naked membrane vesicles do not carry out oxidative phosphorylation while containing the enzymes of the respiratory chain, whereas the reconstituted membranes (with knobs added back) are again active in the synthesis of ATP.

The interaction between cells

While the idea of self-assembly is able to account for a number of features of the formation of a particular cell, it is unreasonable to expect it to extend to all facets of development. For example, the development of a tissue-

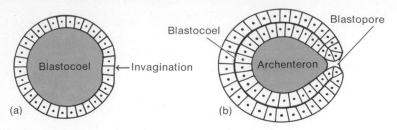

Figure 18-3 Early stages in the development of *Amphioxus*. (a) The blastula; in formation of the gastrula, invagination occurs at arrow. (b) Gastrula; the former blastocoel is now a thin space between the two layers surrounding the archenteron.

forming part of an embryo includes changes in a large number of cells, and some of these changes appear to involve interactions between different cells in nearby regions. A number of such interactions are known (but, frequently, little understood), but we shall concentrate on one example, the *induction* of formation of neural tissue in a number of vertebrate embryos.

If one follows the early development of an amphibian, two phases are of interest to us here, the *blastula* and the later *gastrula*. Cell division beginning with the fertilized ovum leads rapidly to the blastula stage, where the embryo is a hollow ball of cells with walls a few cells thick enclosing a central space, the *blastocoel*. Subsequently, invagination occurs as shown in Figure 18-3, producing a second hollow space, the *archenteron*, which is later to give rise to the gut. The channel remaining where the original invagination occurred is called the *blastopore* and is usually partially or completely plugged with cells that store yolk. Now, the upper layer of cells, above (dorsal to) the archenteron is fated to become the neural tissue of the developing embryo, and the question that we ask is: What stimulates these particular cells to abandon, rather suddenly, their nondifferentiated state and commence development into the highly specialized and organized system of cells that comprise the nervous system? The reader is reminded that the task of producing a nervous system is even more complicated than that of providing a population of connected neurons able to transmit impulses and store information, since the main nerve chord of vertebrates is hollow. The manner in which the layer of cells external to the upper wall of the archenteron gives rise to such a hollow nerve chord is illustrated in Figure 18-4; what interests us is the stimulus that sends these events on their way.

The first hints about the process of neural tissue formation came from elegant studies of tissue transplantation carried out over fifty years ago. It was found that transplantation of portions of the dorsal lip of the blastopore, at the proper stage, to various regions of a recipient embryo would induce formation of neural tissue, no matter where on the recipient they were placed. Thus, it was concluded that the cells of the dorsal lip induced the development of neural tissue, and such *induction* has been the subject for investigation during the intervening decades. If dorsal lip cells were transplanted to an embryo that was itself past the gastrula stage, induction did not occur, suggesting that there was a finite period during which the embryo was *competent* to be induced.

Figure 18-4 Formation of the neural tube in amphibians. (a) A cross section through the dorsal wall of the gastrula showing the two layers and the remnant of the blastocoel between them. (b) The outer layer thickens to form the neural plate. (c) The borders of the neural plate are raised and begin to surround a space that will form [in (d)] the cavity of the neural tube. Thus, the neural tube, which develops into the spinal cord, originates in the outer layer of the gastrula, the ectoderm.

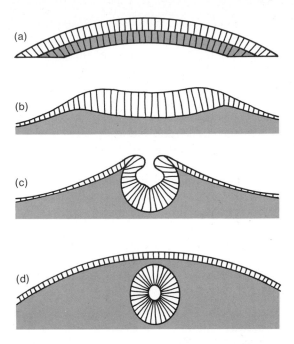

While certain early workers in the field thought that induction might reflect the action of some type of "vital force" able to organize cells into neural tissue, this view was replaced by a growing conviction that some chemical substance was produced in the dorsal lip cells and diffused to the upper layer of cells where neural formation took place. This latter view was supported by experiments in which dorsal lip tissue was killed by boiling (or other means) after which it still exhibited the ability to induce, suggesting that it still contained the active chemical. Moreover, direct contact between cells turned out to be unnecessary, as a layer of inert filter paper, which allowed diffusion of chemicals without cell contact, still permitted induction to take place.

Efforts to extract the active-inducing compound from embryonic tissue have been carried out during the last 40 years, but results have been somewhat sparse. The technical problems imposed by the small amount of tissue available from each embryo have restricted this line of investigation, but some interesting results have been forthcoming. For one thing, ether extracts of dorsal lip tissue, which presumably contain only lipid, are able to induce. Since the active compound in such extracts does not appear to form a salt on the addition of strong base (that is, is not saponifiable), it may well be a steroid. This supposition is in conformity with the experiments on heat inactivation, since steroids are relatively stable when compared to, say, proteins or nucleic acids, both of which otherwise might be thought to be good candidates for the role of inducing substance.

Unfortunately for any unified view of induction, a number of other chemicals (and treatments) also were discovered to induce neural tissue formation in embryos. In some instances, water-soluble extracts of dorsal lip (unlikely to contain steroids) contained active inducer. A number of pure organic acids were also found to induce, the list including such disparate members as fatty acids, DNA, and adenosine monophosphate. It now appears that the action of these acids is trivial, reflecting only an alteration in the pH of the cellular environment. More recently, it has become clear that extremes of pH in either acid or basic range led to the induction of competent embryonic tissue, provided the extreme was not so great as to be lethal.

It is very likely that induction as caused by a pH extreme cannot have much to do with induction by the dorsal lip of the blastopore *in vivo*, and research has proceeded with the strategy of isolating inducing substance from saline extracts of embryonic tissue. During the past few years, several laboratories have succeeded in isolating material by means of ammonium sulfate fractionation followed by chromatography, which is active and which contains, in some cases, mostly RNA and, in others, protein. Furthermore, the inducing ability of other extracts is not abolished by treatment with the enzyme ribonuclease (RNAase) but is abolished by the proteolytic enzyme trypsin. Since the molecular weight of the supposed inducing substance is estimated to be greater than about 40,000, it must be a protein and not a small peptide, a finding somewhat difficult to reconcile with its heat stability mentioned above.

From the foregoing discussion, it is clear that our information about induction is at a very preliminary level. One would like to know not only about the exact chemical nature of the inducer but also something about the early effects of inducing substance on the function of competent cells. The efforts of extremely able investigators during five decades have produced only the most tentative beginnings of answers to such questions.

Studies with cultured cells

In principle, the ideal system for the study of differentiation should be some form of cell culture where conditions should be well controlled and changes measurable. If one could, for example, inoculate a culture medium with a single, undifferentiated cell and follow the differentiation of its progeny until the latter formed a tissue, one would have many opportunities to see what was happening. Unfortunately, only recently has such a tactic begun to yield results, since formidable technical problems have had to be overcome. The central problem is, perhaps, one inherent in much of the animal cell culture, which could be called the unknown identity of the cell being cultured. Thus, if one seeks to culture a liver cell, the cell that actually survives in culture may be any one of the numerous different cell types found in the highly heterogeneous liver tissue. The problem is compounded by changes that often occur in cultured cells as a result of their highly artificial environment. Whether by selection or by direct transformation, cells in culture often bear little resemblance to cells in tissue, apparently losing many of their tissue-specific properties (see Chapter 10). However, one wishes to observe the reverse process in the study of differentiation, namely the acquisition of "tissue-specific" properties by undifferentiated cells, and only recently has this proved possible.

Thus, a suspension of cells from chick embryo muscle may be inoculated into standard cell culture medium and grown as a single layer of cells adhering to the glass of the culture vessel. These cells grow relatively rapidly, at first, with a mean doubling time of the order of 12 hours. Then, after a few days, the rate of cell division (and that of DNA synthesis) becomes slower, and the cell monolayer, up to then undifferentiated, begins to exhibit long fibers. These fibers may be said to have differentiated to the extent that they show the characteristic striations of skeletal muscle fibers and are able to contract spontaneously. It should be said that the cell alterations that occur in such differentiation are of a drastic nature, since not only must there be formation of the proteins of the contractile mechanism, but the overall cell type changes from the primitive (*myoblast*) possessing a single nucleus to fused multinuclear cells characteristic of skeletal muscle tissue.

It is also interesting that cell interaction is required for the successful development of muscle fibers, as described above. If cells are propagated from a single primitive muscle cell (myoblast) (also present in embryonic muscle), so that *fibroblasts* are excluded from the culture, muscle fibers fail to develop. It appears that fibroblasts, which function to lay down connective tissue in the developing muscle, secrete collagen which is required to coat the surfaces of the muscle cells in order that they can fuse into fibers. In the absence of fibroblasts themselves, muscle fibers can develop if the medium in which fibroblasts have previously grown is added to the muscle-cell culture medium.

Dedifferentiation

It is worth remarking that differentiation, whatever its physiological and biochemical basis, is essentially a reversible process. If a uniform population of cells can give rise to muscle fibers or other aggregations characteristic of a particular tissue, organized cell populations can revert to a less-organized state. Such a process forms an essential part of the regeneration of tissues in many organisms, dedifferentiation preceding the formation of a newly differentiated organ or limb.

A second context in which dedifferentiation is of considerable interest is that of *neoplastic diseases*, the so-called cancers. We observed, in the case of developing muscle fibers, a concurrent decline in the rate of cell division and an increase in specific organization related to the function of contraction. The neoplastic condition is, in most general terms, the reverse of this, including a loss of tissue-specific traits and a dramatic increase in cell division. Since this dramatic increase occurs in an organism where, on the average, cell division is under tight control, cells that have escaped this control quickly proliferate to the extent that they threaten the functioning of the whole.

Massive efforts have been underway for a long time directed to

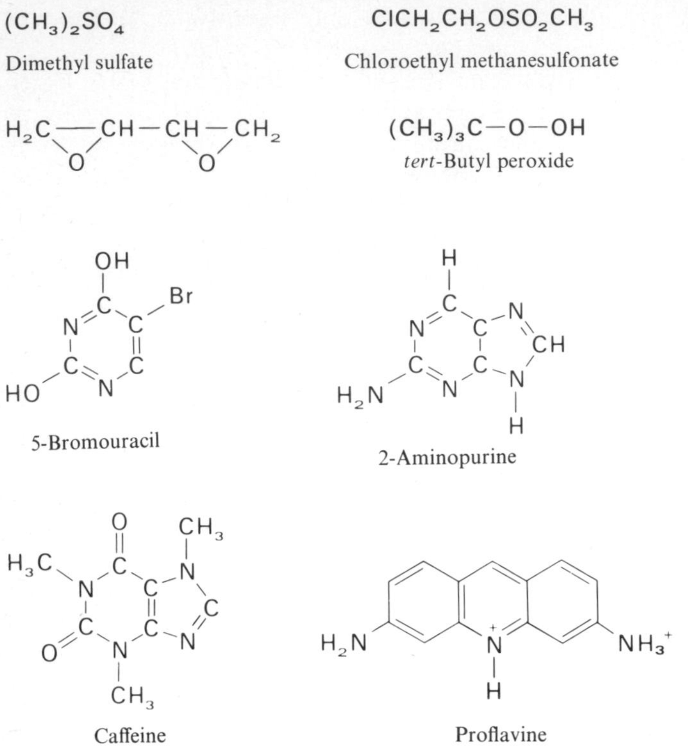

$(CH_3)_2SO_4$

Dimethyl sulfate

$ClCH_2CH_2OSO_2CH_3$

Chloroethyl methanesulfonate

Figure 18-5 Examples of mutagenic reagents.

H₂C——CH—CH—CH₂ (epoxide structure with O)

$(CH_3)_3C-O-OH$

tert-Butyl peroxide

5-Bromouracil

2-Aminopurine

Caffeine

Proflavine

uncovering the primary cause of neoplastic loss of control of cell division and related dedifferentiation, and it is fair to say that we are only on the threshold of understanding. It is clear that a number of seemingly unrelated stimuli give rise to the process, including ionizing (and nonionizing) radiation, oncologic (tumor-producing) viruses, and any one of a large number of chemical mutagens, several of which are inconveniently included in tobacco smoke. The variety of compounds that confer mutagenic effects is illustrated in Figure 18-5.

While agreement as to the nature of the "primary lesion" in neoplastic disease has not been reached, it seems reasonable that a central (if not causally primary) aspect resides in the loss of regulation of cellular DNA replication. Whereas DNA synthesis is suppressed in most cells of an adult plant or animal, occurring at a high rate only in certain specific tissues, such as the meristem of plants or blood-forming cells of higher animals, it occurs at a rapid rate in tumors. This high rate forms the basis for one of the most successful approaches to chemotherapy, that of using nucleic acid poisons to selectively inhibit tumor cells. Thus, a DNA base analog such as 6-mercaptopurine,

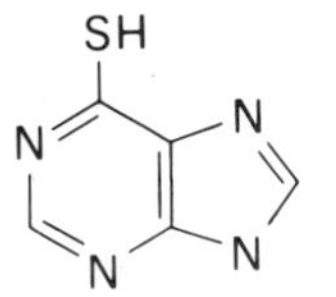

SH (6-mercaptopurine structure)

or 5-fluorouracil,

inhibits those cells most actively engaged in DNA synthesis, and these cells are most likely to be those that have undergone the neoplastic transformation.

An added feature of neoplastic cells that both further complicates the picture and suggests possible avenues of approach is the variety of detectable changes that occur in the structure of chromosomes. Whether the neoplastic state is initiated by radiation, carcinogenic chemicals, or viruses, these cells often exhibit considerable changes in the microscopic appearance of their chromosomes. These changes include deletions of various magnitudes that have often been regarded as causing the transformation and dedifferentiation of the cells, but they may just as well reflect a prior disruption in the functioning of the cell, such as that possibly attendant upon the initiation of very rapid DNA replication.

The role of RNA in differentiation

Since ribonucleic acid (RNA) plays a role intermediate between DNA (the primary information storage) and synthesized protein, it can reasonably be expected to exert a major influence in differentiation. Thus, differentiation involves, among other things, the synthesis of proteins not found in the less-differentiated cell. Synthesis of these proteins must necessarily be preceded by the formation of the specific mRNA molecules corresponding to them, and it is likewise probable that the synthesis of mRNA's associated with protein of the undifferentiated cell is suppressed. Moreover, the concentration of ribosomal RNA in a cell reflects the growth rate of a cell, and, since we have seen that differentiation is often associated with a decline in growth, it is likely to be also linked to a suppression of ribosomal RNA formation.

Recent studies suggest that changes in the level of various types of RNA during embryonic development are the result of alterations in the levels of the enzymes that are involved in RNA synthesis. For example, sea urchin embryos contain three, separable ribonucleic acid polymerases in their nuclei. Of these, one is probably located in the nucleolus and is associated with the synthesis of ribosomal RNA. This form of the polymerase is synthesized at a rate comparable to cell division, so that a constant level of the enzyme is maintained during a considerable period of time. The other two sorts of polymerase are located in the nucleoplasm (that is, not in the nucleolus), and at least one of them is associated with the formation of *DNA-like RNA*, a species of RNA notable for the

433

similarity of its base composition to that of DNA. The level of such DNA-like RNA appears to be highest during the earliest stages of embryonic development and to decrease dramatically by the late gastrula stage. The RNA polymerase thought to be associated with the synthesis of DNA-like RNA is seen likewise to decline as development proceeds. The question as to the role of DNA-like RNA in development as well as the nature of the regulation of RNA polymerase synthesis are presently under active study, as they appear to be central to a fuller understanding of the nuclear control of developmental processes.

Metabolic changes in developing embryos

The purpose of this section is to point out that an increasing body of information is available regarding the physiology of embryos and that this information is likely to be of increasing importance in the attempt to understand developmental processes. It should be said initially that the results described below have often been obtained with considerable difficulty, owing largely to the relatively small quantities of tissue available for study. For this reason, the information that is in hand represents, for the most part, results of experiments performed on a distinctly micro scale.

For example, the first events in embryonic development begin at the stage of a fertilized ovum, usually a fairly small object if one wishes to study it singly. Often one does not want to do so and employs, instead, a large number of ova, all fertilized at the same time, a procedure that presents few difficulties in the case of organisms that lay external eggs. It is from such studies that much of the following information about the role of nucleic acids has been obtained.

Following completion of fertilization, the first major externally recognizable event is *cleavage* of the single cell into, first, two and, then, multiple cells (*blastomeres*). The volume of the embryo remains approximately the same during this period, so that the number of cells per unit volume and, therefore, the number of nuclei per unit volume must increase. This increase is reflected in a comparable increase in the amount of DNA, so that the DNA/nucleus ratio, quite predictably, remains constant. Synthesis of DNA may occur in part by means of mobilization of some of the cytoplasmic DNA contained in the ovum, but the large part of it must be synthesized *de novo* from low-molecular-weight precursors. The rapid synthesis of DNA during this period of development has been confirmed by studies that follow the uptake of radioactive precursors into nuclear DNA.

On the other hand, the synthesis of RNA during cleavage presents a very different picture. Ribosomal RNA does not appear to be made during this period, and it is significant that the nucleolus (see the preceding section), where rRNA is made, is undeveloped or lacking until the formation of the blastula. Following cleavage, rRNA synthesis rapidly reaches a considerable period of subsequent development. In contrast to rRNA, the protein coat, and the intracellular form, the *provirus*, which is effectively both messenger RNA (mRNA) and transfer RNA (tRNA) are synthesized during cleavage, although inhibitor studies mentioned below render their role somewhat unclear during this phase. For one thing, treatment of cells with actinomycin D (Figure 18-6), which inhibits DNA-dependent synthesis of mRNA, fails to prevent cleavage, so that any mRNA produced during cleavage could hardly be called essential. On the other hand, puromycin, an inhibitor of protein synthesis, does stop cleavage, suggesting that protein synthesis employing previously synthesized mRNA must play a vital role. It is interesting that such mRNA must have been present in the cytoplasm prior to fertilization and is therefore likely to be completely determined by maternal DNA.

If little growth is associated with rapid cell division during cleavage, the same comment applies during the following period of gastrulation. Changes that occur in the shape of the embryo are the results of deformation, not of net growth. On the other hand, gastrulation is a period of rapid protein synthesis and consequent synthesis of RNA, which must precede it. Indeed, it is quite clear that proteins synthesized at this stage are new proteins—that is, not identical to any already present in the blastomeres—since they may be detected as newly made antigens. The mRNA formed during the same period are likewise new, as should be expected from their specific role in the synthesis of particular proteins.

The gastrula stage is also characterized by a rapid increase in the rate of oxygen consumption by the embryo, an observation that has been made with sea urchin and amphibians. Since this phase of development includes rapid synthesis of new proteins and since protein synthesis requires prior synthesis of ATP, it is reasonable that oxidative phosphorylation should be carried out in an active manner. Oxygen consumption in gastrulae

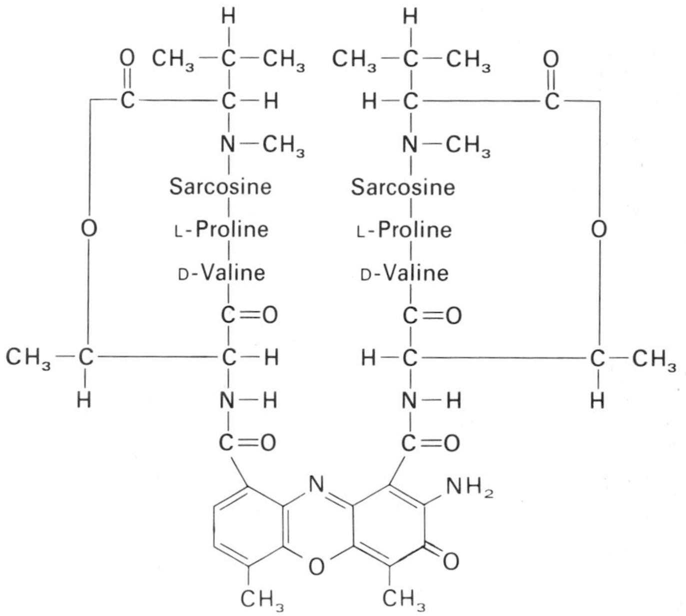

Figure 18-6 Actinomycin D.

has been measured under conditions where the substrate for oxidation was endogenous—that is, already contained in the cells—and it seems probable that the large part of oxygen uptake represents the oxidation of the breakdown products of cellular glycogen. That glycogen oxidation (via glycolysis and the citric acid cycle) is the main contribution to cell respiration in the gastrula stage of development is consistent with the independent observation of a decline in cell glycogen during the same period.

As development proceeds beyond the gastrula phase, additional biochemical and physiological changes are related to the specific organs and tissues that appear. For example, as muscle tissue becomes recognizable, the characteristic muscle proteins *actin* and *myosin* become detectable. Likewise, when red blood cells appear, so does the oxygen-carrying pigment *hemoglobin*. It is interesting that in human beings the hemoglobin that is synthesized in the embryo is of a different sort than that which finally predominates in the adult. Thus, the adult human hemoglobin molecule consists of four polypeptide units, two each of type α and β. Fetal hemoglobin consists of two α chains, while β chains are replaced with two γ chains, which have a primary structure different from the β chain. During fetal life, the circulating red blood cells contain entirely hemoglobin of the fetal sort, and soon after birth the fetal hemoglobin gives way to the adult form of the molecule, although trace amounts of the fetal molecule may persist through life. It should not be surprising that a different hemoglobin functions in the fetus where oxygen comes, not via the lungs, but across the placental barrier. In this connection, it may be significant that fetal hemoglobin is considerably more resistant to denaturation at extremes of pH than is adult hemoglobin, a property that may possibly be related to less developed regulation of the pH balance in fetal blood.

Suggested Reading

Books

Balinsky, B. I., *An Introduction to Embryology*, 3rd ed., W. B. Saunders Co., Philadelphia, 1970.

Bonner, J., *The Molecular Biology of Development*, Oxford University Press, Oxford, 1965.

Davidson, E. H., *Gene Activity in Early Development*, Academic Press, New York, 1968.

Ebert, J. D., and I. M. Sussex, *Interacting Systems in Development*, 2nd ed., Holt, Rinehart and Winston, New York, 1970.

Harris, H., *Nucleus and Cytoplasm*, Oxford University Press, New York, 1968.

New, D. A. T., *The Culture of Vertebrate Embryos*, Academic Press, New York, 1966.

Romanoff, A. L., *Biochemistry of the Avian Embryo*, John Wiley & Sons (Interscience Division), New York, 1967.

Waddington, C. H., *Principles of Development and Differentiation*, The Macmillan Co., New York, 1966.

Weber, R., *The Biochemistry of Animal Development*, vols. 1 and 2, Academic Press, New York, 1967.

Articles

Denis, H., "The role of messenger ribonucleic acid in embryonic development," in *Advances in Morphogenesis*, M. Abercrombie and J. Brachet (eds.), vol. 7, Academic Press, New York, 1968, p. 115.

Gordon, J. B., "Transplanted nuclei and cell differentiation," *Sci. Am.*, **219** (Dec., 1968).

Green, H., and G. J. Todaro, "The mammalian cell as differentiated microorganism," *Ann. Rev. Microbiol.*, **21**, 573, 1967.

Guidice, G., and V. Mutolo, "Reaggregation of dissociated cells of sea urchin embryos," *Advances in Morphogenesis*, M. Abercrombie and J. Brachet (eds.), vol. 8, Academic Press, New York, 1970, p. 115.

Spirin, A. S., "On masked forms of messenger RNA in early embryo-genesis and in other differentiating systems," *Current Topics in Developmental Biology*, **1**, 2 (1966).

Wessells, N. K., B. S. Spooner, J. F. Ash, M. O. Bradley, M. A. Luduena, D. L. Taylor, J. R. Wrenn, and K. M. Yamada, "Microfilaments in cellular and developmental prcesses," *Science*, **171**, 135, 1971.

Wessells, N. K., and W. J. Rutter, "Phases in cell differentiation," *Sci. Am.*, **220** (Mar., 1969).

Part V

The cellular environment

19 Human activities and the cellular environment

Since we are, by necessity, concerned only with cells that live their lives on the planet earth, we cannot avoid notice of the impact of the human species on this environment. Since that species has become notorious for contamination of the finite world with products of his civilization, the sort of impact that we now deal with is largely in the area of pollution. In other words, man produces a large number of materials that are toxic to cells, and in this chapter we consider the nature of some of this toxicity.

The decade of the 1960's has seen an enormous growth in the awareness, on the part of humanity, of the sort of problems that arise from the destruction, especially the chemical destruction, of the environment. Many chemicals have been identified as dangerous, or potentially dangerous, and many environmental questions have entered the area of politics even before they were fully accommodated in the area of science.

For example, large amounts of insecticides have entered the environment owing, largely, to the needs of agriculture and the drives to eliminate insect-borne diseases such as malaria. The amounts of pesticides became easily detectable in the general environment at a time when no one had any idea about the possible effects that these chemicals might have on organisms other than insects. That they should have some toxic effects on noninsect organisms is hardly a wild idea, since they were developed for their toxicity and since the specificity of any antiorganismal chemical is likely to have limits. In any case, these compounds reached relatively high levels in the world before toxic effects were noted, so that the difficulties of reversing the situation were (and are) very great.

There is no longer the slightest question about the generalized toxicity of many insecticides. A number of species of North American birds are likely to become extinct or to be exterminated in much of their range because of the toxic effects of DDT and similar materials. Since a number of these species are spectacular and beautiful (as in the case of the brown pelican or the peregrine falcon), and since one species is, in fact, the national symbol of the United States (which is, by far, the world's leading producer of pesticides), their demise is likely to be heralded by considerable public outcry. The clamor cannot necessarily be expected to influence the

441

course of events, coming rather too late in the process to have rapid effect.

The causal relation between pesticide levels and the decline in the fortunes of a number of species of birds and sea animals is quite clear, and a great deal of work has been carried out as to the mechanism of the toxicity. Thus, it is clear that one feature of the action of DDT and similar compounds on birds is a decline in the ability of the birds to successfully rear their young. This is the result of a defect in the production of egg shells. In some way, the ability of the birds to manufacture insoluble calcium, out of which shells are made, is impaired, and their eggs are thus enclosed in excessively thin and, therefore, fragile shells. Chicks are seldom hatched from such eggs and populations decline rapidly.

We see, then, that at least one toxic effect of at least one class of pollutant is understood to interfere with calcium balance in birds. Thus, the toxicity could be said to be most influential on the level of the organism or even its reproductive system, but this effect must surely reflect an influence of the chemical on the cells that make up the systems involved. From the standpoint of cell physiology, there must be a fundamental cellular toxicity that gives rise to the effects mentioned. As we shall see later in this chapter, the exact nature of the cellular effects is not well understood; there are some clues, however, and the cell physiologist should be aware of them.

The purpose of this chapter, then, is to identify some of the major classes of toxic compounds produced by human culture and to outline what is known of their action. We shall also consider the closely related matter of cellular methods for detoxification of harmful materials, in other words, how toxic substances can be rendered nontoxic. The goal in these discussions, apart from the obvious one of imparting something of the subject matter involved, will be to emphasize the importance of combining a cellular and an ecological approach to biology. The idea that these fields are far apart in method and concept seems to us to be damaging and anachronistic, and we would encourage the reader to view cell physiology in as wide a context as possible.

Cellular detoxification mechanisms

Detoxification mechanisms are sets of enzyme-catalyzed reactions that render toxic substances either less toxic or more readily removed from the region of cells and tissue where they produce a harmful effect. In general, these reactions occur in the microsome fraction of cells, and some tissues appear to be more active in this respect than others. For example, liver tissue is the primary locus in mammals of many of the reactions that have been studied. Detoxification has been most extensively studied with relation to drugs, a choice of material that is of obvious practical significance. If, for example, a drug is rapidly converted to an inactive form in the body, this fact must be taken into account. Moreover, if a given drug is not metabolized at all, it may persist in the body for a certain length of time, a situation that makes its administration highly dangerous.

Relatively little is known about the mechanisms whereby pesticides may be detoxified in nontarget organisms. Indeed, if there were active and recognizable pathways for their inactivation, a number of species would clearly be less endangered than is presently the case. For this reason, it will be convenient for us to consider detoxification in a rather general way and to illustrate some of the tactics that cells employ in ridding themselves of toxic substances.

Detoxification by binding

Perhaps the least specific, but often most important technique for limiting the harmful effects of many foreign compounds is by binding the compound in question so that it cannot act upon the cell. For example, addition of a small amount of a lipid-soluble insecticide, such as DDT, to an intact organism may produce no effect, owing to the binding of the chemical in the fat storage depots of the animal. Subsequent additions of the chemical to the animal may produce no effect—until the fat cells can hold no more, at which time the toxic effects begin to be noted. If the organism is, in fact, capable of metabolizing the compound in question, then the binding may represent an effective way of sequestering it until metabolized. Unfortunately, many pesticides are the problem they are because the rate of metabolism or elimination from the cell is so low that the binding locations are easily overwhelmed. There are also a number of proteins that appear to exist primarily as binding agents. For example, albumins of the circulating blood of vertebrates bind large quantities of fat-soluble chemicals, including many of known cellular toxicity. An important function of such proteins is probably to sequester fatty acids (many of which are quite toxic), keeping their concentration in the unbound form low in the circulating plasma. The binding properties of serum albumins are such that these proteins have long been used in physiological experimentation to reverse the effect of some lipid-soluble inhibitory reagent. For example, mitochondrial or chloroplast electron transport is inhibited by the addition of small amounts of nonyl-hydroxyquinoline-N-oxide,

and this inhibition can be completely reversed on addition of serum albumin, which removes the reagent from solution.

Finally, there is a generalized binding in cells that is probably of importance in reducing the impact of a foreign toxic substance. Thus, there are many compounds that harm cells, because they react with

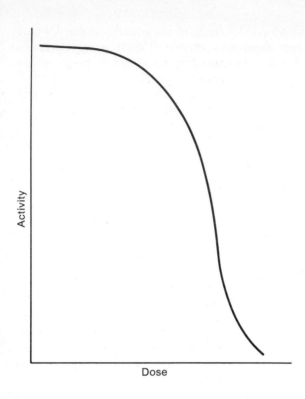

Activity

Dose

Figure 19-1 A sigmoid dose-inhibition curve. Biological activity of whatever sort being measured is unaffected by small additions of a substance, owing to the existence of multiple sites of interaction (see text). Note that other mechanisms can underlie this sort of curve. For example, in some instances, the addition of the first increments of a substance render a biological system more sensitive to subsequent doses. Such an occurrence is often observed in the case of allosteric enzymes (see Chapter 4).

sulfhydryl groups located on essential enzymes. The effect of these compounds, at least at low levels, may be somewhat diminished by the large number of sulfhydryl groups that exist in a cell but do not represent critical activities. If a sizable number of the noncritical binding sites bind more tightly to the inhibitory compound than does the critical binding site, a measure of protection will occur until the concentrations become excessively high. The effect of this sort of multiple-site situation will be to produce a *sigmoid dose-inhibition* curve, whereby the effect of the compound is measured in a quantitative fashion (see Figure 19-1). In the first part of the curve, nothing happens on addition of the substance, since it is being preferentially bound to the noncritical loci; when these are *saturated*, then further addition begins to produce its characteristic effect.

Enzymatic detoxification

Chemical modification of a compound to render it less toxic (or more easily excreted) may occur in a number of ways. The change must be great enough to alter the reactivity of the compound or to change other properties, such as solubility or the ability to traverse membranes, in a major way. For example, the functional group responsible for its harmful effects may be removed, altered, or blocked by any one of a number of chemical means.

Probably the first case of detoxification to become known was the reaction whereby benzoic acid is coupled to the amino acid glycine to form *hippuric* acid (whose name should give some indication where it is to be found):

$$\text{C}_6\text{H}_5-\overset{\overset{\displaystyle O}{\|}}{\text{C}}-\overset{\overset{\displaystyle H}{|}}{\text{N}}-\text{CH}_2\text{COOH}$$

The closely related phenylacetic acid is also excreted as a compound with glycine in many animals, although, in man, it forms the glutamine derivative, instead:

$$\text{C}_6\text{H}_5-\text{CH}_2-\overset{\overset{\displaystyle O}{\|}}{\text{C}}-\overset{\overset{\displaystyle H}{|}}{\text{N}}-\overset{\overset{\displaystyle COOH}{|}}{\text{CH}}-\text{CH}_2-\text{CH}_2-\overset{\overset{\displaystyle O}{\|}}{\text{C}}-\text{NH}_2$$

Thus, the tactic of conjugating toxic substances to amino and other acids is often encountered in cellular self-protection. An additional example is the reaction of phenol with sulfuric acid to form phenylsulfuric acid:

$$\text{C}_6\text{H}_5-\text{OSO}_2\text{OH}$$

A second type of chemical modification that is of importance in detoxification is oxidation. For example, primary alcohols, many of which are pollutants in one sense or another, are detoxified by oxidation to form the corresponding acid, or even, in the case of ethanol, carbon dioxide and water. Methanol ("wood alcohol") is oxidized to formate, which is, itself, toxic and is very slowly metabolized or eliminated. In some cases, detoxification mechanisms backfire, by rendering toxic substances even more harmful. For example, fluoroethanol, which is an industrial solvent, and, hence, a pollutant, is "detoxified" by oxidation to fluoro-acetic acid, which is highly toxic, owing to its inhibition of the aconitase reaction of the citric acid cycle. In this case, the failure of the tactic results from an insufficient specificity of the enzymes involved; the introduction of the fluorine atom into the molecule does not prevent action of the enzyme, but leads to the production of a metabolic inhibitor. Obviously, cells did not evolve under conditions where fluorinated alcohols played an important role. Unfortunately, cells also did not evolve under conditions that rendered it advantageous to carry on active metabolism of a number of compounds, including insecticides and herbicides, which represent the ingenious efforts of synthetic organic chemists.

Hydroxylation reactions appear to be of importance in the detoxification of many substances. The site of such reactions appears to be the endoplasmic reticulum (isolated as the microsome fraction of cells). The

445

mechanisms of these reactions are often complex, requiring enzymes called *hydroxylases*, as well as heme and nonheme iron proteins, and flavin and $NADP^+$ coenzymes. Hydroxylases are also called *mixed function oxidases*, since they utilize one atom of molecular oxygen (O_2) in the hydroxylation and oxidize a reduced coenzyme (for example, $NADP^+$) with the other. Compounds that are oxidized by such enzyme systems include steroids, a number of drugs, and various other aliphatic and aromatic compounds.

Finally, although all the examples of detoxification mechanisms that we have given have involved organic molecules, we should point out that there are instances where inorganic ions and molecules are specifically detoxified by enzymic mechanisms as well. For example, hydrogen peroxide, which is extremely toxic to cells and a mutagen, is broken down in a reaction catalyzed by the iron-heme protein *catalase*:

$$2\,H_2O_2 \;\longrightarrow\; 2\,H_2O + O_2$$

Cyanide and cyanide derivatives are extremely toxic through their inhibition of cell respiration. They are, however, transformed to the much less toxic thiocyanate radical by a replacement reaction between cyanide and thiosulfate, a reaction catalyzed by the enzyme *rhodanase*:

$$SSO_3{}^{2-} + CN^- \;\rightleftharpoons\; SO_3{}^{2-} + SCN^-$$

Effects of some inorganic pollutants on cells

A great deal of attention is presently centered on the occurrence of lead as a major pollutant in liquid and atmospheric spaces surrounding human activity. Lead finds its way into waters as an industrial waste and into the air in the form of tetraethyllead from unburned gasoline. Moreover, lead is included in many paints and is often taken in by children who chew on painted objects. Lead toxicity in humans takes a number of forms, including neuromuscular weakness and gross mental aberrations. In the human body, lead can be regarded, in part, as a calcium analog since it is incorporated in bone in the form of insoluble phosphate salts. Tetraethyllead, which is a major environmental form in industrialized areas of the world, appears not to be toxic itself, but it is converted in the liver to triethyllead, which is very harmful to nervous tissue.

When lead is absorbed in a vertebrate, it is transported to the tissues in close association with the red blood cells; consequently, a great deal of effort has gone into studying its effect on the formation and stability of red blood cells. It turns out that lead poisoning leads to a slight anemia (low red cell level), which is associated with a short life span for red cells. For example, when human red cells are labeled with the radioactive tracer ^{51}Cr, their half-life is a minimum of about 30 days and, in the case of lead poisoning, about 20 days. Isolated red cells appear to be somewhat more fragile than normal when confronted with mechanical stimuli, a defect that may underlie their altered half-life. It also appears that red cell production is inhibited, and this inhibition, in turn, appears to represent the action of lead on several steps leading to the synthesis of hemoglobin.

The action of lead on hemoglobin synthesis turns out to be quite complicated, as there are several steps in heme synthesis that are affected and synthesis of the protein part of hemoglobin, globin, is also diminished. Although we need not go into the details of heme synthesis here, it is possible to divide the pathway into several parts as outlined below, and evidence has been offered suggesting that each part is, indeed, affected by lead intoxication:

$$\text{Glycine} + \text{succinyl}-\text{S}-\text{CoA} \quad \rightarrow \quad \gamma\text{-Aminolevulinic acid}$$
$$\downarrow$$
$$\text{Heme} \quad \leftarrow \quad \text{Protoporphyrin III} \quad \leftarrow \quad \text{Porphobilinogen}$$

Mercury contamination of the environment has also become a serious problem, and recent years have seen certain inland waters become so rich in mercury as to preclude the sale of fish caught there. Mercury enters the environment as the metal and in the form of various organic mercuric compounds. Important sources of contamination include the effluent of a number of industries, notably the paper industry, as well as the use of organic derivatives as fungicides.

Chronic mercury poisoning produces damage in most tissues of the organism but is especially harmful to the nervous system, kidneys, and digestive system of vertebrates. Examination of cells from these regions in cases of mercury intoxication reveals distinct degenerative changes, including extensive formation of vacuoles, often an indication of something amiss with cells.

The basis for the harmful effects of mercury is associated in many instances with its ability to form complexes with the sulfhydryl groups of proteins and other compounds. This binding is very tight in the cases of organomercuric compounds, some of which, including p-chloromercuribenzoate,

$$\text{Cl}-\text{Hg}-\underset{}{\bigcirc}-\text{COO}^-$$

are important reagents in the study of enzymes in their ability to test for the importance of sulfhydryl groups in enzyme mechanisms (see Chapter 4). Since there are large numbers of enzymes in the cell that require a free sulfhydryl group for their function, it is not possible to pinpoint the action of mercuric compounds on one, or a few. The affinity between mercury and sulfhydryl groups also provides a technique for the treatment of mercuric toxicity, since the introduction of such a compound as dimercaptopropanol,

$$
\begin{array}{l}
CH_2-SH \\
| \\
CH-SH \\
| \\
CH_2OH
\end{array}
$$

(also called British antilewisite, and a respiratory inhibitor in its own right) tends to sequester the free mercury in the blood stream and leads to a net removal of the metal from the tissues.

A number of other metals are toxic to cells and are placed into the environment as industrial waste, although in most cases, the exact nature of their toxicity is not fully known. One example of a metal that is toxic and, in some cases occurs naturally at concentrations that are harmful, however, is selenium, which occurs in potentially dangerous amounts in the soil of parts of the central region of the United States. Selenium exerts its effect by becoming incorporated in place of sulfur in amino acids, such as cysteine and methionine, and leading, thereby, to the synthesis of proteins with altered properties. The symptoms of this sort of poisoning can be alleviated by an interesting variation of a natural detoxification mechanism. Striken animals (or people) may be given bromobenzene that is normally detoxified by linkage to the acetylated derivative of the amino acid cysteine. In this case, the cysteine used is the selenium compound, so that the complex that is excreted contains selenium, which is thereby eliminated from the body.

Effects of insecticides and herbicides

Insecticides are, of course, designed to be toxic to insects, so that any effect that they have on other forms of life should not be surprising. Indeed, the chemical industry that manufactures these materials is correctly fond of pointing out that many of these compounds are remarkably specific in producing death to insects at concentrations where they do not produce noticeable harm to other forms. The problem is, of course, that some of the harm to other species may require time to become noticeable. For instance, the effects of DDT on certain species of bird did not become clearly recognized until the birds were on (or beyond) the brink of extinction. Furthermore, many insecticides are novel enough in structure as to be degraded only very slowly by microorganisms, so that they tend to build up in the environment until present in amounts dangerous to organisms other than insects. Finally, most of these compounds are fat-soluble, which means that they tend to become concentrated in the lipid of organisms rather than being randomly distributed in the environment. This being the case, they also tend to be especially concentrated in the case of predatory animals whose intake of contaminated fats is necessarily high. Thus, among birds, the hawks are more likely to be harmed than are birds whose diets are largely seed. Likewise, flesh-eating marine animals exhibit high levels of lipid-soluble insecticides, a notable example being (alas) the lobster.

448

One of the most effective and inexpensive of insecticides is DDT, 2,2-bis(*p*-chlorophenyl)-1,1,1-trichloroethane, whose structure is shown below:

$$Cl \overbrace{\hspace{1em}}^{} \quad \underset{|}{\overset{H}{C}} - \underset{|}{\overset{Cl}{C}} - Cl$$

This chlorinated hydrocarbon is highly lethal to a wide spectrum of insects and has played an immense role in the increase of food crops and the eradication of insect-borne diseases, such as malaria. The exact nature of its mechanism of action is not yet understood, but the picture that seems to be emerging is that the compound acts at the level of cellular membrane systems. This conclusion is consistent with the high-lipid solubility of DDT, which enables it to become concentrated in the lipid-rich membranes of cells. Thus, in insects, one aspect of DDT toxicity appears to be the inhibition of normal function of the excitable membranes of nerve cells. In addition, DDT has been shown to interfere with oxidative ATP synthesis in mitochondria isolated from insects, and it will be recalled that such phosphorylation is closely associated with the mitochondrial inner membrane. Finally, there is some evidence that DDT, at least at higher concentrations than normally employed to kill insects, prevents photosynthetic ATP synthesis in green plants, a function associated with chloroplast membranes.

The harmful effects on nontarget organisms are, likewise, not fully understood on the cellular level. It is possible that the failure of birds to form normal eggshells may reflect an aberration in the transport of calcium across cellular membranes (a process seen in Chapter 6 to be closely associated with oxidative ATP synthesis). It is also possible that the decline in reproductive efficiency that has been noted in the case of a number of species can be attributed to some defect in membrane function, although there is presently only the most indirect kind of evidence relating to this.

In fact, both the effects of DDT on calcium mobilization and on the efficiency of reproduction appear to be related also to the action of steroid hormones. For example, birds to whom DDT has been administered exhibit a lowered concentration of the steroid *estradiol*, which functions both in the regulation of calcium deposition and in maintaining the condition of the sex organs. One possible consequence of this alteration is the observation that activity of the enzyme *carbonic anhydrase* is also lowered in oviducts when birds are fed DDT or related compounds. This lowering may account for the inability to form normal eggshells. The reaction catalyzed by the enzyme is

$$CO_2 + H_2O \rightleftharpoons H_2CO_3$$

which is believed to function in eggshell formation by providing carbonate used in the deposition of insoluble calcium carbonate. The action of the

449

pesticides appears to be inhibition of the enzyme (and not prevention of its synthesis), since they also inhibit carbonic anhydrase from human erythrocytes.

Even low levels of DDT (and related compounds) administered to rats give rise to changes in the liver cells which seem to be specific for such insecticides. For example, dosages of DDT that produce no detectable physiological harm nonetheless lead to an increase in the smooth endoplasmic reticulum of liver cells as well as to the occurrence of small lipid-inclusion bodies in the cells. These effects, which are visible in the electron microscope, appear to be related to the preparation on the part of the cells to carry out detoxification reactions. It is significant that similar doses of insecticides also lead to the induction of several enzymes located in the endoplasmic reticulum and associated with drug detoxification reactions.

Detoxification of DDT appears to fall into three categories, the reactions belonging to them having been demonstrated in a variety of organisms. In the first place, cells from rat liver, yeast, and bacteria can carry out the *reductive dehalogenation* of DDT to yield a molecule in which one of the chlorine atoms is replaced by hydrogen. Second, enzymes have been described in DDT-resistant insects that carry out *dehydrohalogenation* of DDT, resulting in the removal of one chlorine atom, one hydrogen atom, and the resultant formation of a double bond. This reaction is carried out by a number of other organisms, but at rates several hundred-fold lower than in the case of DDT-resistant insects. The enzyme responsible for dehydrohalogenation appears to occur in cells unlinked to the cellular membranes. Finally, mammalian liver cells detoxify DDT by a *hydroxylation* reaction that forms an alcohol:

2,2-bis(*p*-chlorophenyl)-1,1,1-trichloroethane $\rightleftharpoons$

$\qquad$ 2,2-bis(*p*-chlorophenyl)-1,1,1-trichloro-2-ethanol

This reaction is carried out by a mixed function oxidase from the endoplasmic reticulum of the sort described earlier in this chapter.

A second class of important insecticides that exert known effects on cells is that of the organic phosphates. Of these, an important example is TEPP (tetraethyl pyrophosphate),

$$CH_3CH_2 \diagdown \underset{\displaystyle \underset{O}{\diagup} \; \underset{\displaystyle \overset{\|}{P}}{\diagdown}}{O} \; O \quad CH_2CH_3$$

Compounds of this sort have an important advantage when compared to the chlorinated hydrocarbons, in that they are fairly rapidly degraded by the action of microorganisms in the environment. They are thus classed as *nonpersistent* and their use is, thus, less prejudicial to the

450

ecological future. On the other hand, these compounds are extremely toxic to animals with nervous systems and, although insects are particularly sensitive, higher concentrations can be lethal to many other organisms, including man. Indeed these compounds are closely related to other agents that have been developed specifically for their toxicity to man; they are called *nerve gases*.

Organic phosphate insecticides (and related military nerve gases) act on the cellular level by inhibiting a class of enzymes called esterases, most importantly, *choline esterase*, which plays a vital role in nerve transmission at synapses. The action of these inhibitors is the result of the formation of a covalent enzyme-substituted phosphorus compound, which is hydrolyzed only very slowly. The formation of this compound is, in a sense, analogous to the formation of the enzyme–substrate complex, with which it interferes.

The action of some herbicides

A final category of chemicals that have been recently introduced into the world ecosystem in large amounts is the herbicides, used mainly as "weed-killers" and (unfortunately) as military defoliants. In view of the dependency of all life on earth upon the green plants, any major increase in the levels of herbicides in the environment should be (and is) viewed with alarm. From the human point of view, a major harmful effect of some of these compounds is the induction of abnormal development in human fetuses, and some of these compounds (as well as some insecticides) have also been suggested as possible carcinogens and mutagenic agents.

The mechanism of action of a number of herbicides appears to be understood, and, in several cases, they are specific in killing plants by virtue of their inhibition of a process unique to plants, chloroplast photosynthesis. For example, diuron (dichlorophenyl-1,1-dimethylurea)

is effective as a herbicide through its inhibition of noncyclic photophosphorylation in chloroplasts (see Chapter 7). This compound also exhibits a number of related effects, including lowered formation of carbohydrates, diminished ability to transport phosphate, and inhibition of electron transport. In the case of plants that are relatively insensitive to this compound, a detoxification mechanism occurs, which leads to a demethylation of diuron to yield the relatively harmless dichlorophenylurea,

A number of other herbicides also interfere with photosynthetic reactions. In addition, some are apparently specific inhibitors of respiration in plant mitochondria, an example being Picloram (4-amino-3,5,6-trichloropicolinic acid). Finally, a herbicide that is very widely used and that produces a bewildering array of effects on cellular processes in sensitive organisms is 2,4-D (2,4-dichlorophenoxyacetic acid),

This compound, together with a number of closely related ones, has been shown to inhibit glycolytic enzymes, uncouple oxidative phosphorylation, and stimulate protein synthesis in sensitive plants. The latter stimulation may be most fundamental to the action of this compound, which appears to represent a synthetic analog of naturally occurring plant growth hormones (*auxins*), of which a well-known example is indoleacetic acid:

It appears that the destructive effects of a synthetic auxin such as 2,4-D may represent the stimulation of unregulated growth in sensitive plants, which, in turn, leads quickly to their death. Binding of 2,4-D, as well as natural auxins, to cytoplasmic and nuclear proteins and to chromosomes has been demonstrated. The stimulation of protein synthesis noted with this compound appears linked to a stimulation of the synthesis of messenger RNA as well as ribosomal RNA, which, in turn, may represent a release of control mechanisms that normally place bounds on these important processes.

Suggested Reading

Books

Edwards, C. A., *Persistent Pesticides in the Environment*, Chemical Rubber Co., Cleveland, 1970.

Harris, J. W., *The Red Cell: Production, Metabolism, Destruction—Normal and Abnormal*, Harvard University Press, Cambridge, Mass., 1963, Ch. I.

Miller, M. W., and G. G. Berg (eds.), *Chemical Fallout: Current Research on Persistent Pesticides*, Charles C Thomas, Springfield, Ill., 1969.

O'Brien, R. D., *Insecticides*, Academic Press, New York, 1967.

O'Brien, R. D., and I. Yamamoto, *Biochemical Toxicology of Insecticides*, Academic Press, New York, 1970.

Phillips, J., *Environmental Health—A Paradox of Progress*, William C. Brown Co., Dubuque, Iowa, 1971.

Report of the Subcommittee on Environmental Improvement, American Chemical Society, *Cleaning Our Environment—The Chemical Basis for Action*, Washington, D.C., 1969.

West, E. S., W. R. Todd, H. S. Mason, and J. T. van Bruggen, *Textbook of Biochemistry*, 4th ed., The Macmillan Co., New York, 1966, Ch. 32.

Articles

Ames, P. L., "DDT residues in the eggs of the osprey in the Northeastern United States and their relation to nesting success," *J. Appl. Ecol.*, **3**, 87 (1966).

Bitman, J., H. C. Cecil, and G. F. Fries, "DDT-induced inhibition of avian shell gland carbonic anhydrase: a mechanism for thin eggshells," *Science*, **168**, 594 (1970).

Butler, P. A., "Pesticides in the Marine Environment," *J. Appl. Ecol.*, **3**, 253 (1966).

Hickey, J. J., and D. W. Anderson, "Chlorinated hydrocarbons and egg shell changes in raptorial and fish-eating birds," *Science*, **162**, 271 (1968).

Peakall, D. B., "*p,p'*-DDT: effect on calcium metabolism and concentration of estradiol in the blood," *Science*, **168**, 592 (1970).

Coda

20 Cell investigations in the future

It is hoped that the reader will have come from the previous 19 chapters possessed of the idea that cell physiology (1) has a considerable way to go in the understanding of cellular and subcellular events and (2) will probably employ a very wide array of techniques in approaching and understanding these events. It should be fairly clear that cells are difficult to study because of their small size and because of the complexity of what goes on in them. For this reason, students of cellular affairs have always been very open-minded about using any methods that have any possibility of advancing their cause. Students of the cell should not (and generally do not) worry about whether they are employing the techniques of the traditional physiologist, those of the biochemist, or those of the biophysicist. Things can be difficult enough already, without worrying about spheres of influence.

In this chapter we have set out for ourselves the formidable task of predicting something about the ways in which the field will develop over the next period of years. This is a fool's errand in the sense that we cannot really have much idea about the future, so rapidly are things changing. It is, on the other hand, a very important exercise in the sense that we can thus emphasize the unfinished nature of our collective task and can suggest the wide basis of preparation that is probably required for a student *now* so that he can be an effective physiologist a decade, or more, *hence*.

An important difficulty in prediction is the lack of information about the techniques that will be available in the future. Obviously, our present techniques and apparatus will be continuously refined, and new approaches will branch off from existing ones. On the other hand, completely new methods are continuously falling into the lap of the physiologist from unlikely sources, and only the prepared mind is often able to grasp their utility and promise. For example, military research on radar during the Second World War could not have been predicted to lead to benefits in the study of the cell, but in fact it was of great indirect importance in leading to the exploitation of magnetic resonance techniques through the development of microwave electronics.

Not only do new methods of answering questions become continuously available, but the sorts of question being asked become constantly refined and altered. In some instances, the sorts of questions about the cell that make sense and lead to useful experimental work change abruptly. For example, the introduction of important theories has often completely changed the nature of areas of interest. The increasing certainty that nerve transmission was based on ionic movements across membranes led to a complete revision of one's thinking about the whole field, and physiologists who were unable to adapt to the new approach were simply left behind. Likewise, the postulation of the *chemiosmotic* approach has quite changed the nature of creative thought about energy coupling in the cell. Those who have been unable to understand the implications of this point of view appear doomed to increasing triviality and isolation from the theory in their own field. Thus, one can only exhort students of cell biology to keep an open mind and retain as many options as possible.

With regard to options that should be kept open, it seems worthwhile to note that cell physiologists have not gone very far in the direction of exploring many of the experimental options that exist. Most of our knowledge about the operation of the cell has come from relatively few types of cell in relatively few sorts of organism. The reason why a physiologist works with a much studied organism should be clear enough: He can usually get directly to the subject of interest to him without having to obtain too much background information. Thus, when one looks for an experimental animal, the laboratory rat is terribly tempting, since so much is known about it already. One is much less likely to turn to a more obscure animal (the bandicoot, for example) because it presumably presents many more areas where little knowledge is available. It may also be considerably more difficult to obtain. Our knowledge of mitochondria has come from livers and heart of rats and cattle, largely because the former can be obtained or bred in a simple manner and the latter obtained at an abattoir. Little or nothing is known about mitochondria from the liver of the grizzly bear (because the bears do not give up their livers easily), but that is no reason to believe that their livers are less interesting than those of rats.

For this reason, it may be argued that future developments in cell physiology can be expected to come, in part, from the study of a wider variety of cells. There are undoubtedly many surprises ahead for physiologists who examine a wider selection of material. The variety of the selection is itself important in providing a wider basis for what might be termed *comparative cell physiology*. We are clearly at the threshold of interesting developments in this area but have only begun to touch the surface. One can imagine that the next decades will see much of physiology providing information about evolutionary matters much in the way anatomy supported (and continues to support) the study of evolution. It is interesting that, while there have been cell physiology and biochemistry journals around for a century or so, journals of comparative physiology and biochemistry are recent developments.

As cell biologists develop more catholic tastes in their choice of experimental material, so are they ranging more widely in their experimental tactics. We have said that one should await and adapt eagerly to new techniques. It is also true that more and more investigators have become freed from any one technique or type of technique and have ranged as widely as productivity requires. For example, more and more biologists have stopped worrying about whether they were morphologists or physiologists and have addressed problems from as wide a standpoint as possible. No more do many people worry about whether they are physiologists or biophysicists, being mindful perhaps that these terms are such a matter of fashion that last year's biophysicist is today's physiologist. When one encounters investigators who use the term *molecular biologist* as one of praise or condemnation, the chances are that they are more interested in scientific politics than in science itself. The common experience is that life is too short to worry about identifying with a particular label. It is also too short to allow one to identify too strongly with a particular technique, and the trend, as we have said, seems to be rather in the direction of freedom to follow one's inclination and curiosity.

We promised at the beginning of this chapter to discuss areas of cell physiology that appeared to us to be particularly susceptible to major advances in the future. We have already mentioned comparative cell physiology as one example of this sort of thing. In adding to the list, it is important to note first that predictions are a matter of taste and that these predictions reflect the prejudices of the present author only. Any other writer would, no doubt, suggest a different list, and the suggestions made below should not be taken as gospel. We also note that one's ability to predict this sort of thing is strictly limited by the natural tendency for scientific areas to develop in a somewhat noncontinuous manner, while the insights of one or several people sometimes rapidly alter the whole direction that the field takes. For example, in the history of chemical thermodynamics, men such as Gibbs and, more recently, Onsager obtained insights of such importance that they changed the whole direction of the field. Similarly, in cell and molecular biology the names of Watson and Crick (DNA), Bernstein (nerve), Hodgkin and Huxley (muscle), and Hill (also, muscle), all come to mind. So, in contemplating future directions in cell physiology, let us keep in mind the likelihood that investigators such as the above will upset them anyway.

With these disclaimers in mind, we can argue that the field of cell physiology is at the threshold of important advances with regard to the relationship between the cell and the action of hormones. The discovery of cyclic AMP as an amplifier of hormonal messages when they reach the cell surface (see Chapter 8) has led to a rapid development of understanding of the action of a number of hormones. It is safe to say that we are in the beginning phase of this study and that important developments lie ahead. This area is an example of the more general topic of the relationship between the cell, considered in isolation, and the cell in the intact multicellular organism. This whole area appears to be one where relative ignorance prevails and where we now have the basis for important advances. The ability to study mammalian cells under culture conditions is proving to be of great importance in understanding the interactions between cells. For example, the regulation of growth in a multicellular organism appears to be reflected in contact inhibition, where cultured cells stop dividing when they make sufficient contact with each other. In other cases, we have begun to understand the process whereby nerves transmit impulses (although it must be said that our understanding is somewhat fragmentary, especially with respect to the gating mechanism). We are, however, very little able to conceive of the manner of operation of the human brain and central nervous system as well as the action of drugs on nervous tissues and most pathological conditions. We are, therefore, presently unable to make the transition from the nerve cell to the collection of cells that operate as a system. It is interesting (and highly promising) that a number of physiologists have turned to the study of very primitive, invertebrate nerve nets. In these relatively simple systems, there appears to be the possibility of coming to understand the intercellular interactions that underlie the whole of neuronal coordination.

We also appear to be at the point of beginning to understand a number of diseases that act at the cellular and molecular levels. For instance, a few human diseases have come to be understood during the last decade or so as resulting from congenital defects in the synthesis of specific enzymes. Such metabolic diseases include the porphyrias, glycogen storage diseases, and phenylketonuria. These diseases, when correctly understood, are susceptible to treatment that consists of minimizing the impact of the defect by special diets and other tactics. There has been a great deal of noise in the public press about the possibility of repairing such genetic defects by somehow adding the normal gene (in the form of DNA) to the defective cells. While the press appears to consider such "molecular surgery" as imminent, it would appear to be a much more remote possibility but, nonetheless, a possibility. In the near future, we may expect to see the list of diseases that are understood on a cellular and molecular level become much more extensive, possibly including some pathological conditions of much more widespread occurrence. As we said in an earlier section (Chapter 14), the increasingly successful application of cell biology to human problems will, no doubt, turn up information of value in the study of "pure" biology as well.

It has become popular to speak of molecular biology as a field where all the great questions are answered. In this instance, we use the term *molecular biology* in its most narrow context, that is, the study of nucleic acids and their action. Indeed, with the discovery that DNA is the genetic material, together with the actual "cracking" of the genetic code in terms of a mapping relationship between nucleotide triplets and amino acids, one might feel that the big questions were answered and that only details remain. Apparently, many investigators who call themselves molecular biologists feel much the same way, as they are presently abandoning the field in droves. Many have migrated to the study of nerve systems, where they, with good reason, feel that their talents can be applied to "big" questions. Be that as it may, there appear to be questions of sufficient importance left in molecular biology to keep everyone busy and happy. For instance, the discovery of the very small RNA bacterial viruses has led to the possibility of a complete understanding of a single genetic system, a prospect engendering considerable excitement.

While the physical chemistry of solutions of DNA is quite well understood, a number of features of the role of DNA *in vivo* remain unclear. For instance, the mechanism by which double-stranded DNA unwinds prior to replication is not clear, especially with respect to the energy required to break the very large number of interchain hydrogen bonds. The relationship of DNA to the eukaryotic chromosome is, likewise, not completely understood, and molecular biologists are actively investigating this matter. There appears to be no shortage of interesting questions even in the much picked-over field of nucleic acid chemistry and biology.

One could extend the list of active and soon-to-be active areas of cell biology. We are clearly a short distance in time from beginning to make important strides in the understanding of membranes. Recent experiments on the reconstitution of membranes that retain some of their biological properties appear to be pointing the way to significant advances. When more information is in hand regarding the structure and function of biological membranes, we will be better able to understand their regulatory role, which is presently regarded as important but which is not very well understood.

Thus, there appears to be no end to the thresholds across which the cell physiologist may step if he so chooses. The only limitations are the vigor of his imagination and the clarity of his understanding of the subject matter at hand. The only pitfalls are a failure of courage in advancing new explanations and a failure of candor in rejecting ideas that are not consistent with observations. In all of this, the rewards are immense, since so much beauty exists in the cellular world. One need only ask the question:

"Always the beautiful answer who asks a more beautiful question"

(*E. E. Cummings*)

Index

A

A-band, 319
Absorption spectra, 136–137, 141, 250
Accessory pigments, 189
Acetabularia, 394
Acetaldehyde, 148
Acetate, 60
Acetic acid, 60
Acetoacetate, 117
  in bacterial fermentation, 236
Acetoacetyl—S—ACP, structure, 218
Acetolactate, structure, 221
Acetylcholine, structure, 314
Acetyl—S—CoA, 149
  from fatty acid oxidation, 183
  in gluconeogenesis, 211, 213
  from glycolysis, 148–149
  in glyoxylate cycle, 155–156
  Krebs cycle breakdown, 150–152
  in lipid biosynthesis, 217
Acid bath experiment, 198
Acids, 59–62
  Lewis, 64–65
ACP, *see* Acyl carrier protein
Acrasieae, 422
Actin, 324, 326, 436
Actinomycin D, 242, 435
  inhibition of DNA synthesis, 242
  inhibition of mitosis, 360
Action potential, 310–313
  biphasic, 311
  ionic basis, 313–315
  in nerve axon, 311
Active transport, 270, 278–280
  energy source, 280–282
Actomyosin, 327
  ATP binding, 321
Acylcarnitine, 183
Acylcarnitine shuttle, 183
Acyl carrier protein, 217–218
Acyl—CoA, 182, 183
Adenine
  in base pairing, 371
  structure, 370
Adenosine diphosphate
  from ATP hydrolysis, 146, 179
  in ATP synthesis, 165–167, 174–175

phosphofructokinase stimulation, 160–161
  in respiratory control, 176–177
  structure, 165
Adenosine monophosphate, 165
  in amino acid activation, 383
  cyclic, *see* Cyclic AMP
  in flavin adenine dinucleotide, 171
  in gluconeogenesis, 212
  in luciferin oxidations, 203
  from myokinose reaction, 166
  in the Pasteur effect, 160–161
  in urea cycle, 157
Adenosine-3′,5′-monophosphate, structure, 392
Adenosine triphosphatase, 179
  in the chemiosmotic hypothesis, 291–296
  in cytoplasmic motion, 339
  in flagella contraction, 332–334
  knobs, 264, 267, 298, 427
  quaternary subunits, 103
  sodium–potassium, 287, 302, 304, 310
  transport linked, 286–287
Adenosine triphosphate, 145–146
  in active transport, 279, 280–281
  in actomyosin ATPase reaction, 321, 327
  in amoeboid movement, 341
  in biosynthesis, 207–208
  and chemiosmotic coupling, 292–295
  control of synthesis, 176–177
  in cytoplasmic motion, 338–340
  as energy source for contraction, 328–330
  from fatty acid oxidation, 183
  feedback inhibition, 143, 160–161
  in flagella contraction, 333
  in formation of action potential, 313
  free energy of hydrolysis, 90
  in gluconeogenesis, 210–212
  from glycolysis, 143, 146–147, 148, 150, 235–237
  as high energy compound, 145

hydrolysis, 89, 90, 146, 179, 327
  from Krebs cycle, 152–153, 169
  in luciferin oxidation, 203
  measurement, 140–141
  mitochondrial synthesis, 169
  in muscular contraction, 320–330
  in nitrogen fixation, 205
  oxidative synthesis, 174–176
  in photosynthetic CO_2 fixation, 200
  in resting potential formation, 310
  in reversal of respiratory chain, 178
  structure, 144, 165
  synthesis, 165–167
  synthesis by coupling factors, 179–180
  synthesis by photosynthesis, 188, 195–199
  in urea cycle, 157
Adenyl cyclase, 215–216, 392
ADP, *see* Adenosine diphosphate
ADP/ATP ratio, 161, 163
Alanine, 94, 95, 220
  synthesis, 220
  transport, 305
Alanyl-tRNA, yeast, 376–377
Alcohol dehydrogenase, 103
Aldolase, 140, 143
  quaternary subunits, 103
Alga, blue-green, 226–227, 232, 318
All-or-none event, 311
Allosteric enzymes, 119
Allosteric inhibition, 119, 222
  in amino acid synthesis, 221–225
  in the Pasteur effect, 160–161
Alloxan, structure, 214
Amino acid activation, 383
Amino acids, 94–98
  biosynthesis, 218–225
  essential, 219–220
  structures, 95–97
  synthesis of aromatic types, 222
Aminoacyl-tRNA, structure, 243
γ-Aminobenzoic acid, structure, 239
γ-Aminobutyrate, structure, 314
γ-Aminolevulinic acid, 447
Aminopterin, structure, 239

463

Index